FLORA OF DERBYSHIRE

JACOB'S LADDER (*Polemonium caeruleum*); *Taddington Dale.*

Flora

of

Derbyshire

A. R. Clapham, MA, PhD, FRS, FLS
Formerly Professor of Botany, University of Sheffield

Ashbourne Editions

ISBN 1 873775 08 3

First published 1969
Reprinted 1992 by Ashbourne Editions

This reprint published by:
Ashbourne Editions, Clifton, Ashbourne, Derbyshire

Printed in the UK by:
The Cromwell Press Ltd
Broughton Gifford, Wilts

CONTENTS

PHOTOGRAPHS

FIGURES IN THE TEXT

PREFACE

THE publication of a county flora involves a great deal of preparatory work of two main kinds. In the first place there is the collection of the basic field-data: the identification of the various species and the recording of their distribution within the county, and also the intensive searching required for the confirmation or otherwise of old records. A large number of field botanists, named in the Introduction, have contributed to this Flora of Derbyshire, sending us records from all parts of the county and over long periods. Without their help our task would have been impossible.

The second kind of preparatory work is that of assembling the records checking them and listing or summarizing them in the most appropriate manner for publication. This is no less necessary but a good deal less enjoyable than work in the field, and we have been fortunate in finding volunteers who have been willing to devote many long hours to it. For the initial planning we owe much to Mr. (now Councillor) F. W. Adams, of Sheffield, whose system has been retained almost unchanged throughout. After this valuable start the work was continued and completed by Mr. A. L. Thorpe, F.M.A., Curator of the Derby Museum and Art Gallery, Mr. F. Crosland and Miss K. Hollick, to whom we are very deeply indebted. It has been extremely valuable to have this important activity centred on the Derby Museum, where all the original records will continue to be housed and where they can be consulted.

It is appropriate at this point to acknowledge the indispensable help given us by certain taxonomic experts in difficult plant-groups. Mr. Peter Sell, of the Herbarium of the Cambridge University Botany School, and Professor J. N. Mills, of Manchester University, have made possible a modern account of Derbyshire hawkweeds, and Mr. E. S. Edees similarly gave generous information and advice on our brambles, as did Dr. P. F. Yeo on eyebrights. Dr. Franklyn Perring, of the Biological Records Centre at Monk's Wood Experimental Station, Huntingdon, has always been ready to send distributional information collected for the Atlas of the British Flora and its recent Critical Supplement, and his help is also very gratefully acknowledged.

We take this opportunity, too, of expressing our thanks to Professor Alice Garnett and Professor R. S. Waters of Sheffield University, and to Mr. W. H. Wilcockson, who contributed the prefatory sections on Climate, Geomorphology and Geology respectively; and to Dr. J. G. Dony, to whom we naturally turned for advice on all matters connected with the actual publication of a county flora and who was every bit as helpful as we had confidently expected him to be.

A special word of appreciation and gratitude is due to Dr. T. T. Elkington, of the Botany Department of Sheffield University, for all the work he has undertaken so unselfishly and carried out so competently. Dr. Elkington not only wrote the prefatory sections on the Vegetation of Derbyshire but has also acted as executive editor on behalf of the Flora Committee.

Finally we are glad to acknowledge generous financial assistance from the Publications Fund of the Royal Society towards the costs of publication and in particular of the preparation of the pre-publication brochure; and to express our very sincere thanks to the Museum Committee of the Derby County Borough Council for consenting to publish this new Flora of Derbyshire.

A. R. CLAPHAM

Chairman of the Flora Committee

Chapter 1

INTRODUCTION

'In 1903 justice was done to Derbyshire by the publication of an excellent *Flora* of the county, written by the late Rev. W. R. Linton, M.A.'. None would wish to disagree with this judgment expressed by E. & H. Drabble shortly after Linton's death in 1908, and it is perhaps not difficult to understand why his *Flora of Derbyshire* has been neither revised nor replaced in all the sixty-five years since it appeared. Much botanical field-work has nevertheless been carried out in the county during that period, and it has resulted in an impressive body of scientific publications.

Mention must first be made of Dr. Eric Drabble (1877-1933), who came of an old Derbyshire family and was born at Herne House near Chesterfield. He was educated at Chesterfield Grammar School, University College, Sheffield, and the Royal College of Science, South Kensington, where he obtained the London B.Sc. degree with First Class Honours in Botany. He became Lecturer in Botany, first at St. Thomas' Hospital Medical School, then at the Royal College of Science, and finally at the University of Liverpool. In 1908 he was made Head of the Botanical Department at the Northern Polytechnic, and in that same year was awarded the D.Sc. degree of Sheffield University.

In the first of their Notes on the Flora of Derbyshire, quoted above, Dr. and Mrs. Drabble referred to Linton's request for lists of plants found in various parts of Derbyshire so that he might incorporate them in his Flora, and to the fact that they continued to send such lists after the appearance of the Flora. When Linton died they felt it necessary to publish their new records, and the Notes appeared in the Journal of Botany at intervals from 1909 to 1916. A further paper entitled Additions to the Flora of Derbyshire was published in the Journal of the Derbyshire Archaeological and Natural History Society, Vol. XXXIX (1917), and a second in Vol. II of the New Series of the same Journal (1929) reverted to the former title Notes on the Flora of Derbyshire. These papers contain many important new records and express views on the taxonomic status of a number of interesting Derbyshire plants. It should be added that the volume in which the last of the Notes appeared contained also an article by Dr. Drabble on Derbyshire Pansies, a group in which he was the national expert.

In more recent times many local botanists have made their contributions to the steady increase in our knowledge of the Derbyshire flora and of the changes which it has undergone. Special mention should be made of F. T. and R. H. Hall, father and son, who worked assiduously for many years in the north-western part of the county, round their home in Buxton. They published their valuable Notes on the Flora of Buxton and District in the Report of the Botanical Society and Exchange Club of the British Isles for 1939-40. Meanwhile Miss K. M. Hollick, of Ashbourne, had been equally assiduous in the south-west and she followed Dr. Drabble's example in publishing her Botanical Record for Derbyshire, at frequent intervals from 1941 onwards, in the Journal of the Derbyshire Archaeological and Natural History Society. Many other workers have from time to time contributed notes or new plant records in various national and local publications.

The period immediately following the publication of Linton's *Flora* saw the initiation of ecological studies in this country, and a great deal of the early work was centred on the southern Pennines, including the Peak District of Derbyshire. At the turn of the century Dr. W. G. Smith was on the staff of the Biological Department of the Yorkshire College at Leeds, and his students and disciples included C. E. Moss as well as W. B. Crump, W. M. Rankin and T. W. Woodhead. Moss made a special study of the Peak District, and when that classic of British plant ecology, Tansley's *Types of British Vegetation*, appeared in 1911, Moss was one of the main contributors. He wrote sections on the vegetation of the older limestones and of the older siliceous soils and, with F. J. Lewis, on the Upland Moors of the Pennine Chain. In all these accounts descriptions and photographs of Derbyshire vegetation figure largely. In 1913 his studies of Derbyshire vegetation were brought together in his well-known book *Vegetation of the Peak District*, still an extremely valuable source of information about the plant communities of the county and their interrelationships in space and time. It has full lists of species found in the various communities and these have proved valuable in revising Linton's *Flora*.

In 1938 W. H. Pearsall became Professor of Botany in Sheffield University and his appointment initiated a long period of active ecological research which has continued to the present day. Pearsall's excellent book *Mountains and Moorlands* draws heavily on his intimate knowledge of the southern Pennines. That work, and the scientific publications of other Sheffield research workers from Dr. Verona Conway and Miss Olive Balme to Dr. (now Professor) C. D. Pigott and Dr. D. J. Anderson and then on to Drs. Ian Rorison, Philip Grime and Philip Lloyd of the Nature Conservancy's Research Group in Sheffield University, constitute a very considerable volume of information about Derbyshire vegetation whch we have found most valuable.

Soon after the war, when it became possible to resume botanical field-work, it was suggested that the time had perhaps come to revise Linton's *Flora*, taking advantage of the new information already available and enlisting the help of the many field botanists throughout the county and on its borders. There was clearly a sufficient number of competent and determined volunteers to embark on the task, and a Committee was set up to organize and supervise the work. The Committee met first in 1949, the original members being Dr. R. Abercrombie, F. W. Adams, J. Brown, Miss R. Carey, Professor A. R. Clapham, F. Crosland, Miss M. Dawson, F. T. and R. H. Hall, Miss K. Hollick, A. L. Thorpe, Mrs. E. Wain and C. B. Waite. Two of this number, Dr. Abercrombie and F. T. Hall, have, sadly, died before they could see the task completed. Others have joined the Committee at various subsequent times, these including Dr. D. J. Anderson, R. Brown, E. Caulton, Dr. T. T.

Elkington, Miss S. J. Herriott, A. Mather and Dr. C. D. Pigott. Meetings of the Committee, apart from a few held in the field, have taken place in the Prince Charlie Room of the Derby Museum and Art Gallery, where all decisions have been taken as to the collection and recording of the field data and the format of the final publication.

Members of the Committee have assumed responsibility for the accuracy of all the records published without comment. The great majority of these records were made by members of the Committee, but some have been supplied by other workers known to one or more of us. The following list includes most of these, but some will inevitably have been omitted: R. H. Appleby, Mr. and Mrs. P. Ball, A. Beaumont, Mr. and Mrs. R. A. Carr, Dr. J. G. Dony, V. Gordon, W. H. Hardaker, Miss M. C. Hewitt, S. Hill, J. Mills, D. McClintock, A. R. Proctor, Miss M. Shaw, W. H. Somers, Mrs. Torry, M. M. Whiting and D. P. Young. To all of these we are deeply indebted, as also to Mr. E. S. Edees, Professor J. N. Mills and P. Sell, already mentioned in the Preface.

4

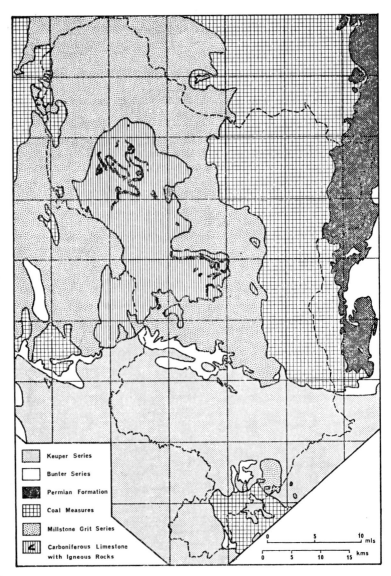

Keuper Series

Bunter Series

Permian Formation

Coal Measures

Millstone Grit Series

Carboniferous Limestone
with Igneous Rocks

Fig. 1.   Solid Geology of Derbyshire.

Chapter Two

GEOLOGY

TOPOGRAPHY

THE county of Derbyshire lies at the southern end of the Pennine Chain and is composed for the most part of Carboniferous rocks, though there are outcrops of Permian on the east side and Triassic in the southern part of the county. Superficial deposits cover the 'solid' rocks in many places. The general succession of rocks exposed is as follows:—

Pleistocene and Recent	Superficial Deposits
Triassic	{ Keuper Marl. Waterstones. Bunter Sandstone.
Permian	Magnesian Limestone and Marls.
Carboniferous	{ Coal Measures. Millstone Grit Series. Carboniferous Limestone.

The Carboniferous Limestone outcrops in a rather irregular dome occupying about 180 square miles, about 25 miles from north to south and 10 miles from east to west and is also exposed in two anticlinal inliers at Ashover and Crich. Altogether some 1,700 feet of limestone is exposed with an additional 870 feet proved in the boring at Woo Dale. On the north, east and west it is encompassed by continuous outcrops of the Millstone Grit Series, which, on the east and west are succeeded by the Coal Measures. Along the southern margin of the dome the massive grits are absent and the limestone and the succeeding shales sink beneath the unconformable Triassic rocks which extend almost to the county boundary, only interrupted by the northern margin of the South Derbyshire Coalfield, the Millstone Grit area about Melbourne and the Limestone inliers of Ticknall and Calke Abbey.

On the outcrops of the Carboniferous Limestone there is little surface drainage and the country takes the form of a high undulating plateau dissected by steep sided valleys, the dales. In some of the dales rivers may be seen at the surface but many are dry. The erosion of the dales must have been accomplished either when the rocks were frozen or when the water table was higher than at present. Even with a higher water table the drainage of the plateau would still be underground and the sides of the dales remain steep from lack of surface erosion. The Millstone Grit Series which rests on the limestone is composed of beds of sandstone or gritstone alternating with soft shales and the same may be said of the succeeding Coal Measures. At the base of the former series there are about 700 feet of shales, the Edale Shales, that give a strip of low-lying clay land around the limestone which is especially well seen on the north-east side of the dome around Castleton and Bamford. In places, where the shales have been eroded, the margin of the limestone plateau is steep and cliff-like as at Treak Cliff near Castleton. Above the Edale Shales come the alternating sandstones and shales of the Millstone Grits which give the plateaux of Kinder Scout and Bleaklow and the prominent cuestas on both sides of the dome. The Coal Measures exhibit a similar, though more subdued, topography. On the eastern side of the coalfield, between the Yorkshire border and Pleasley the Permian Magnesian Limestone stands up as an escarpment above the Coal Measures. It has a wide dip slope with typical limestone topography. In the south of the county the Keuper Marl makes low-lying country over a wide area between Derby and the River Dove. The underlying Bunter Sandstone provides rather more elevated topography on the line between Ashbourne and Derby and, in the south, between Repton and Ticknall.

THE PRE-CARBONIFEROUS ROCKS

No rocks older than the Carboniferous Limestone are known at the surface, but a boring at Woo Dale about two miles south-east of Buxton penetrated more than 100 feet of altered volcanic rocks below the limestone. These rocks were said to be similar to the pre-Cambrian of Charnwood Forest and it is possible that such ancient rocks are widely present beneath the Carboniferous of Derbyshire.

THE CARBONIFEROUS FORMATION

Carboniferous Limestone. In the northern part of England rocks of this age occur in two facies, the 'massif' or 'standard' facies and the 'basin' facies. The limestones of the massif facies are usually light coloured and massive in character with a fauna of corals and brachiopods and make up the whole of the plateau region. The basin facies is made up of dark limestones and shales with goniatites and lamellibranchs. The latter are not extensively developed in Derbyshire. They have been proved in borings beneath the shales at Edale and Alport and have wide outcrops in the Dovedale area. They may also be present in the south about Bradbourne and Kniveton.

In the plateau region the limestones lie with gentle folds and it may be that it is underlain by a stable block like that of the North of England. Round much of the margin steeper dips off the plateau can be observed and in the southern region in the valleys of the Manifold and Dove and in an area north-east of Ashbourne, the rocks exhibit quite extensive folding. Over much of the plateau the limestones are normal bedded rocks, but around the margins, they often assume what is known as the 'reef' facies, a variety with a rich fauna of brachiopods and crinoids and frequently masses of algal growth, which in places appear to have been deposited as steeply inclined 'apron reefs' or 'fore reefs' along the cliff-like edge of the limestone massif in a similar manner to modern barrier reefs.

Dolomitic Limestones. Over much of its outcrop the limestone is free from magnesium, but dolomitisation has taken place in several areas. The lower limestones in the neighbourhood of Buxton are dolomitic as are the rocks of the basin facies around Dovedale. In the Matlock district almost every division of the limestone shows signs of dolomitisation and this change has also taken place near Winster and Elton, between Arbor Low and Gratton Dale and along the south boundary of the limestone from Rainster Rocks to Hopton. In some places the dolomitisation has been almost complete and there the rocks tend to weather into steep crags and upstanding tower-like formations known as 'tors'.

Sand Pockets. South-west of a line from Arbor Low to Hopton there are deposits of sand and clay occupying steep-sided pockets in the limestone. These appear to be solution cavities, into which remnants of beds which once were present over the whole limestone area were let down and so preserved. As would be expected they lie with steep dips, often away from the edges of the pockets. The top few feet are often filled up with boulder clay.

Volcanic Rocks. These are found in several parts of the limestone region. They are basaltic in character and comprise both lavas and tuffs and are also found occupying the throats of old volcanoes. There are also several large intrusive masses of dolerite at Water Swallows, Peak Forest, Calton Hill, Ible and elsewhere. The volcanic rocks are impervious to water and have affected the underground drainage.

Mineral Veins in the Carboniferous Limestone. This formation is traversed by numerous mineral veins containing galena, blende, fluorspar, barytes and calcite together with a number of other minerals. The deposits occur as 'rakes', veins filling vertical fissures, and 'flats' and 'pipes' both formed by replacement of beds of limestone. The rakes mostly trend in an east and west direction and, in the northern part of the massif are grouped on the broad anticlines. The ores were deposited from solutions that rose along the fissures and the richest concentrations of mineral are found beneath the overlying shales or even under beds of toadstone where the solutions were trapped. Of the gangue minerals fluorspar is found along the eastern margin of the limestone outcrop, barytes and calcite further west. The ores have been worked for lead from the time of the Romans and now the fluorspar, which the old miners discarded, is being recovered from mines and old waste heaps.

The Fauna of the Limestone. In the massif and reef facies the fauna consists mainly of brachiopods and corals and with their aid a series of life zones have been established. In the basin facies the fossils are mostly of lamellibranchs and goniatites and from the latter group another series of zones have been worked out. Correlation between the two systems is not easy because the two faunas are rarely found in the same beds but a very general correspondence has been arrived at as follows:—

Coral-brachiopod fauna	Lamellibranch-goniatite fauna
	P_2
D_2	P_1
D_1	B_2
S_2	B_1
S_1	Pe_3
C_2	Pe_2
C_1	Pe_1

The first section of the limestone to be described in detail was that along the Wye Valley between Buxton and Monsal Dale and this will be taken as a type. The rocks exposed are as follows:

		Approximate thickness
P2	The Ashford Beds, dark limestones with chert and shaly partings	80 feet
D2	The Monsal Dale Beds, dark thin bedded limestones with chert	400 feet
D2	Priestcliffe Beds, massive pale limestones with chert	100 feet
	Upper Lava	0–100 feet
D2	Station Quarry Beds, thinly bedded dark to black limestones with chert	30 feet
	Disconformity	
D1	Miller's Dale Beds, thick bedded light coloured limestones	100 feet
	Lower Lava	0–80 feet
D1	Chee Tor Beds, thick bedded and massive pure light coloured limestones	300–400 feet
S2	Daviesiella Beds, dark, often dolomitised limestones	300 feet

The succession of the rocks is roughly similar throughout the region north of the Wye Valley. The greater part of the area is underlain by normal massif type limestones, but along the northern and north-eastern borders of the limestone outcrop, reefs which are often richly fossiliferous and with steep outward dips occur. To the north the limestones dip down under the rocks of the Millstone Grit series and pass into the basin facies. To the south of Castleton a group of thin bedded cherty limestones with occasional reefs, the Eyam Limestones, come on at the summit. They appear to represent an incursion of the basin facies onto the massif and can be traced along the margin of the limestone as far as Hassop. The massif type rocks of D1 and D2 age extend to the western side of the limestone and are there covered by the Edale Shales. Further south near Earl Sterndale the D1 limestones form the edge of the massif and are fringed on their western side by reefs of B2 age. These dip steeply outwards and make the spectacular outcrops of Chrome and Parkhouse Hills.

The middle and south-eastern region, including Wirksworth, Matlock and the country as far west as Monyash, shows a similar succession. The greater part is composed of normal facies type limestones with two or more lava flows, volcanic necks at Grangemill and intrusive sills at Bonsall and Ible. At the summit there is a variable group of limestones with reefs and shales, the Cawdor Limestones. Dolomitisation is widespread in this area. South of a line between Wirksworth and Tissington the Carboniferous rocks are, with the exception of an inlier of limestone near Bradbourne, a shaly group now referred to the lower part of the Millstone Grit Series.

The south-western region lies on the east side of the River Dove and is a transitional area between the massif and the basin. Here all three facies, massif, reef and basin have been recognised. There is considerable folding and faulting and the outcrops of the different facies are somewhat intermingled. North-east of a line from Hartington to Parwich the rocks belong to the massif facies while to the south-west they are of reef and basin types. The upper division of the massif facies, the Alsop Moor Limestone, occupies the plateau country from near Hartington to Parwich, while the lower divisions occur on either side of the Dove between Beresford Dale and Iron Tors. Of the Reef Beds only the lower division, the Dovedale Limestone, makes an outcrop of any size. It consists of one large reef mass and forms the sides of Dovedale below Dove Holes, where it is responsible for most of the spectacular crags. The lower divisions of the basin facies crop out round Milldale and the upper division, the Gag Lane Limestone, covers a wide area on the plateau south of Alsop-en-le-Dale. It includes bedded limestones and dolomite both cherty and crinoidal. The latest section of the limestone series, the Hollington End Beds, are only found as a narrow strip faulted into the Gag Lane Beds.

The Millstone Grit Series. This series represents a great change in sedimentation. The majority of the rocks were laid down under deltaic conditions but contain occasional beds of shale with marine fossils, the 'marine bands', which provide marker horizons for purposes of correlation. The series has been subdivided into four groups with characteristic goniatites.

The Rough Rock Group with Gastrioceras	Zonal letter G_1
The Middle Grit Group with Reticuloceras	Zonal letter R_2
The Kinder Scout Group with Reticuloceras	Zonal letter R_1
The Edale Shale Group (Upper) with Homoceras	Zonal letter H
The Edale Shale Group (Lower) with Eumorphoceras and Cravenoceras	Zonal letter E

The Edale Shales are a group of dark marine mudstones with occasional beds of muddy limestone and siltstone and many fossiliferous beds with abundant lamellibranchs and goniatites. They occupy the low ground of the Edale and Hope valleys and the lower slopes of the Millstone Grit hills throughout the region. They rest on the limestones with unconformity and are often banked against the old cliffs of the marginal reefs. The lower beds of the Kinder Scout Group are of a similar series of shales.

The middle portion of the Kinder Scout Group is made up of the Mam Tor Sandstone and the succeeding Shale Grit Series. The Mam Tor Sandstones are soft and thinly bedded with shale intercalations. They form the ridge from Lose Hill to Mam Tor. The Shale Grit Group is similar to the last but with rather more sandstone. These two groups weather readily and make rounded shoulders on hills below the Kinder Scout Grit. Where the impervious shales below them have given way they give rise to extensive landslips as at Mam Tor, Alport Castles and elsewhere. The upper part of this group is composed of the Kinder Scout Grit, the first of the coarse, massive grits. This is a thick rock divided into an upper and lower portion by a band of shale. The lower leaf is usually massive and coarse grained and the upper thinner bedded and finer in grain. Over the northern extension of the Derbyshire Dome the flat-lying grits make the plateau country of Kinder Scout and Bleaklow. On either side of the dome, where the dips are away from the axis of the fold, the bed makes prominent escarpments such as the Derwent and Bamford Edges. The lower leaf dies out near Bamford, but the upper continues to the south. It makes the high ground of Eyam Moor but finally fails between Chatsworth and Matlock.

The Middle Grit Group contains several gritstone horizons but only two are important. These are the Chatsworth or Rivelin Grit near the top and the Ashover Grit near the bottom. The Ashover Grit is first seen near Calver and increases in thickness to the south until near Matlock and Ashover it is the most prominent of the Grits and makes the escarpment of Riber and the high ground of Cromford Moor. In this area it is the most persistent bed and has a continuous outcrop as far as Little Eaton near Derby. In the northern and central parts of the county the succeeding Chatsworth or Rivelin Grit is the most important bed. It makes an almost continuous escarpment from Moscar in the north to the south of Belper where it dies out. The scarps are often crowned by crags affording excellent rocks for climbing. It is often coarse with abundant fresh felspar and maintains a rough surface under wear, for which reason it was much sought after for millstones, thus giving the name to the series. On the top of the bed there is a persistent coal, the Baslow or Ringinglow Coal. The succeeding Rough Rock Group contains two grit horizons, the Redmires Flags in the middle and the Rough Rock at the top. Neither of these is very persistent. Both are present to the west of Sheffield but the Rough Rock fails not far from Baslow and the Redmires Flags do not continue as far as Ashover.

The Millstone Grit Series is well developed in the north-west part of the county about Glossop and Chinley. Here the succession is similar to that to the east of Kinder Scout, the main Grits being represented. On the west slopes of Kinder Scout the Shale Grit and the Kinder Scout Grit make a wide outcrop as do also the Chatsworth Grit and Rough Rock further west. To the south, with steeper dips, the outcrops are narrower, the most obvious being that of the Chatsworth along Cracken Edge. About Chinley beds equivalent to the Ashover Grit appear and the complete succession continues till its outcrop passes out of the county at the south end of Axe Edge which is an escarpment of the Roaches Grit the representative of the upper Ashover Grit of the east.

The Coal Measures. These rocks are found on both sides of the Derbyshire Dome and also in the South Derbyshire Coalfield. The largest area is on the east, the Derbyshire Coalfield. On the west small thicknesses of Lower Coal Measures are preserved in synclines. In the north-east of the county the whole of the exposed coalfield belongs to Derbyshire from the outcrop of the Millstone Grit to that of the Permian rocks in the east, but south of Chesterfield the county boundary is coincident with the Erewash River and the eastern part of the exposed coalfield is in Nottinghamshire. Lithologically the Coal Meaures are similar to the Millstone Grits, but the sandstones are rather fewer and finer in grain and make less imposing features. Like the Millstone Grits the Coal Measures are of deltaic origin and the fauna is, for the most part, non-marine, mainly lamellibranchs. There are a number of marine bands which provide marker horizons, but the broad palaeontological subdivisions of the measures have been determined by assemblages of non-marine lamellibranchs. These zones are usually named after a species of a characteristic lamellibranch.

5. Upper similis-pulchra.
4. Lower similis-pulchra.
3. Modiolaris.
2. Communis.
1. Lenisulcata.

For purposes of mapping the Geological Survey divide the series as follows:

	Approximate Thickness
Middle Coal Measures with several marine bands and important coals in the lower part. These beds belong to the similis-pulchra zones and the Upper Modiolaris zone	1,700 feet

Clay Cross Marine Band.

Lower Coal Measures with important coals in the upper part and marine bands near the base. This subdivision belongs to the Lower Modiolaris, Communis and Lenisulcata zones.	1,800 feet

Pot Clay Marine Band.

Millstone Grit.

The lowest part of the Coal Measure sequence has much in common with the Millstone Grits and contains a high proportion of sandstone. Some of these are topographically important and persistent. The first sandstone above the Pot Clay Coal is known as the Crawshaw Sandstone. This persists from the neighbourhood of Sheffield almost to the south of the coalfield. In places it is a thick rock and was earlier mistaken for the Rough Rock. It also occurs on the western side of the Pennines under the name of the Woodhead Hill Rock. Another sandstone or sandstone series which lies some 400 or 500 feet higher in the measures is known in Derbyshire as the Wingfield Flags. It is a series of sandstones with interbedded shales and is found under different names in all the coalfields round the southern end of the Pennines.

Structure of the Coalfield. The general dip of the Coalfield rocks is to the east, but they are disturbed by folds and faults. Along the Yorkshire border the folds strike about west-north-west to east-south-east. But about Dronfield an important fold, the Brimington Anticline, turns from an east to west direction to north to south and again near Chesterfield to north-west to south-east, finally passing under the Permian towards Mansfield. Further south the measures are folded along an anticline striking nearly north and south along the line of the River Erewash. The strike of the folds is rather irregular, but the predominant direction is north-west to south-east or at right angles to this direction.

These structures and the uprise of the Derbyshire Dome were the result of the Armorican Earth-movements which took place at the end of Carboniferous times. These movements were followed by a long period of erosion during which the region was reduced to a peneplain.

THE PERMIAN SYSTEM

Over the post-Carboniferous peneplain the Zechstein (Upper Permian) sea advanced and deposited the rock series known by that name. In Derbyshire only the lowest beds are exposed, the higher beds having been eroded. The local succession is:

	Approximate Thickness
Middle Permian Marls	
Lower Magnesian Limestone	120 feet
Lower Permian Marl	30 feet
Basal Sands	0–20 feet

The Basal Sands. These are uncemented sands filling hollows in the old land surface. In Nottinghamshire they are replaced by a thin breccia.

Lower Permian Marl. An impersistent group of dolomitic marls of variable thickness, thin or absent on the Yorkshire border, but rather thicker to the south.

Lower Magnesian Limestone. A dolomitic limestone sometimes approaching theoretical dolomite. The lower beds have less magnesia and contain a considerable number of fossils, mainly brachiopods and lamellibranchs with occasional masses of polyzoa. The limestone is thick bedded and well jointed and has yielded large size building blocks. It is now quarried for basic refractories. Owing to its low dip it makes a wide outcrop, attaining a width along the eastern border of the county of four or five miles.

Middle Permian Marls. These have been almost entirely eroded and only appear as one or two small patches about Clowne and Shirebrook.

THE TRIASSIC SYSTEM

The rocks of this system in Derbyshire are confined to the south of the county where they rest unconformably on the Carboniferous and lie in a syncline bordered on the north on an irregular line between Ashbourne and Quarndon and on the south by the South Derbyshire Coalfield and the Millstone Grit inlier around Melbourne. The general succession is as follows:—

> Keuper Marl.
> Keuper Waterstones.
> Bunter Pebble Beds.
> Bunter Lower Mottled Sandstone.

Bunter Beds. The Lower Mottled Sandstone is often absent. The Pebble Beds, a series of red sandstones with abundant quartzite pebbles, are persistent over the whole of the area and make restricted outcrops on both sides of the syncline. In the north they vary between 100 and 130 feet in thickness. In the south around Repton they are about 200 feet thick and thicken considerably at the southern county boundary.

The Keuper Waterstones. Beds of white and pink sandstone with marl bands and a breccia at the base. They crop out on both sides of the syncline and vary in thickness from 50 to 100 feet.

The Keuper Marls. A series of marls with thin sandstones and bands of gypsum. Near Chellaston the gypsum bands become thicker and have been a source of alabaster for statuary purposes. The marls are soft rocks. They cover a wide area and are responsible for the low lying country enclosed by the Rivers Dove and Trent.

THE SUPERFICIAL DEPOSITS

Drifts of Pleistocene age. Apart from the Trias the Mesozoic and Tertiary formations are absent in Derbyshire. Glacial deposits are widely distributed and are divided into an Older and Newer Drift. The Older Drift has largely been eroded, but remnants are found as scattered erratics and small patches of boulder clay over the surface of the Derbyshire Dome. The erratics include rocks from local sources and from the Lake District and Southern Scotland and from their distribution it may be inferred that ice came from the west, through the gap at Dove Holes and continued in a south-easterly direction down the valleys of the Wye, Derwent and Dove. Erratics found further north in the upper Derwent valley suggest also a more northerly invasion by the older glaciers. Probably also belonging to the Older Drifts are small thicknesses of boulder clay preserved in the tops of the sand pockets in the limestone and more extensive spreads around Matlock, the northern part of the Ashover Dome, and around Crich and Belper. Some small patches of boulder clay around Barlborough and Clowne on the Magnesian Limestone may also belong here. The evidence of the later glaciation is much more distinct, though its ice does not appear to have invaded the Derbyshire Dome nor even the outcrop of the Magnesian Limestone in the east. On the west side of the county it made significant advances. Around Glossop and Chinley erratics are found up to a height of 1,000 to 1,300 feet and considerable spreads of boulder clay cover the Millstone Grit and Coal Measures between the Etherow Valley and Buxton. During the retreat stages of this ice-sheet a system of lakes with accompanying overflow channels were developed along the western slopes, but these are all outside the county. In the south drifts are widespread. The outcrops of the Trias are largely covered by boulder clay as are the also Carboniferous rocks as far as the line between Tissington and Wirksworth.

Head. This is a periglacial deposit resulting from solifluxion and accumulated on many slopes usually below gritstone outcrops. It is a clayey or sandy loam crammed with blocks of sandstone of all sizes up to 20 feet across. It extends up to and merges with the ancient screes and extends down the slopes by flow, being thinner on spurs and thicker in depressions. It has been able to move on slopes of 3 deg. or less and in places extends onto the more gently sloping ground as in the Edale Valley and the shale flats below the Ashover Grit at Ashover. The thickness is usually only a few feet but it has been measured as much as 30 feet. Head is best developed in the Millstone Grit country in the east. It is known on both Coal Measures and the Permian, but there is very little on the west side of the Derbyshire Dome.

Peat. Over much of the high moors on the Millstones Grit thick deposits of peat have accumulated up to a thickness of as much as twenty feet. It is now undergoing erosion. The deposits are intersected by deep gullies and from large areas the protecting vegetation has been removed and destruction is going on apace.

Alluvium. The rivers of Derbyshire are mostly fast, flowing through deep and narrow valleys so that there are no very extensive alluvial flats. Spreads of alluvium are seen in the valleys of the Wye and Derwent above Matlock, in the Derwent Valley from Belper to its junction with the Trent, the lower valley of the Dove and in the Trent Valley. In the coalfield it is also developed in the valleys of the Rother and the Erewash. The Trent Valley alluvium contains much gravel which is now being actively exploited.

REFERENCES

Arnold-Bemrose, H. H. (1907). The Toadstones of Derbyshire, their Field-relations and Petrography. Quart. Journ. Geol. Soc., 63, 241–281.

Cope, F. W. (1933). The Lower Carboniferous Succession in the Wye Valley of North Derbyshire. Journ. Manch. Geol. Assoc., 1, 125–145.

Cope, F. W. (1965). The Peak District, Derbyshire. Geol. Assoc. Guides., No. 26.

Dalton, A. C. (1957). The Distribution of Dolerite Boulders in the Glaciation of N.E. Derbyshire. Proc. Geol. Assoc. 68, 278–285.

Eden, R. A., Stevenson, I. P. & Edwards, W. (1957). Geology of the County around Sheffield. Mem. Geol. Surv. (Sheet 100 N.S.).

Eden, R. A., Orme, G. R., Mitchell, M. & Shirley, J. (1964). A Study of Part of the Margin of the Carboniferous Limestone 'Massif' in the Pin Dale Area of Derbyshire. Bull. Geol. Surv., 21, 73–118.

Ford, T. D. (1963). The Dolomite Tors of Derbyshire. East Midland Geographer., 3, 148–153.

Green, A. H. and others (1887). The Geology of North Derbyshire. Mem. Geol. Surv.

Hudson, R. G. S. & Cotton, G. (1945). The lower Carboniferous in a Boring at Alport, Derbyshire. Proc. Yorks. Geol. Soc., 25, 254–330.

Jowett, A. & Charlesworth, J. K. (1929). On the Glacial Geology of the Derbyshire Dome and the Western Slopes of the Southern Pennines. Quart. Journ. Geol. Soc., 85, 307–334.

Neves, R. & Downie, C. (1967). Geological Excursions in the Sheffield Region. Sheffield.

Parkinson, D. (1947). The Lower Carboniferous of the Castleton District of Derbyshire. Proc. Yorks. Geol. Soc., 27, 99–124.

Parkinson, D. (1950). The Stratigraphy of the Dovedale Area, Derbyshire and Staffordshire. Quart. Journ. Geol. Soc., 105, 265–294.

Shirley, J. & Horsfield, E. L. (1940). The Carboniferous Limestone of the Castleton–Bradwell Area, North Derbyshire. Quart. Journ. Geol. Soc., 96, 271–317.

Shirley, J. & Horsfield, E. L. (1945). The Structure and Ore Deposits of the Carboniferous Limestone of the Eyam District, Derbyshire. Quart. Journ. Geol. Soc., 100 (for 1944). 289–306.

Shirley, J. (1959). The Carboniferous Limestone of the Monyash–Wirksworth Area, Derbyshire. Quart. Journ. Geol. Soc., 114, 411–429.

Sibley, T. E. (1908). The faunal Succession in the Carboniferous Limestone (Upper Avonian) of the Midland Area. Quart. Journ. Geol. Soc., 64, 37–53.

Smith, E. G., Rhys, G. H. & Eden, R. A. (1967). Geology of the Country around Chesterfield, Matlock and Mansfield. Mem. Geol. Surv. (Sheet 112 N.S.).

Wolfenden, E. B. (1958). Palaeoecology of the Carboniferous Reef Complex and Shelf Limestones in North-West Derbyshire, England. Bull. Geol. Soc. Amer., 69, 881–898.

Chapter Three

GEOMORPHOLOGY

INTRODUCTION

DERBYSHIRE owes its scenic variety to its geological diversity and to spatial and temporal variations in the nature and effectiveness of the geomorphic processes which have given visible topographic expression to its diverse rocks and structures.

The close dependence of the surface form on the nature of the underlying rocks is revealed by a sub-division of the county according to the character of its terrain, defined in terms of height above sea-level, amplitude of relief from interfluve crest to adjacent valley floor, and surface texture as expressed in assemblages of landforms. Many of the morphological subdivisions thus delimited are, in essence, lithologically determined tracts of land broadly coincident with the pattern of outcrops on the geological map. From the high Millstone Grit plateau of Bleaklow in the north to the Swadlincote hill country on Coal Measures in the south, the areal extent, trend and altitudinal range of each of the major morphological units reflects the disposition, attitude and lithology of the rock strata from the Carboniferous Limestone to the Keuper Marl.

Currently all of these morphological subdivisions are being affected by a common suite of geomorphic processes appropriate to the mid-temperate, humid region of which Derbyshire now forms a part. Mechanical and chemical weathering, mass movements on slopes and fluvial processes are everywhere moulding the land surface, albeit at varying rates depending on height above base-level, land slope, drainage density and discharge and the nature of the earth materials which are being fashioned. Virtually all the minor morphological features, like the gullies fretting the slopes of the Rushup Edge–Lose Hill ridge, owe their origins to these current processes. But during the cold phases of the Pleistocene, tundra-like bioclimatic conditions prevailed over large parts of the county while other parts, particularly in the south, were invested by ice from both east and west of the largely ice-free Pennine upland. Uniformity of geomorphological development was thus precluded, and any interpretation of the land form must take account of the two alternating morphogenetic systems which have affected the county during the last one million years—the fluvial/glacial alternation in the south and the fluvial/periglacial alternation over the remainder of the area. The consequences of these diverse Pleistocene geomorphic regimes are writ large in the landscape. Thus the terraces flanking the Dove and Trent between Uttoxeter and Long Eaton are dissected and, in places, reworked remnants of widespread deposits

of fluvio-glacial outwash containing flints and other 'foreign' pebbles; whereas
in the north the terraced floors of Edale and Hope Valley are built of spreads
of ill-sorted, locally derived solifluction debris.

The recognition of such polygenetic landforms is relatively straightforward.
They display clearly features diagnostic of an initiation in a bioclimatic
environment which no longer exists. But the interpretation of other inherited
elements in the landscape is less securely based through lack of unequivocal
evidence on the date and manner of their initiation. This applies to the
planation surfaces, like the gently undulating interfluves of more limited
extent on the Triassic outcrop between Ashbourne and Derby. Their testimony
is of value nonetheless. If it can be demonstrated that closely spaced groups
of such 'flats' originated as base-levelled surfaces, that is, as valley floors or
footslopes developed in relation to the level of a permanent stream or other
manifestation of the water table, they may be accepted as indicators of well-
defined stages in the history of denudation. Base-levelled surfaces, recognized
as such by their discordant relations with underlying structures or their
uniform development across rock outcrops of contrasted lithologies, have been
widely identified in the major drainage basins of Derbyshire and used as the
bases of the county's geomorphological history.

MORPHOLOGICAL SUBDIVISIONS

Derbyshire is situated astride the boundary between highland and lowland
Britain, and in terms of relief and surface texture may be divided by the
generalised 800 ft. (240 m.) contour into an upland zone with relatively
subdued interfluves and deep-cut valleys and a lowland zone of more dissected
plateau and hill country and weaker relief (Figs. 2 and 3).

1. **The Limestone Plateau.** This upland unit is clearly separated from the
surrounding grit and shale country by a well-defined topographic discontinuity.
In places the edge of the limestone itself is steep and scarp-like, as on the
southern side of the Hope valley, on Longstone Edge, between Winster and
Matlock, or near Parwich and Brassington; elsewhere the steep feature is
provided by gritstone scarps overlooking the limestone, as between Axe Edge
and Mam Tor or Bradwell and Eyam. Its steep western edge overlooking the
lowland between Longnor and Hartington has been interpreted as a composite
fault-line scarp.

Within this well-defined boundary the plateau is morphologically homo-
geneous. Its gently undulating surface lies at 1,000 ± 100 feet (300 ± 30 m.)
in the south and rises gradually northwards over a distance of 25 miles (40 km.)
to 1,200 ± 200 feet (400 ± 60 m.). The northern half is the more accidented
and is surmounted by low hills reaching c. 1,500 feet (500 m.). Even in its
more subdued southern portion, which exhibits the largest single patch of
unbroken plateau on the crest of the Derwent–Dove interfluve, it is differen-
tiated by low rounded swells rising to over 1,200 feet (400 m.).

Although the varying width of the plateau reflects the broad structural
disposition of the limestones, its surface shows no consistent relations with those
structural elements. Its widest extent in the south (10½ miles; 18 km.) occurs
on a broad area of doming which extends westward from Matlock, where

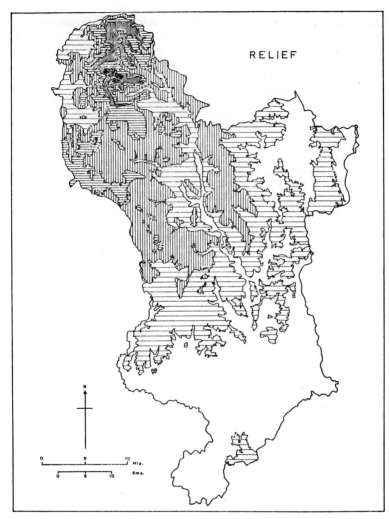

Fig. 2. Relief of Derbyshire.

the fold is transected by the Derwent, to beyond the Dove in Staffordshire. Here the surface is discordant to structure and bevelling of the upfold has exposed toadstones and underlying older limestones. Three miles to the north the plateau is at its narrowest—a mere 5 miles (8 km.) across between Youlgreave and Hartington—over a synclinal tract; but its approximately constant surface elevation gives no indication of the downfold. A second broad area of doming accounts for the increased width of the plateau between Calver and Buxton (11 miles; 20 km.). This complex structural feature does have topographic expression: its eastern part forms the broad swell of Longstone Moor; farther

west in the Upper Wye basin above Monsal Dale the upfold has been deeply
eroded and inverted such that the oldest limestones are exposed in the Wye
valley in the core of the dome and discontinuous inward-facing scarps overlook
the broad shallow basin below which the river is incised. Toadstones outcrop
in the face of the escarpment which runs from Taddington by Chelmorton to
Harpurhill on the southern side of the inverted dome, 'but are only inter-
mittently present on the northern side where the scarp is represented by Bee
Low and Eldon Hill. All the higher eminences of the limestone outcrop are
included in the scarp rim of this inversion'. (Linton, 1956).

The valleys which coarsely dissect the plateau show striking morphological
contrasts between their upper and lower portions. They head in broad
depressions, as at Monyash, with smooth, grass-covered sides and dry floors,
so shallow as to be barely distinguishable from the general undulations of the
surface. But lower down, as in Lathkill Dale, they become narrow and gorge-
like, with sides steep and wooded or near-vertical and bare, sharply separated
from the flat interfluve surface by angular breaks of slope. Indeed it was the
deep incision of the gorges that lowered the nearly flat water-table in the well-
jointed limestone, left the valley heads dry and precluded further fluvial
dissection of the plateau. At the end of the eighteenth century a further
lowering of the water-table by the lead miners, who drove in long 'soughs'
at water level from the river Derwent, left all but the main valleys dry. Nowhere in the limestone is there a record of permanent or flowing water more
than a few feet above the 1,000 feet (300 m.) contour (Fearnsides, 1932). Some
of the most impressive dry dales, like the Winnats, trench the steep margins of
the plateau.

It is towards the margins of the plateau too, that the deep water-table,
commonly 300 to 500 feet (90 to 150 m.) below the surface, has encouraged
the development of extensive cave systems, as near Castleton, Eyam and
Matlock. Surface karstic phenomena are, however, relatively undeveloped;
swallow holes are rare save in the north west where they are fed by streams
draining Rushup Edge, and bare limestone surfaces are absent. The absence
of limestone pavements is a consequence of the area lying outside the limit
of the last glaciation (Pigott, 1962).

2. **The Shale and Gritstone Upland.** The morphological diversity of the
shale and grit country stands in marked contrast to the uniformity of the
limestone plateau that it encircles. Moreover, within the outcrop of the
Millstone Grit Series, variations in the attitude, thickness and lithological
character of the grits and in the trend and amplitude of the many folds which
affect them are reflected in the land form. Thus the limestone plateau is overlooked on three sides—west, north and east—by an impressive series of
inward-facing escarpments, buttressed by edges of dark, grey grit standing
above long boulder- and block-strewn shale slopes; but on its southern margin
the shales and less resistant sandstones of the Millstone Grit Series form part of
the adjacent undulating lowland.

Within the high Black Peak itself the relations of surface form to underlying
structure are variable. Deep erosion and inversion of shallow eastward-
trending flexures have created the broad vales of the Hope Valley and Edale,
cut some 1,000 feet (300 m.) below the narrow, steep-sided, Mam Tor–Lose

Hill synclinal ridge between them. North of Edale the broad and very subdued 2,000+ feet (600 m.) plateau of Kinder Scout with its imposing outward-facing escarpments is determined by a shallow synclinal: farther north the the plateau bevels the major structural element in the area, the Alport dome, maintaining, as on Featherbed Moss, a consistent elevation of 1,750+ feet (525 m.) on the Shale Grit brought up by the fold; and another shallow synclinal on the northern flanks of the dome brings in the Kinder Scout Grit once more and the plateau rises to 2,000+ feet (600 m.) on Bleaklow. Over this extensive region from Edale to the Etherow valley dips are gentle

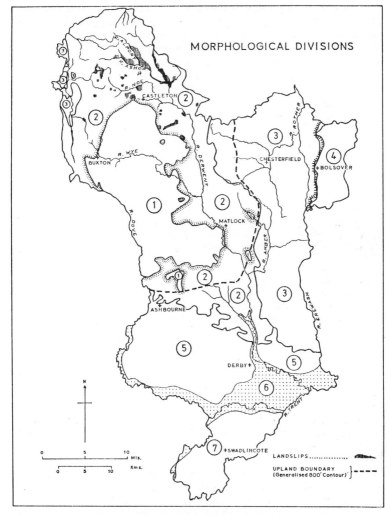

Fig. 3. Morphological divisions of Derbyshire.

and the grits support broad tabular interfluves that separate the deep-cut valleys with benched sides of the headwaters of the Derwent, Goyt and Etherow. Landsliding and cambering are characteristic of this area where the alternation of thick beds of shale and massive flat-bedded grits have provided conditions favouring slope-failure (Fig. 3).

From this high plateau in the north the surface of the grit and shale country declines steadily southwards on either side of the limestone plateau. Dips are steeper on the west than on the east and the superposition of grits on shales is there expressed in compact cuestas like Black Edge and Axe Edge. It is on the east however that the edges or free-faces in gritstone form such distinctive elements in the morphology. North of Bamford the low-dipping Kinder Scout and Chatsworth grits build broad and lofty cuestas buttressed by strong edges (Bamford, Derwent, Stanage and Strines) which overlook the Derwent valley from the east. But farther south, as D. L. Linton has emphasized, the innermost series of inward-facing escarpments lies west of the Derwent. 'It is commonly implied that from here on the horseshoe is continued by the fine series of edges that overlook the Derwent from the east. In reality that river runs, except at Matlock, wholly behind a forward series of scarps which runs in zigzag fashion along the eastern margin of the limestone plateau, the zigzags resulting from the minor folding which carries the gritstones westward in synclines to build the bold salients of Bradwell Edge and Eyam Edge in the north, of Hassop and Bakewell in the centre, and Harthill Moor and Stanton Moor in the south. The bases of these salients are truncated by the Derwent valley in slopes hardly less steep in places than those which lead up to the bold if composite rock rim which bounds the opposite side of the valley' (Linton, 1956). Below Matlock where it cuts deep into the limestone as it transects the eastern end of two steeply pitching anticlines, the river breaks right across the whole of the Millstone Grit sequence until it reaches the Trias just south of Duffield.

To the east of the Derwent valley and paralleling its course from Grindleford to Rowsley lies the bevelled upland of East Moor with its subdued surface cut across various members of the Millstone Grit series and even passing on to sandstones of the Lower Coal Measures. But farther south the upland surface becomes more accidented. At Ashover and Crich strong folds in the grits have been unroofed to expose limestone cores surrounded by shale vales and inward-facing gritstone scarps. Elsewhere in the southern part of the Millstone Grit country, although dip-slopes attest to the partial adjustment of surface form to underlying structure, many discordances still exist and surfaces bevel alike outcrops of grit, sandstone and shale.

3. **The Coal Measure Lowland.** Although sandstones and shales of the Coal Measures rise high on the flanks of the Pennine upland to underlie part of the East Moor plateau most of the outcrop in Derbyshire lies below the 800 feet (240 m.) contour and gives rise to some strongly accidented lowland relief. West of the Rother long stepped spurs descend from the upland edge to 400 feet (120 m.) between close-set, east-flowing streams. It is in this area that the rocks are affected by strong folds, the Norton anticline, the Dronfield syncline, and the Holmesfield–Brimington anticline. Near the streams these folds have been brought into relief by the removal of the shales and the

consequent emergence of the sandstone outcrops as hogback ridges cut across by narrow watergaps. But at higher levels on the interfluves surface bevellings attest to an earlier planation of the folds, subsequent to which a measure of adjustment to structure has occurred.

Similar hill-top bevels have been identified in the area south of Chesterfield, particularly around Ripley. In this southern part where strong easterly dips concentrate the outcrops, differential erosion has given a distinct north–south grain to the country, as between Clay Cross and Pentrich, by bringing the sandstones into relief as long ridges. Elsewhere although many local examples of structural control occur in sandstone scarps and shale-floored valleys the more resistant members of the Coal Measures fail to impose order on an irregular pattern of relief much disturbed by mining and industrial development.

4. **The Magnesian Limestone Cuesta.** A small part of this striking landscape unit occupies the north-eastern corner of the county. From Hardwick Hall to just north of Barlborough it is fronted by an impressive west-facing escarpment built largely of Coal Measure shales with only a thin capping of limestone, 550 feet (165 m.) high at Hardwick Hall. The crest attains its maximum elevation of 613 feet (184 m.) two miles farther north. It is here some 300 feet (90 m.) above the Doe Lea, an interval which is maintained as it declines northwards to 500 feet (150 m.) beyond Barlborough. The open tabular dip-slope is diversified by a network of dry valleys, or grips. These commonly lead down to stream-occupied gorges which like Creswell Crags traverse secondary cuesta features on the backslope of the limestone.

5. **The Keuper Marl Lowland.** South of the generalised 800 feet (240 m.) contour bounding the upland between the Derwent and Dove the outcrop pattern is less clearly expressed in the landscape. Moderate to well dissected hill country with an amplitude of relief of some 200 feet (60 m.) extends southwards from a line between Ashbourne and Belper across various formations to the Dove–Trent confluence. Even to the north of that line there is a penetration of lowland on weak members of the Millstone Grit series between the upland outliers of Hognaston Winn (Carb. Lst.) and Callow (Shale Grit). Farther south the interfluves decline step-like from 700 feet (210 m.) to a little below 300 feet (60 m.) across the broad Keuper Marl outcrop towards the lower Dove and the Trent and give a certain morphological unity to this broad belt of foothills. The chief contrast is between its northern part where the streams flow in relatively narrow valleys, or dumbles, and its southern portion of much weaker relief. Patches of outwash gravel and till indicate that this area in common with the whole of southern Derbyshire experienced several glaciations, the effects of which have further masked or erased the influence of underlying structure on surface form.

6. **The Trent Floodplain and Terraces.** Although narrow and discontinuous strips of floodplain and terrace deposits are present along the middle and upper reaches of all the major rivers, and their significance in the geomorphological history of the area has been widely appreciated, the areal extent of these features is too limited to warrant their distinction as landscape units. However along the Trent and on the Derwent below Duffield and the Dove below Rocester they do constitute a mappable unit consisting of three elements: Hilton Terrace, Beeston Terrace and Floodplain.

The Hilton Terrace extends from Uttoxeter on the Dove to Long Eaton on the Trent. It has two distinct parts, the upper at above 90 feet (27 m.) above the surface of the floodplain and the lower from 40 to 60 feet (12 to 18 m.) above the alluvium. Its gravels (at Hilton itself: 70 per cent Bunter pebbles, with flint as the next most frequent constituent) have the characteristics of fluvio-glacial outwash. Torrent bedding in the gravels of the upper Hilton Terrace suggest aggradation by heavily laden meltwater streams. Those in the upper part of the lower Hilton Terrace are evenly bedded suggesting an amelioration of climate. The Beeston Terrace is also an important morphological feature. It forms a wide bench in the Dove valley and along the Trent and lower Derwent valleys at up to 30 feet (9 m.) above the present alluvium. Although its gravels have been ascribed to various dates by different authors its topographic relations suggest that it clearly postdates the deposition of both Hilton Terraces. The present floodplain is the result of the deposition of river alluvium below the remnants of the three Pleistocene terraces during the post-glacial aggradation.

7. **The Swadlincote Hill Country.** South of the Trent is another well-dissected area of bevelled interfluves rising to some 600 feet (180 m.) in the south-east and relatively narrow valleys draining to the trunk stream. Its landforms are comparable with those of the Keuper Marl Lowland and like that area its general morphological unity belies its diverse geological basis (Keuper Sandstone and Marl, Coal Measures and Millstone Grit).

GEOMORPHOLOGICAL HISTORY

As indicated above, although many elements in the land form reflect the nature and attitude of the underlying rocks others, like the surfaces of low relief, truncate geological structures and pass from one outcrop to another of different lithology with no change in level or morphological character. These discordant elements have been identified and mapped throughout Derbyshire and interpreted as remnants of planation surfaces developed in relation to base-levels of erosion up to 1,000 feet (300 m.) and more above present sea-level (Table I).

On the Pennine upland two extensive surfaces are recognized: the **Summit Surface** (S) on the gritstone country north of the Hope valley and the **Upland Surface** (U) on the limestone of the Low Peak and the gritstone of East Moor and represented farther north by the uppermost valley stage in the valleys of the Derwent headstreams (Linton, 1956). If these upland plains are but little-modified expressions of base-levelled surfaces they would appear to have been fashioned by sub-aerial processes during the Tertiary period and to have developed in relation to two eastward-flowing drainage systems.

Linton suggests that the converging headstreams of the Derwent (Fig. 4), the uppermost Derwent, Westend, Alport, Ashop and Noe (in Edale), united with two now-disrupted tributaries coming from the south (one along the line of Bradwell Dale and the Noe above Hope, the other along the Derwent valley between Win Hill and Bamford Moor) to form a trunk river which flowed through the col at Moscar (1,182 feet; 385 m.) and along the line of the Rivelin valley. This reconstruction of an eastward-flowing Upper

Table 1. Base-levelled surfaces in Derbyshire

PENNINE UPLAND			EASTERN FLANKS

On Carboniferous Limestone	On Millstone Grit	800 ft. (240 m.) Contour	EASTERN FLANKS
	Kinder Scout and Bleaklow residuals 2,000+ ft. (600 m.)		
	Summit Surface (S) 1,650–1,800 ft. (495–540 m.)		**Apperknowle Surface** 650–700 ft. (195–210 m.)
Knolls up to 250 ft. (75 m.) higher on divides			Developed over an area of 40 sq. miles (100 km²) between Sheffield and Chesterfield.
Upland Surface (U) 1,250 ft. (375 m.) in N. & N.W. 1,000 ft. (300 m.) in S. & S.E.	= Uppermost stage in valleys of Derwent headstreams.		(Lewis, 1954)
	On East Moor 930–1,100+ ft. (280–330 m.)		
	Valley-in-valley forms many stages in valleys of Wye, Derwent, Amber.		
	(Linton, 1956)		

SOUTHERN FLANKS

Dove	700–800 ft. (210–240 m.) 580–600 ft. (175–180 m.) 460–495 ft. (138–148 m.) 360 ft. (108 m.) 320 ft. (96 m.) 280 ft. (84 m.) Upper Hilton Terrace to 90 ft. (27 m.) above alluvium Lower Hilton Terrace to 60 ft. (18 m.) above alluvium Beeston Terrace to 30 ft. (9 m.) above alluvium	Derwent	Erewash
	Trent	(Clayton, 1953b)	

Derwent system at the Upland Surface stage accounts satisfactorily for the isolation on their northern sides of the summits of Lose Hill, Win Hill and Crook Hill which rise sharply from a general 1,200 feet (360 m.) plain. Since the hills encircling the reconstructed basin belong to the Summit Surface and are devoid of any signs of abandoned, older valleys, it follows that the reconstructed drainage was inherited from that surface. Thus in the High Peak both the Summit Surface and the upper valley stage representing the Upland Surface were produced before the postulated eastward-flowing drainage system was disrupted and the Moscar col abandoned.

A similar reconstruction of an eastward-flowing Wye on the Upland Surface has been made. The trunk followed a line from Buxton to Monsal

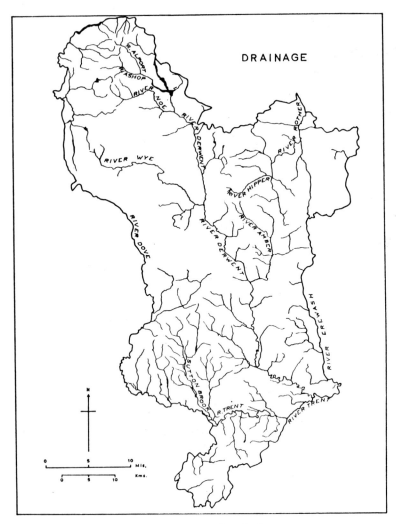

Fig. 4. Drainage of Derbyshire.

Head, thence over the site of Baslow and across East Moor by the Leash Fen col (930 feet; 280 m.).

It is generally agreed that the Summit and Upland Surfaces were formed during the Tertiary, probably after the main mid-Tertiary earth movements. But satisfactory evidence for their age is lacking. They are certainly pre-Pleistocene insofar as their production required relatively long periods of stable base-levels and their dissection is demonstrably Pleistocene. Prior to this dissection the postulated eastward-flowing Wye and Upper Derwent systems would have been disrupted and the main outlines produced of the existing integrated Derwent–Wye drainage pattern. The river captures necessary to effect the disruptions are considered to have been associated with the rejuvenation of the Trent and the extension headwards of the Lower Derwent along shale outcrops on the Upland Surface to divert the Wye at Monsal Head, and subsequently to capture the Upper Derwent system (Linton, 1956; Straw, 1968).

Since the production of the Upland Surface the Southern Pennines appear to have experienced relatively rapid but intermittent uplift, as a consequence of which base-level has fallen through some 900 feet (270 m.). This is expressed by the succession of valley-in-valley forms related to the main drainage lines (Fig. 5). Open, high-level valleys (UV) mark the first stage of incision of the

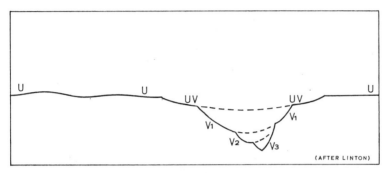

(AFTER LINTON)

Fig. 5. Elements in the denudation chronology of the limestone upland of the Peak.

present valleys below the Upland Surface (U); thereafter rapid downcutting produced the sharp breaks of slope bounding the main valleys (V1) which were subsequently deepened in stages indicated by valley-side benches and valley-floor terraces (V2, V3). All of these valley forms date from the Pleistocene; their relations with glacial and periglacial (solifluction) deposits have been elucidated in the Wye valley below Monsal Dale and in the Noe valley (Waters and Johnson, 1958; Straw, 1968).

Comparable suites of surfaces recur on the Pennine flanks where they bevel strong folds in the Lower Coal Measures (Lewis, 1954), and to the south of the upland (Clayton, 1953b). Here K. M. Clayton recognized a complex pattern over the Dove–Derwent interfluve and less extensively between the Derwent and Erewash (Table I). The highest, a little above 700 feet (210 m.) forms a bench around the upland, broad between the Dove and Derwent and narrow between the Derwent and Amber. Below it a sequence of base-levelled surfaces ranges from a height of nearly 600 feet (180 m.) down to the Trent terraces. Formed by an interrupted fall of base-level, the surfaces slope with the present river valleys. Their disposition suggests that the Trent–Lower Dove shifted southward with each lowering of base-level. Southward shift of the trunk stream seems to have been responsible for the north-south alignment of its left-bank tributaries which extended their courses with each

c

movement. This lateral movement of the main river is undoubtedly an example of simple uniclinal shifting towards the centre of the Stafford–Needwood syncline. As on the upland so on its eastern and southern flanks the lower surfaces have been produced during Pleistocene time, the lowest of them being intimately associated with glacial and fluvio-glacial deposits.

It is generally agreed that three glaciations have affected the county, the second being the most significent geomorphologically. Evidence of the earliest, the Lowestoft (East Anglia), Ante-Penultimate (Zeuner) or Elster (Northern Europe) according to the terminology used, is in the form of outwash gravels at $c.$ 700 feet (210 m.) near Blackwall north-west of Derby and till beneath more recent drift farther south. This ice may have moved across the southern part of the area from the north-west, but little is known of its morphological effects. The second ice advance i.e., the last advance of the Older Drift ice, correlated with the Gipping, Penultimate or Saale, was responsible for most of the till in Derbyshire and, during its retreat, for the fluvio-glacial gravels of the upper Hilton Terrace and outwash in the lower Wye valley. Ice moved southwards both east and west of the Southern Pennines and spread across southern Derbyshire, that from the east taking flint drift to the Dove and beyond. Although much of the upland escaped glacierization, an ice lobe did extend eastward down the Wye valley to Rowsley and thence down the Derwent valley depositing till containing Lake District erratics and outwash on the most extensive and highest terrace, the Pilsley Terrace, in the Great Longstone–Bakewell–Youlgreave area (Straw and Lewis, 1962; Straw, 1968). The third ice advance, the major Newer Drift advance in the last glaciation, did not cover southern Derbyshire; but outwash from Irish Sea ice to the west appears to grade into the gravels of the Beeston Terrace which may therefore date from the retreat phase of the Newer Drift. The Beeston gravels have however been ascribed to various dates between the last (Eemian) interglacial and the New Drift interstadial (Clayton, 1953a, Posnansky, 1960). In the Derwent basin, gravel spreads near Derby and Belper and widespread solifluc-tion deposits are also referred to the Newer Drift glaciation.

During the Pleistocene cold phases large parts of Derbyshire were clear of ice and their landforms show the effects of modification by cryo-nival processes. Even during the Saale glacial phase ice on the upland seems to have been confined to the Wye and lower Derwent valleys, and no ice penetrated any part of the county during the Newer Drift phases. Consequently the effects of frost processes are virtually ubiquitous. Bedrock exposures were broken down by gelifraction, the products of which are seen as angular fragments in superficial deposits and particularly in the cemented screes in the limestone dales (c.f. Prentice and Morris, 1959); and there was a great re-distribution of waste by solifluction and related geliturbation processes. Soliflual deposits or head are known from every part of the area, though the drift sheets of the 1:63360 geological map give little indication of their extent. Nor is information available in the literature in respect of the number of soliflual spreads at any one site, their nature and composition, or their age. Therefore the following tentative conclusions are based on observations connected with research that is currently being carried out.

On the gritstone plateaux and the long shale slopes below the edges that overlook the terraced floor of the Derwent valley between Baslow and Hathersage several heads have been distinguished. Two heads are common on the lower portions of the gentle valley sides and even on the terrace treads: a lower head of brown, fine, weathered material containing a small but variable proportion of medium to small, sub-angular pieces of Millstone Grit; and an upper head of grit blocks and boulders. The source of the upper head is obvious; the blocks and boulders can be followed up to the foot of the edge which shed them and in places, as on Bar Brook, they form a veritable block field. The sheet of lower head on the other hand wedges out upslope. Wherever it is examined and sampled it exhibits no detectable variations in the degree of weathering from top to bottom, but has the appearance of a regolith that has moved after being weathered. It is interpreted as part of a pre-existing weathering profile which presumably formerly overlay the now exposed free face. Its solifinal transfer downslope obviously preceded the exposure of an incipient edge from which the blocks in the upper head were derived. Its removal would also release the rounded boulders or corestones which form so distinctive an element in the upper head. These observations lend support to Linton's views on the evolution of a Pennine edge (Linton, 1964).

Periglacial modification of interfluve areas and valley sides occurred during each of the cold phases well into Late-Glacial times. In the valley of the Burbage Brook an Allerød soil horizon (dated $11,590 \pm 360$ years B.P.) separates heads belonging to Zones I and III. On the evidence of the solifinal material it appears that the periglacial metamorphosis was accomplished through the mass transfer of pre-existing waste by solifluction and the exposure of bedrock surfaces at the base of the zone of pre-periglacial weathering. Once exposed, jointed rocks would undoubtedly be subjected to selective frost weathering, but the very small proportion of sound, frost-broken material in the solifinal spreads would seem to indicate that the amount of bedrock lowering was limited. However the total effect of the mass stripping in each periglacial (cold) phase of regoliths prepared in the interglacial or interstadial (warm) phases undoubtedly contributed to the considerable wasting of interfluves during the whole of the Pleistocene. The vertical extent of a tor is a measure of the minimum amount of lowering of the surface from which it rises. Consequently the definition of high-level planation surfaces within narrow altitudinal limits, particularly on the gritstone plateaux, has no validity. Existing nearly flat interfluves, though initiated near base-level, are not the terminal surfaces of pre-Pleistocene planations. They are more appropriately interpreted as surfaces at the base of the zone of weathering, stripped of their mantles of waste; and their very low gradients are indicative only of the minimal slopes across which the solifinal transfers were effected.

It is becoming increasingly apparent that throughout Derbyshire the end-Tertiary landform of extensive planation surfaces was very subdued indeed and its dissection by successively rejuvenated streams was initiated by Pleistocene changes in the relative levels of land and sea. The steep slopes that were created by stream incision encouraged land slipping and other mass movements. But the progressive adaptation of surface form to underlying structure over entire interstream areas would appear to be the result of differential areal degradation. This was effected by selective bedrock decomposition during mild, interglacial phases, with the production of deeply weathered regoliths on the more susceptible outcrops like the Millstone Grit and, it must be presumed, relatively thin waste mantles on sedimentary formations of stable secondary minerals, and by mass removal of pre-existing debris by soliflual processes in the tundra-like bioclimatic conditions of the cold phases.

REFERENCES

Clayton, K. M. (1953a). 'The glacial chronology of part of the Middle Trent Basin', Proc. Geol. Assoc., 64, 194–207.

(1953b). 'The denudation chronology of part of the Middle Trent Basin', Trans. Inst. Brit. Geogr., 19, 25–36.

Fearnsides, W. G., et al., (1932). 'The geology of the eastern part of the Peak District', Proc. Geol. Assoc., 43, 152–191.

Lewis, G. M. (1954). 'Evolution of the Drainage in the Don Basin', unpublished thesis, University of Sheffield.

Linton, D. L. (1956). 'Geomorphology', in Sheffield and its Region (ed. D. L. Linton), British Association (Sheffield), 24–43.

(1964). 'The origin of the Pennine tors', Zeitschr. fur Geomorph., 8 (Sonderheft), 1–24.

Pigott, C. D., (1962). 'Soil formation and development on the Carboniferous Limestone of Derbyshire', J. Ecol., 50, 145–156.

Posnansky, M. (1960). 'The Pleistocene succession in the Middle Trent basin', Proc. Geol. Assoc., 71, 285–311.

Prentice, J. E., and Morris, P. G. (1959). 'Cemented screes in the Manifold valley', East Mid. Geog., 2, 16–19.

Straw, A. (1968). 'A Pleistocene diversion of drainage in north Derbyshire', East Mid. Geog., 4, 275–280.

Straw, A. and Lewis, G. M. (1962). 'Glacial drifts in the area around Bakewell, Derbyshire', East Mid. Geog., 3, 72–80.

Waters, R. S. and Johnson, R. H. (1958). 'The terraces of the Derbyshire Derwent', East Mid. Geog., 2, 3–15.

Chapter Four

CLIMATE

BASICALLY the climate of Derbyshire should be identified in relation to factors dependent both on the regional position of the county with respect to Britain as a whole and its own local physical diversities. Though not coincident with the broadest sector of Britain, nonetheless the county occupies a central position giving rise to less maritime characteristics than are present in more peripheral coastal regions. The ranges of temperature, both seasonal and diurnal, are more pronounced, and for the county as a whole the period liable to air frosts is extended from mid October to early or mid May—

Table 2. Days of Fog at 09.00 h. G.M.T.

	Jan.	Feb.	Mar.	April	May	June	July	Aug.	Sept.	Oct.	Nov.	Dec.	Year
Chapel-en-le-Frith													
1960	3	2	2	0	0	0	0	0	1	3	1	4	16
1961	3	2	2	3	0	0	0	0	0	0	3	4	17
1962	4	1	2	0	0	0	1	0	0	4	5	5	22
1963	2	2	1	3	0	0	0	1	0	2	2	0	13
1964	1	6	1	0	0	0	1	–	0	4	1	4	–
1965	6	3	4	0	0	0	1	2	2	2	5	–	–
Buxton													
1960	8	7	10	2	0	2	0	4	1	10	5	7	56
1961	13	7	1	6	0	0	0	0	6	1	8	16	58
1962	6	4	7	2	1	0	3	0	5	12	11	10	61
1963	18	14	12	18	3	8	4	4	8	9	11	8	117
1964	11	14	9	4	1	1	2	4	2	9	11	8	76
1965	5	5	11	2	1	0	2	2	9	12	6	6	61
Chesterfield													
1960	1	3	3	1	0	0	0	0	1	4	3	6	22
1961	2	2	4	1	0	4	4	2	11	7	13	13	63
1962	5	3	12	4	5	1	4	0	8	17	9	13	81
1963	4	12	11	3	1	1	1	0	9	9	–	–	51
1964	12	8	10	0	0	0	0	0	2	7	7	3	49
1965	0	0	2	0	0	0	0	2	0	11	3	3	21
Derby													
1960	6	6	0	1	0	0	0	0	3	6	8	12	42
1961	4	3	2	1	0	0	0	0	1	7	13	14	45
1962	5	0	6	0	0	0	0	0	2	9	8	10	40
1963	7	12	5	0	0	0	0	0	1	0	7	5	37
1964	9	4	3	0	0	0	0	0	3	9	5	6	39
1965	0	0	5	1	0	0	0	0	0	8	1	4	19

features shared with the Midlands generally, and with increasing severity in the Pennine sector of north Derbyshire. The county falls within a sector of Britain not well endowed with sunshine. An annual average of about 80–90 days without sunshine and only 30–40 days with prolonged sunny weather characterizes the region generally, and this may be contrasted with the conditions along, for example, England's southern fringe which reveal an average of only 60 days without sunshine, and 60–70 days of prolonged sunny weather (Climatological Atlas, 1952). The higher incidence in Derbyshire of days with obscurity can be accounted for by both hill fog coincident with low cloud base, and high cloud frequency in the northern Pennine sector, and industrial haze and fog in more low-lying terrain, particularly near urban areas in the south and east. Table 2 (days with fog at 9.0 a.m.) emphasises the local importance of hill fogs within the highland zone throughout the year. Buxton shows no month with freedom from fog for all the years for which records are given and in 1963 at this station there was fog on very nearly a third of the days of the year. Lowland stations, such as Mackworth and Derby, on the other hand record a strong seasonal incidence with no fog recorded in any year for the four months May to August.

Table 3. Smoke and Sulphur Dioxide Air Pollution ($\mu g/m^3$) April 1966–March 1967

(a)=daily mean: Smoke. (c)=daily mean SO_2 – =insufficient number
(b)=extreme reading: smoke. (d)=extreme reading SO_2 of readings

		April	May	June	July	Aug.	Sept.	Oct.	Nov.	Dec.	Jan.	Feb.	Mar.
Derby ..	(a)	163	59	39	34	46	135	217	257	215	291	164	108
	(b)	269	164	83	80	93	348	422	605	687	508	442	369
	(c)	178	105	89	56	66	167	214	241	201	273	200	119
	(d)	312	167	148	86	123	418	354	516	594	451	430	353
Ilkeston ..	(a)	128	94	56	52	61	167	256	277	193	290	185	90
	(b)	263	167	131	128	183	531	542	809	659	557	738	229
	(c)	114	108	118	81	68	136	171	205	178	227	168	113
	(d)	214	161	178	169	171	295	297	485	540	419	415	217
Morley ..	(a)	–	–	58	48	53	149	–	–	–	207	131	–
	(b)	162	137	99	122	88	439	248	339	535	477	355	167
	(c)	–	–	107	75	86	177	–	–	–	224	161	–
	(d)	289	269	184	129	188	472	190	461	354	402	379	175
Belper ..	(a)	133	95	78	–	79	143	190	234	188	257	161	109
	(b)	210	175	152	124	151	245	302	699	719	442	338	191
	(c)	130	99	89	–	84	106	147	172	160	193	142	99
	(d)	188	160	142	108	179	174	196	342	502	348	281	177
Ripley ..	(a)	138	124	98	60	43	136	178	153	136	166	138	69
	(b)	274	336	182	124	116	342	313	528	651	347	603	223
	(c)	–	154	96	73	79	136	206	195	175	246	180	89
	(d)	231	668	202	135	152	306	416	541	531	427	336	173
Alfreton ..	(a)	161	87	80	56	79	181	235	259	163	278	138	65
	(b)	244	167	156	122	191	513	536	764	951	625	738	247
	(c)	94	77	60	73	73	87	51	55	73	58	62	90
	(d)	149	148	100	140	142	144	92	185	284	192	158	174
Chesterfield	(a)	143	50	44	44	53	110	161	199	134	201	113	58
	(b)	359	131	81	105	118	269	292	716	822	524	634	246
	(c)	145	101	87	79	90	106	146	170	134	208	150	85
	(d)	306	198	132	120	155	178	249	357	478	340	424	193
Bolsover	(a)	60	46	37	36	35	77	108	60	–	–	–	44
	(b)	148	101	83	95	128	245	189	219	–	–	230	158
	(c)	104	69	63	76	71	41	168	103	–	–	–	48
	(d)	326	355	239	307	290	103	553	226	–	–	1291	127

Local variations as between rural and urban land use can be important contributory factors in accounting for some very local climatic diversity, and air pollution has significance in this regard. Whilst there is an absence of pollution sampling stations for a large part of the county, records over the eastern sector (*see table* 3) are indicative of a broad zone of polluted air coincident with the industrial lowlands. Conditions in the rural uplands to the west and north cannot so easily be assessed, though, locally, some farmers record the unduly rapid corrosion of fences and farming implements, even in the heart of the High Peak, in country that is very thinly populated but is set midway between the industrial areas of south Yorkshire and Lancashire. An index of the differences in pollution can be gauged by a comparison of statistics for Buxton and Derby recorded for example, for March, 1967. In Buxton the average daily smoke sampling for the month was only 10 μg/m³ and the highest daily reading was only 18 μg/m³. Average sulphur dioxide sampling was 39 μg/m³ and the highest daily reading 59 μg/m³. The equivalent values for a site in Derby for the same month gave a daily average and extreme daily value for smoke at 108 μg/m³ and 367 μg/m³ respectively, and for sulphur dioxide of 19 μg/m³ and 352 μg/m³ (data compiled from The Investigation of Air Pollution, 1967). Particularly during winter months the industrial lowlands may be seriously affected by fog and haze that forms in the shallow valleys that are set in the subdued lowland relief and encouraged in this by the higher winter air pollution (notably of sulphur dioxide). But, important though these influences may be a comparison of the monthly differences in the number of hours of sunshine received at stations such as Morley, Buxton and Derby (*see tables* 3 *and* 4), highlights the overall disadvantageous position of the highlands as compared with the lowlands in response to more general climatic controls. Thus Table 4, giving means of the monthly hours of sunshine for December (1956-60) shows that the upland site of Buxton had only one third of the sunshine recorded at Morley, and the industrial site at Derby had only three-quarters. In three of the five years 1956-60, only 5·5–6·9 hours of sunshine were in fact recorded at Buxton during the month of December.

Table 4. **Mean number of hours of sunshine to nearest hour 1956-60**

	Jan.	Feb.	Mar.	Apr.	May	June	July	Aug.	Sep.	Oct.	Nov.	Dec.
Morley	45	63	81	141	197	209	161	152	114	90	49	30
Derby	39	57	86	40	195	210	159	150	112	83	42	23
Buxton	22	51	73	118	175	196	128	121	110	79	30	10

From most points of view relief is the dominant influence in modifying the climate locally. Differences of altitude alone—in a range of from 2,000 feet in the High Peak to less than 200 feet in the Trent basin in the south of the county—give rise to a substantial general diminution of temperature across the county from south to north-west. As the Tables indicate, southern Derbyshire has warmer summers than the north and somewhat warmer winters, but during extreme weather conditions less favoured localities in the south can be as cold as localities at much higher altitudes in the north. In 1959 temperatures at Derby ranged from a monthly minimum of —2°C. (28°F.) in January to a maximum of 23·1°C. (73·7°F.) in July. For the same months

Table 5. Hours of Sunshine

DERBY   Grid Ref. SK 359367.   Altitude 159 feet (48 m.)

	Jan.	Feb.	March	April	May	June	July	August	Sept.	Oct.	Nov.	Dec.	Total
1960	21·5	54·8	48·5	141·6	165·8	266·4	121·8	142·6	100·6	44·1	45·4	44·2	1197·3
1961	27·9	51·8	131·1	67·6	187·8	216·4	157·8	176·8	121·1	107·3	54·2	36·1	1335·9
1962	53·5	66·9	109·5	138·4	151·1	221·0	117·6	145·3	106·2	85·4	40·2	38·9	1274·0
1963	47·4	53·6	87·2	107·3	176·3	208·2	178·1	117·2	120·7	67·9	40·0	39·7	1243·6
1964	27·8	49·3	53·6	107·0	195·1	123·3	214·2	199·4	179·6	115·1	52·3	40·1	1356·8
1965	58·9	33·4	125·8	167·6	169·7	201·1	113·5	181·6	96·3	73·9	75·3	54·6	1351·7

MORLEY   Grid Ref. SK 385401.   Altitude 350 feet (107 m.)

	Jan.	Feb.	March	April	May	June	July	August	Sept.	Oct.	Nov.	Dec.	Total
1960	25·6	58·6	55·9	147·9	169·8	268·5	127·6	144·0	109·6	47·6	46·9	56·7	1258·8
1961	31·2	52·2	138·6	66·4	183·7	218·3	160·6	179·2	120·8	111·3	64·9	52·3	1379·5
1962	70·6	69·2	114·3	150·8	150·8	191·8	97·2	133·5	108·2	95·0	43·8	47·5	1272·7
1963	50·9	61·7	91·6	105·8	190·6	206·0	168·2	110·3	131·8	66·4	39·9	40·9	1264·1
1964	30·0	50·6	55·3	98·7	194·5	116·1	193·4	199·6	175·7	115·2	53·6	48·3	1331·0
1965	61·2	29·2	124·6	159·8	156·5	186·7	105·2	175·1	92·8	78·0	72·0	50·8	1291·9

BUXTON   Grid Ref. SK 059738.   Altitude 1,007 feet (307 m.)

	Jan.	Feb.	March	April	May	June	July	August	Sept.	Oct.	Nov.	Dec.	Total
1960	9·4	52·7	47·5	128·0	155·9	263·9	115·0	115·9	110·4	49·1	30·4	12·6	1090·8
1961	10·9	35·9	121·8	74·1	150·0	180·9	140·7	157·3	119·5	92·8	32·1	11·4	1127·4
1962	24·2	46·4	102·5	151·0	131·2	173·5	93·4	130·0	90·9	68·9	19·7	10·8	1042·5
1963	28·3	55·2	73·8	88·8	168·9	199·2	179·3	102·3	101·0	64·1	23·9	10·9	1095·7
1964	14·7	33·1	34·3	72·2	152·5	82·3	—	—	125·4	83·7	17·6	15·0	—
1965	18·0	12·8	98·3	100·0	104·7	123·5	88·5	165·0	87·1	86·0	56·7	24·8	965·4

Fig. 6. Climatological stations and reference points.

at Buxton the range was from —3°C. (26·9°F.) to 19·6°C. (67·3°F.) (i.e. a difference of only 1°C. (1·1°F.) in the winter extreme, but of 3·5°C. (6·4°F.) in the summer).

A noteworthy peculiarity of the climates of all highland areas in Britain is the increasingly pronounced rapid deterioration in temperature that occurs with increasing altitude (see Manley, 1945). This feature is in fact well exemplified in north Derbyshire, particularly where, in the High Peak, altitudes approach 2,000 feet. Some continuous records of temperature were obtained for stations at 2,000 feet (610 m.), 1,700 feet (518 m.), 1,000 feet (305 m.) and 700 feet (213 m.) in the neighbourhood of the Kinder Scout plateau and the adjacent Ashop valley, immediately to the north. Whilst these records represent no more than a sample evaluation only taken over the period of one year they are nonetheless indicative of the rate of change that locally may occur in such terrain. For each of these stations the percentage of time was assessed during each month when temperatures were below both, 5·6°C. (42°F.) (i.e. growing threshold temperatures) and 0°C. (32°F.) (i.e. with air frost). The year to which these readings related was not one of unduly severe or 'atypical' weather.

The monthly percentages of time when air temperatures were below the growing threshold 5·6°C. (42°F.) at 2,000 feet were very pronounced. At this altitude time losses ranged from 11–25% during the summer months (June to September) to 50–55% in May and October and more than 80% from December to April—the maximum loss of over 90% of growing time being recorded during January and February. At 1,700 feet the winter loss was slightly less, though still severe, exceeding 80% from December to February, 35–40% in May and October and 60–50% in April and November; the loss was however negligible during midsummer months. At the valley sites at 1,000 feet or 700 feet below Kinder Scout, the loss of growing threshold time in winter was appreciably less, being no more than 60–70%, with spring and autumn losses only 20–30% (i.e. roughly half the time loss at 2,000 feet). The percentage of time lost under conditions of air frost followed a slightly different pattern. No site during this year recorded frosts between June and September, and winter months showed expected increases in frost time with altitude. The monthly maxima which occurred in February ranged from 30% of time with frosts at the lowest sites to over 40% at 1,700 feet and over 50% at 2,000 feet. In December on the other hand, frost incidence occupied 32% of time at 2,000 feet as compared with 27% at 1,700 feet but no more than 7% at 1,000 feet and 9% at 700 feet. A slight increase in the length of frost time at the lowest site might seem paradoxical, but in fact, for the year in question was a feature also characteristic (often in more marked degree) of the months March, April and May, whilst in November 19% frost time was recorded at 700 feet and this exceeded that at 1,700 feet (only 11%) and all but matched that at 2,000 feet (20%). These facts are reminders of the complexity of meso-temperature patterns in regions of bold relief, where there may be unexpected prolongation of periods of low temperature, and even of severe killing frosts, in valleys, though at the same time the general condition (as assessed by growing threshold temperature losses) is markedly less severe in the valleys that at summit altitudes.

The above seemingly contradictory incidence in fact reflects another important control by relief and one which is frequently to be observed under some weather conditions, when stagnant air circulation favours local air drainage into valleys and areas of lowest relief. Particularly in association with anticyclonic weather, these conditions give rise to inversions of temperature in all parts of Derbyshire, but their local effects can be very conspicuous and prolonged in the more broken and higher relief of the north-west of the county (Garnett, 1951, 1956). Continuous temperature recording at a number of sites in the High Peak has revealed how prolonged and severe these valley inversions may be and how they vary in intensity and duration from season to season with changing sun altitudes. The records also have shown how far there may be differences in the temperature regimes in adjacent valleys according to the varying local conditions of the physical environment present in each case. Equally the diurnal temperature changes and inter-relationship

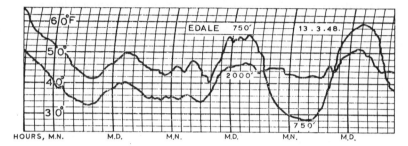

Fig. 7. Cyclonic and anticylonic regimes on successive days in Edale and the Ashop valley.

of summit and valley sites can show very marked contrasts day by day according to the changing weather controls. Some examples as recorded in the High Peak illustrate these seeming anomalies. Thus as Fig. 7 illustrates, during a day in March of normal cyclonic weather, temperatures in the vale of Edale at 750 feet reached 10°C. (50°F.) when at 2,000 feet on Kinder Scout they were no more than 4·4°C. (40°F.)—a difference indicative of a rate of fall of temperature with altitude that is severe. Yet on the following day, under anticyclonic weather, with calms and clear skies temperatures at the valley floor fell at night to —2·2°C. (28°F.), whilst at 2,000 feet these exceeded 5·6°C. (42°F.) continuously. The diurnal pattern of temperature inter-relationships over summit and valley were thus reversed on successive days.

In the Ashop valley, near Kinder Scout an assessment was made of the duration and severity of inversions at different seasons of the year, with respect to sites at 700 feet and 1,700 feet. During the summer months inversions did not persist by day, for normal lapse was then possible and valley temperatures were usually about 2·2-3·3°C. (4°-6°F.) higher than at the summit station. At night however between the hours of about 8 p.m.–8 a.m. during summer, inversion conditions gave rise to night minima in the valley 3·3-4·4°C. (6°-8°F.) lower than that at 1,700 feet. With waning sun altitudes in the autumn (e.g. in October), the inversion period lasted from 6-7 p.m. until 10-11 a.m. and night minima in the valley at 700 feet fell to about 12·7°C. (23°F.) lower than temperatures at the higher station. By the end of November, and continuing through the midwinter months, sun altitudes were so low, insolation so weak, and hill shadows falling across the valley so deep and continuous, that inversions remained unbroken by day and night and persisted continuously for several days—so long as anticylonic weather occurred (Garnett, 1951, 1956). In these circumstances very prolonged and severe valley frosts developed with valley-minima falling to —7·7°C. (18°F.) whilst temperatures at 1,700 feet were continuously above 4·4°C. (40°F.). By the spring (e.g. during mid March and April) inversions were once more broken by day, approximately between the hours of 8 a.m. until 4 p.m., though the ensuing night inversion was then still quite severe, with valley temperatures 6·6°C. (12°F.) lower than at the higher station, and registering 4·4°C. (8°F.) of frost in March and 1·1°–2·2°C. (2°–4°F.) of frost in April.

Comparisons of records for sites in adjacent valleys showed that under the same weather regime appreciable local modifications of temperature conditions can occur. Thus at 800 feet in the Ashop valley night minima were often 2·2°C. (4°F.) lower than at a site at the same altitude in the adjacent valley beside the Derwent valley reservoirs. Again, a comparison of conditions on the same occasions in Monsal Dale, the vales of Hope and of Edale showed interesting variations. From the standpoint of local climate Monsal Dale depicts conditions in a very narrow well screened limestone dale fed by air drainage from a limestone plateau at 1,000–1,500 feet. The Hope valley on the other hand presents a widely opened plain rimmed by sharply rising relief both of shale-grits and limestone. Edale, a somewhat narrower valley in shale-grit country is enclosed by lofty ridges and plateaux rising to 2,000 feet. Under anticyclonic weather the daily minima at Monsal Dale were always 2·2°–3·3°C. (4°–6°F.) warmer than in the other two valleys and often escaped night frosts that persisted elsewhere. Edale was persistently the coldest of the three valleys, though at times very severe killing frosts developed on the widely open vale of Hope, well exposed to night radiation. Thus a record in February, during a prolonged phase of temperature inversion with continuous night frosts occurring on the plain, gave evidence of minima of —7·7°C. (18°F.) and —8·3°C. (17°F.) on three successive nights when, only 270 feet upslope on the flanks of Lose Hill, lowest temperatures in the range —1·1°– 0°C. (30°–32°F.) were the exception, occurring only for very short spells. In Derbyshire, therefore, the surface form of the land and its characteristics are of no little importance in local climatological study, and at times may be of greater significance than the factor of altitude, as can, finally, be indicated by the following data. A comparison was made of temperatures recorded at two sites in the Ashop valley; both were located at 1,000 feet, but one was on the floor near the head of the valley, whereas the second was high on the valley flanks where the floor lay at 600–700 feet. The site on the valley floor was consistently 3·3°–4·4°C. (6°–8°F.) colder than that upslope, during anticyclonic and other weather conditions inducing stagnant circulation—though both stations were identical in altitude.

Through its influence on both the patterns and intensities of insolation, as has been noted above, relief can cause indirect modifications of temperature conditions through the prolongation of inversions in well shaded valleys into which direct sunlight cannot readily penetrate. These restrictions can however also have more direct importance, particularly with respect to rates of soil warming and drying, and ground temperatures generally. In the bold hilly country of north Derbyshire, the relation of sunlight to aspect and degree of slope, and to the pattern of hill shadows that the relief induces, becomes a matter of increasing importance both at higher altitudes and where valleys are orientated from west to east. Thus, in midwinter, the admission of direct sunlight to the slopes of the south wall of the vale of Edale is all but precluded by hill shadows. Insolation over most of the valley floor is severely restricted, and not until above heights of about 800–900 feet on the north flank of the vale can sunlight freely reach the surface during favourable weather conditions. Surfaces situated well upslope and facing south are thus at a great advantage with respect to potential sunlight incomes—both duration and intensity during the winter months (see Garnett, 1939). In this regard it is of interest to record that in the Hope valley, land adjacent to the hamlets of Aston and Thornhill—located upslope on a south face that escapes hill shadows—is regarded with favour since (as a local farmer has described to the writer), the land there is 'a topcoat warmer' than in the valley, and 'where the sun never sets'.

The local variations in insolation are not without importance in relation to soil temperatures, as can be seen with dramatic effect with the advent of sunshine after a spell of frost or snowfall, leading in the High Peak to patterns of thawed surfaces precisely dictated by aspect, degree of slope, landform, and relief. The effects at the higher altitudes, above 1,500 feet can persist for some weeks as snow-melt on northerly aspects is retarded whilst southerly aspects are warmed and dried. Even at a depth of one foot below the surface the effect of aspect is still evident as the daily records of earth thermometers have shown when maintained at stations at 800 feet (244 m.) on either side of the Ashop valley, where north and south aspects are sharply defined and juxtaposed. Over the period of a year it was of interest to note that not only were maximum and minimum temperatures higher by at least 2·7°–2·2°C. (5°–4°F.) on the southward face, but also there was an appreciable increase in the number of days with temperatures above the 'growth threshold' of 5·6°C. (42°F.). On both slopes, however, the extreme monthly minima at one foot depth (recorded in February) were above freezing temperatures (i.e. 1·1°C. (34°F.) and 2·2°C. (36°F.) respectively) despite the several records of very severe killing air frosts that were obtained four feet above ground level close by in the Ashop valley over the same period.

The lengthening on the southward slope of the period of growing threshold temperatures at one foot depth amounted to about 45 days (225 as compared with 180 days on the northward facing slope). These results may be compared with the assessment of 235 days for lowlands in the Midlands more generally. On the basis of calculations from air rather than soil temperatures, using monthly means in relation to the threshold *air* temperature of 5·6°C. (42°F.), the duration of the growing season for upland *versus* lowland areas in Derbyshire has been assessed as of the order of ±200 days in the High Peak (Buxton 202 days) and 230–240 days in the lowlands (Belper 238 days), indicative of a potential increase in the growing season of more than one month at the lower altitudes.

In terms of temperatures, the High Peak country has, therefore, a severe climate, and this is well summarised in the long period data for Buxton, as given in Table 6, typifying conditions at about 1,000 feet in a region where most of the surface is appreciably above this altitude. In each of six months at least 11·1°C. (20°F.) of frost has been recorded, and in each of three months the absolute minimum temperature has fallen below −17·7°C. (0°F.). Only in the month of August is there no record of frost occurrence. Yet, even at 1,000 feet, in summer, quite hot spells may occur. The means of the monthly maxima at Buxton exceed 21·0°C. (70°F.) throughout the summer (May to

Table 6. Buxton records, altitude 1,007 feet (from Garnett, 1956)

	Jan.	Feb.	Mar.	April	May	June	July	Aug.	Sept.	Oct.	Nov.	Dec.
Mean Monthly temperatures (°C) (1926–50)—												
	2·2	2·3	4·0	6·2	9·2	12·3	14·2	13·8	11·6	8·2	4·9	3·0
Mean of the daily maxima (1926–50)—												
	4·5	4·7	7·2	9·8	13·4	16·4	18·0	17·5	15·0	11·1	7·3	5·1
Mean of the daily minima (1926–50)—												
	0	0	0·7	2·7	5·0	8·2	10·5	10·1	8·2	5·3	2·6	0·8
Means of the monthly extreme maxima (1869–1954)—												
	9·3	12·1	13·2	17·1	21·5	24·4	25·0	23·8	21·0	16·6	12·1	9·9
Means of the monthly extreme minima (1869–1954)—												
	−9·4	−8·3	−6·6	−3·9	−1·1	1·6	4·4	3·8	0·5	−3·9	−5·0	−8·3
Absolute maximum recorded (1869–1954)—												
	13·2	14·8	19·9	22·7	27·3	28·8	31·0	31·0	29·9	25·0	17·1	15·6
Absolute minimum recorded (1869–1954)—												
	−17·8	−23·5	−17·2	−13·3	−7·1	−1·3	−0·6	0·5	−2·2	−7·7	−11·1	−20·0
Mean monthly totals of bright sunshine (hours) (1921–50)—												
	22	42	89	120	161	176	154	145	111	76	36	14
Mean monthly rainfall (1881–1915) (inches)—												
	4·5	3·8	4·1	2·9	3·1	3·2	3·9	4·4	3·2	4·9	4·7	5·7
Wettest month (inches) (1928–54)—												
	12·1	9·5	5·8	5·5	5·2	5·7	8·7	7·3	7·2	10·1	11·4	9·7
Driest month (inches) (1928–54)—												
	2·2	0·3	0·2	0·5	0·5	0·8	1·7	0·6	0·7	0·9	0·9	0·5
Mean number of rain days (1881–1918)—												
	19	18	19	16	16	14	16	19	15	19	19	21
Mean number of days with thunder—												
	0·2	0·3	0·2	0·9	2	2	2	2	>1	0·3	0·2	0·2
Mean number of days with snow—												
	8	8	8	3	1	0	0	0	0	>1	3	6

September), with absolute extremes for each month ranging from 27·3°–31°C. (81°–88°F.). Not without relevance is the fact that at 1,400 feet on the plateau between Castleton and Chapel, oats can be ripened, which is possibly at the highest altitude in Great Britain.

For the rest of Derbyshire temperature records are less dramatic though it should be noted that on occasion, absolute minima in the winter months may be more extreme at the lowland rather than the highland stations (*see table 7*). Thus in January, 1959, the minimum recorded at Belper (203 feet) was −3·3°C. (26·0°F.), at Chesterfield (298 feet) −2·5°C. (27·4°F.), and at Buxton (1,007 feet) −2·8°C. (26·9°F.). Widespread *regional* inversion of temperature in some depth may at times characterise the lowlands generally, above which the highlands stand as an 'island' of comparatively warmer conditions. Such conditions in the lowlands can bring serious problems in valleys within or adjacent to sources of air pollution which, during such inversions is held low in the atmosphere, spreading extensively horizontally rather than vertically, over the land, and often closely associated with fog.

Table 7 Monthly maximum and minimum temperatures 1960-1965 for selected stations in Derbyshire (In degrees Centigrade.)

DERBY

Grid Ref. SK 359367.   Altitude 159 feet (48 m.)

Monthly maximum temperature

	Jan.	Feb.	Mar.	April	May	June	July	Aug.	Sep.	Oct.	Nov.	Dec.
1960 ..	7·5	7·1	9·1	13·8	17·9	22·2	19·7	19·5	17·0	12·8	10·1	6·9
1961 ..	6·3	10·0	13·4	13·9	15·4	20·2	19·5	20·1	19·9	14·4	9·6	5·3
1962 ..	7·7	7·9	6·9	11·8	14·4	19·1	18·9	18·6	17·0	14·3	8·6	5·3
1963 ..	1·6	2·6	9·6	12·4	15·3	20.0	20·4	18·1	17·4	14·3	10·8	5·6
1964 ..	5·6	7·3	6·5	12·8	18·8	18·0	20·6	20·5	19·3	12·8	10·5	6·5
1965 ..	6·2	6·1	9·7	12·6	16·1	19·4	17·7	19·7	16·3	14·7	7·2	7·3

Monthly minimum temperature

	Jan.	Feb.	Mar.	April	May	June	July	Aug.	Sep.	Oct.	Nov.	Dec.
1960 ..	1·4	0·9	3·1	4·9	8·1	11·2	11·3	10·6	9·0	6·9	4·2	1·5
1961 ..	0·7	4·1	3·6	5·9	6·5	9·8	11·4	11·3	10·8	6·9	2·8	0·8
1962 ..	1·2	1·7	-1·3	3·8	6·3	8·8	11·2	10·7	9·1	6·3	2·5	-0·9
1963 ..	-4·4	-3·1	2·3	4·9	6·7	10·7	10·8	10·7	9·0	7·8	4·9	0·2
1964 ..	0·1	1·7	1·6	5·2	8·9	10·3	12·3	11·3	8·7	4·9	4·7	0·1
1965 ..	1·0	1·5	0·7	4·3	7·8	10·7	10·2	10·7	9·4	7·2	1·9	1·6

BELPER

Grid Ref. SK 346468.   Altitude 203 feet (62 m.)

Monthly maximum temperature

	Jan.	Feb.	Mar.	April	May	June	July	Aug.	Sept.	Oct.	Nov.	Dec.
1960 ..	6·5	7·1	9·1	14·1	18·0	21·7	19·7	19·4	17·7	13·1	10·6	7·2
1961 ..	6·1	9·8	13·7	14·0	15·8	19·9	19·8	19·9	19·7	14·7	9·3	4·7
1962 ..	7·4	7·6	6·8	11·7	14·7	18·7	18·7	17·7	16·6	14·2	8·4	4·7
1963 ..	1·2	2·1	9·6	12·5	15·3	20·0	20·1	17·9	17·6	13·9	10·8	5·9
1964 ..	5·6	7·3	6·7	12·7	18·9	17·6	20·7	20·1	18·1	12·2	10·3	6·3
1965 ..	5·7	6·2	9·5	12·8	16·1	19·6	17·7	19·9	15·9	14·8	7·3	7·4

Monthly minimum temperature

	Jan.	Feb.	Mar.	April	May	June	July	Aug.	Sept.	Oct.	Nov.	Dec.
1960 ..	1·1	0·2	2·6	4·5	7·7	10·5	10·6	10·3	8·6	6·5	3·1	0·7
1961 ..	-1·3	3·4	2·8	4·3	5·4	8·8	9·8	11·2	9·5	6·6	2·6	1·2
1962 ..	1·2	0·9	-1·9	3·6	5·7	7·9	10·6	9·8	8.0	5·3	1·7	-2·4
1963 ..	-5·4	-4·1	1·5	3·8	5·9	9·8	9·4	9·4	9·0	6·2	4·1	-0·3
1964 ..	-0·4	1·0	1·0	4·1	7·5	9·3	10·9	10·6	7·6	3·0	3·5	-0·4
1965 ..	0·1	1·1	-0·2	3·7	6·8	9·9	10·1	9·3	8·3	6·8	1·3	1·1

CHAPEL-EN-LE-FRITH

Grid Ref. SK 056809.   Altitude 732 feet (223 m.)

Monthly maximum temperature

	Jan.	Feb.	Mar.	April	May	June	July	Aug.	Sept.	Oct.	Nov.	Dec.
1960 ..	5·1	5·5	7·6	12·0	16·3	20·2	17·6	17·6	15·4	11·8	8·8	5·3
1961 ..	4·7	8·6	11·3	12·7	13·9	17·7	17·3	17·7	18.1	13.1	7·8	3·7
1962 ..	6·1	5·9	5·4	10·1	12·8	17·0	17·1	16·7	15.2	12·7	7·3	4·5
1963 ..	0·3	1·1	7·9	10·7	13·3	18·0	18·3	16·4	15·9	13·2	9·4	4·6
1964 ..	4·9	5·8	5·7	11·4	16·9	16·1	18·1	18·8	17·2	11·7	9·3	5·5
1965 ..	4·7	4·9	8·2	11·1	14·6	18·4	16·2	18·2	14·6	13·8	6·2	5·6

Monthly minimum temperature

	Jan.	Feb.	Mar.	April	May	June	July	Aug.	Sept.	Oct.	Nov.	Dec.
1960 ..	0·3	-0·3	2·3	3·2	7·1	9·3	9·6	9·2	8·0	6·0	3·2	0·5
1961 ..	-0·4	3·1	3·5	4·8	4·4	7·8	9·3	10·1	9·8	6·2	2·2	-2·1
1962 ..	-0·2	0·3	-2·4	2·3	5·1	7·1	9·4	9·2	7·9	5·7	1·4	-2·2
1963 ..	-4·8	-4·4	1·4	3·8	5·2	9·5	9·3	9·2	7·5	6·4	3·9	-0·8
1964 ..	0·1	0·5	0·2	4·1	7·9	8·7	10·9	8·9	7·7	3·4	3·4	-1·1
1965 ..	0·1	-0·3	-0·6	2·7	6·3	8·6	8·3	9·4	7·6	6·6	-0·1	-0·2

Rainfall, like temperature, shows marked variation over the county as a
whole, with respect to both amounts received and their seasonal incidence
in the northern as compared with southern and eastern areas. Generally
speaking, there is a progressive diminution in the mean annual total with
decreasing altitude, i.e. from north-west to south and south-east. In the

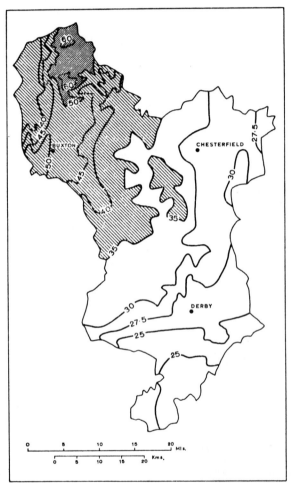

Fig. 8. Derbyshire; Mean annual rainfall in inches.

north of the county (see Fig. 8) in the High Peak, the mean annual rainfall
exceeds 60 inches in the neighbourhood of Kinder Scout and Bleaklow—the two
highest localities—with extensive areas of the adjacent high plateau and
upland surfaces receiving at least 45–50 inches, southward to Buxton and its
neighbourhood. Southward and eastward beyond this, rainfall diminishes

R. H. Hall

1. *Miller's Dale from Priestcliffe Lees; part of the Wye valley with a very rich limestone flora*

R. H. Hall

2. LILY-OF-THE-VALLEY (*Convallaria majalis*); *Chee Dale*

D

R. H. Hall

4. MOUNTAIN MELICK (*Melica nutans*); *Lathkill Dale*

R. H. Hall

3. *Winnats Pass, Castleton*

to less than half these amounts (25 inches ±) near the Trent basin, with 27–30 inches maintained widely over the eastern low plateaux within the county. West of Derby the reduction in rainfall from 30 to 25 inches is sharp but elsewhere in the east the graduation to lower rainfall is slight over the more subdued though still varied relief. In the highland sector however there is much local diversity that is masked in any generalised map, as a comparison of Figs. 8 and 9 will show. As in the case of temperature studies, complex inter-relationships between relief, altitude and rainfall are evident, and again are a reminder of the dangers of climatic generalisation in such environments. In general, at equivalent altitudes, as would be expected, the eastern leeward slopes of the Pennines show, altitude for altitude, a lower rainfall than the west facing windward flanks. Thus within Derbyshire, the western face of the Kinder Scout mass at 1,500 feet receives 56 inches, but at the same altitude on its extreme eastern flank overlooking the Ladybower reservoir, rainfall is only 51 inches, whilst beyond this, and again at 1,500 feet on the eastern flanks of the Derwent valley, rainfall in places is no more than 45 inches.

Paradoxically however, the reverse trend can be traced very locally within this regional pattern, particularly in the areas receiving the heaviest rainfall (Barnes, 1966; Garnett, 1956). Thus at Bleaklow and Kinder Scout, where relief attains 2,000 feet, a mean annual rainfall of 64 or 63 inches respectively is attained. But the precise areas of rainfall maxima in both cases 'overrun' the summit areas and extend well downslope, on the leeward side, to give the heaviest rainfall to leeward and not windward slopes, and to valley heads rather than summit zones. On Kinder Scout this gives rise to quite appreciable differences reversing the trend described earlier. The plateau exhibits remarkably level relief, at 2,000 feet; over the south-west however the rainfall is no more than 54 inches, whereas on the northern margin it reaches 63 inches. The adjacent leeward valleys likewise show unexpected increases; thus the upper Ashop valley floor, at 1,000 feet, receives more rainfall than the western and southern side of Kinder Scout which lies at twice the altitude, and on the more exposed windward side. In the Ashop valley the down valley rainfall gradient (60–48 inches) matches that in Edale of not more than 50–45 inches. Once again, the importance of relief and the details of landform in relation to climatological study is very evident.

The seasonal distribution of rainfall follows the pattern generally applicable to wider regions in Great Britain. In the High Peak and highland sector it is well distributed throughout the year, with pronounced autumn and winter maxima, as the monthly mean values for Buxton demonstrate. Here about 44% of the total rainfall occurs during the months October to January, with spring the driest season and May the driest month. The lowland sector of the county on the other hand shows an increase in the proportion of rainfall during the summer rather than the winter half of the year—a tendency found generally present in passing eastward across the Midlands. The gradation from the highland to the lowland seasonal incidence is however only slight in Derbyshire when related to half yearly percentages; in the High Peak about 40–46% of rain falls in the summer six months as compared with 46%± in central Derbyshire and 50% in a limited sector of the south, in the Trent basin. Table 8 reveals not only how varied may the seasonal pattern be, month by month over individual years, but also how marked may be the

Fig. 9. Rainfall distribution in the High Peak in inches.

Table 8. Rainfall 1960–65 for Selected Stations in Derbyshire (in inches).

DERBY

	Jan.	Feb.	March	April	May	June	July	August	Sept.	Oct.	Nov.	Dec.	Total (")	Total (mm)
1960..	4·88	1·93	1·46	1·33	1·45	2·26	3·13	4·16	4·49	5·11	2·92	3·13	36·25	920·7
1961..	2·91	1·80	·37	2·92	1·35	1·21	2·46	2·63	1·44	2·26	1·38	2·82	23·55	598·2
1962..	2·12	·73	·59	1·77	1·80	·46	1·79	3·09	3·04	·76	1·15	2·03	19·33	491·0
1963..	·90	·48	2·57	2·35	·83	3·91	1·32	3·17	2·47	1·43	4·05	·41	23·89	606·8
1964..	·84	·67	3·56	1·74	1·94	1·95	1·89	2·33	·47	1·67	1·21	2·68	20·95	532·1
1965..	2·68	·83	2·44	1·60	2·06	2·52	2·49	2·31	4·99	·74	2·91	6·32	31·89	810·0

BELPER

	Jan.	Feb.	March	April	May	June	July	August	Sept.	Oct.	Nov.	Dec.	Total (")	Total (mm)
1960..	5·53	2·21	1·91	1·63	1·93	2·20	3·25	4·08	4·12	6·23	3·50	3·37	39·96	1015·0
1961..	2·76	1·85	·34	3·33	1·46	1·06	2·33	2·74	2·02	2·58	1·52	3·21	25·20	640·1
1962..	2·66	1·09	1·21	2·72	2·16	·45	1·80	3·07	3·24	1·08	1·43	2·03	22·94	582·7
1963..	·75	·31	3·58	2·33	1·00	2·97	2·96	3·40	3·43	1·56	4·76	·66	27·71	703·8
1964..	·74	·86	4·45	2·29	2·09	1·99	2·06	2·03	·51	1·48	1·73	2·89	23·12	587·2
1965..	3·29	·89	3·05	1·68	2·41	3·08	2·47	2·10	6·25	·89	3·62	7·44	37·17	944·1

CHAPEL-EN-LE-FRITH

	Jan.	Feb.	March	April	May	June	July	August	Sept.	Oct.	Nov.	Dec.	Total (")	Total (mm)
1960..	6·63	2·84	1·63	2·09	1·92	1·65	7·01	4·27	4·44	5·75	5·63	4·78	48·64	1235·5
1961..	5·78	3·28	2·33	3·51	1·83	2·14	4·50	6·44	2·33	5·72	3·42	2·61	43·89	1114·8
1962..	5·03	3·11	1·39	4·26	3·53	·80	2·20	5·83	5·11	2·40	1·93	5·21	40·86	1037·8
1963..	·55	·61	3·52	3·23	3·72	4·07	1·92	5·72	4·68	3·35	5·92	·59	37·88	962·1
1964..	1·66	1·20	3·90	3·10	2·58	3·41	4·51	3·91	2·10	3·12	2·95	5·83	38·27	972·1
1965..	4·73	·96	2·47	3·47	3·86	2·67	4·39	3·73	8·98	1·14	4·23	11·77	52·40	1331·0

incidence of midsummer rainfall in the lowland sector, where, again, early months in the year may be very dry. This is particularly, but not invariably true of the months February and March. The more pronounced midsummer rainfall is largely associated with increased convective rains at that season, for much of the southern sector of the county lies within or adjacent to a belt across Britain noteworthy for high thunderstorm frequency. It is not therefore surprising to find that, coincident with these general seasonal tendencies, in the lowlands there also occurs a similar seasonal maximum in the occurrence of days with heavy rains. The long period (37 years) records for Belper show that only on four occasions was the daily rainfall more than 2 inches, and that these all occurred in the four months May to August. Of seventeen occasions when it exceeded 1·5 inches, ten were concerned with days in the same summer period, with five alone occurring in July.

A comparison on the other hand of the monthly mean incidence of days with rain shows that for most of the year only a modest difference occurs as between the highland and lowland sector. Table 9 shows that an annual total of 209 raindays at Buxton is matched by 180 at Belper—the excess at Buxton being due mainly to increased frequency through the winter six months. Midsummer (June) shows no difference between the two stations. A total of at least 200 raindays in the year is that generally regarded as applicable to the north-west Pennines.

Table 9.   Mean annual number of raindays, 1881-1915

	Jan.	Feb.	Mar.	April	May	June	July	Aug.	Sep.	Oct.	Nov.	Dec.	Total
Buxton ..	19	18	19	16	16	12	16	19	15	19	19	21	209
Belper ..	17	15	15	14	13	12	14	15	13	17	17	18	180

The table of monthly rainfall (8) indicates how variable in different years month by month, the rainfall may be at different stations. This variability, year by year, increases markedly in the highland sectors as can be seen from the extreme ranges of rainfall that have been recorded month by month at Buxton (see table 6), particularly during the midwinter season. In the localities experiencing heaviest rainfall the range of variability becomes very pronounced, as can be seen from the records of four stations in the Kinder Scout–Bleaklow area during the years 1915-54. (Table 10.) Two of the stations exemplify valley sites and two are near summit areas, but it may be noted that the valleys do not necessarily provide either the wettest or driest extremes. The very wide ranges in the extremes at all four sites are significant and are well exemplified by the records for January, June and October.

Table 10.   Extreme Range of Rainfall, 1915-54, at four stations in the Kinder Scout-Bleaklow Area.

Station	Altitude	January	June	October
Blacklow Cote ..	970′ (296 m.)	14·0″–2·1″	7·1″–0·1″	13·8″–0·8″
Birchin Lee ..	870′ (245 m.)	13·9″–2·3″	6·4″–0·1″	12·5″–1·2″
Ladyclough ..	1,720′ (524 m.)	15·0″–2·3″	7·8″–0·05″	12·4″–1·4″
Alport Moor ..	1,790′ (546 m.)	14·8″–2·0″	7·9″–0·05″	13·3″–1·5″

Table 11. Monthly totals of days with snow or sleet falling and days with snow lying at 09.00 hours G.M.T.

	Jan.	Feb.	March	April	May	June	July	August	Sept.	Oct.	Nov.	Dec.	Year
Chapel-en-le-Frith													
1960	12/10	10/14	3/0	1/0	—	—	—	—	—	1/0	1/0	7/0	35/24
1961	5/1	2/1	1/0	1/1	—	—	—	—	—	—	—	10/10	19/13
1962	3/5	8/3	10/3	3/1	—	—	—	—	—	—	5/4	4/7	33/23
1963	14/31	17/28	1/4	4/0	1/0	—	—	—	—	—	1/0	2/0	40/63
1964	1/2	5/5	4/7	1/0	—	—	—	—	—	—	3/3	5/3	19/20
1965	12/5	6/1	5/4	2/0	—	—	—	—	—	—	0/4	—	25/14
Buxton													
1960	14/11	10/16	2/0	—	—	—	—	—	—	—	—	8/6	34/33
1961	9/3	3/1	1/0	1/0	—	—	—	—	—	—	—	9/10	23/14
1962	2/6	4/4	11/9	—	1/0	—	—	—	—	—	3/5	6/9	26/33
1963	15/31	15/28	3/0	4/1	—	—	—	—	—	—	—	4/2	42/62
1964	5/0	3/8	8/6	—	—	—	—	—	—	—	4/3	5/5	25/22
1965	9/11	6/4	9/11	7/0	—	—	—	—	—	—	8/5	6/9	45/40
Chesterfield													
1960	11/8	10/12	3/0	—	—	—	—	—	—	—	—	2/0	26/30
1961	3/0	3/0	—	1/0	—	—	—	—	—	—	—	10/5	17/5
1962	2/5	6/4	8/6	—	—	—	—	—	—	—	5/0	4/1	25/16
1963	13/29	15/28	3/5	2/0	—	—	—	—	—	—	1/0	3/0	36/62
1964	2/2	5/4	6/5	—	—	—	—	—	—	—	1/0	3/3	17/14
1965	10/6	7/0	6/8	—	—	—	—	—	—	—	6/4	2/3	31/21
Belper													
1960	8/7	2/6	2/1	—	—	—	—	—	—	—	—	—	12/14
1961	1/0	—	—	1/0	—	—	—	—	—	—	—	7/5	9/5
1962	0/8	3/2	6/1	—	—	—	—	—	—	—	3/1	5/6	17/18
1963	13/31	11/28	—	1/0	—	—	—	—	—	—	—	1/1	26/60
1964	1/1	3/0	5/4	—	—	—	—	—	—	—	1/0	1/3	11/8
1965	6/2	2/1	2/7	1/0	—	—	—	—	—	—	4/4	1/2	16/16
Derby													
1960	4/7	5/4	—	—	—	—	—	—	—	—	—	—	9/11
1961	1/0	—	—	1/0	—	—	—	—	—	—	—	5/4	7/4
1962	1/4	5/2	2/1	—	—	—	—	—	—	—	3/3	2/5	13/15
1963	12/29	9/11	—	1/0	—	—	—	—	—	—	—	2/0	24/40
1964	1/1	2/2	4/2	—	—	—	—	—	—	—	1/0	3/3	11/8
1965	7/3	3/2	2/5	—	—	—	—	—	—	—	6/1	1/1	19/12

The monthly median values for these four stations show that in general March is the month experiencing least rainfall in the annual cycle, with lower median values occurring at the two higher sites despite their higher extreme rainfall records. Both May and June are months likely to receive surprisingly low rainfall with September also a drier month at all altitudes. In this area of the High Peak oscillations in the mean annual totals from year to year range from 80–155% of the mean values, and within areas with an average total rainfall of 60 inches, a wet year may bring over 80 inches. In 1954 at Blackden Brook (1,070 feet) in the Ashop valley, 85·56 inches was recorded.

With the possibility in the north of such high winter precipitation in conjunction with low temperatures, it is to be expected that an increase north-westward in the frequency of snowfall will be evident. Table 11 indicates that the number of days with snow or sleet is halved in the south (compare Derby and Belper with Chapel and Buxton). For the years 1960–65 there was a mean frequency of 34 days with snow lying at 9 a.m. at Buxton as compared with 25 at Chesterfield and 15/16 at Morley, Mackworth and Derby. No snow data exist for the highest and wettest areas in Derbyshire, but on the basis of estimates made for Britain as a whole by Manley (1939, 1949) (involving an increase in days with snowfall from 20 a year in the lowlands to more than 30 at altitudes exceeding 1,000 feet, 50 above 1,600 feet, and 70–75 at 2,000 feet) much of north Derbyshire should experience days with some snowfall for periods equivalent to 6–7 weeks in the year, with an unfavourable aspect locally causing snow to lie long on the ground at high altitudes.

The effective value of precipitation in relation particularly to soils is closely related to water loss by evaporation. Soils that are likely to be waterlogged continuously can dry out and become aerated only for very limited periods of the year, and will tend to be associated with very marked soil leaching or even bog formation. It has been assessed by Pearsall (1968, page 64), that in the southern Pennines and their eastern foothills, the annual water loss by evaporation is roughly equivalent to 18 inches of rainfall, distributed seasonally.

Table 12. Monthly Evaporation (in inches).

	Jan.	Feb.	Mar.	April	May	June	July	Aug.	Sep.	Oct.	Nov.	Dec.
Evaporation (ins.)	0·1	0·4	0·9	1·8	2·8	3·3	2·9	2·2	1·8	0·9	0·3	0·2

A comparison of rainfall with these estimated values shows that an appreciable part of north Derbyshire is likely to have an average monthly rainfall that, even in summer, is in excess of evaporation, involving therefore a liability to bog formation. A rough index for this condition is generally associated with the isohyet for 50 inches mean annual rainfall (see Fig. 8), but the long term monthly means for Buxton show that even with a rainfall less than this (48·4 inches) a permanent excess of rainfall over evaporation month by month is likely to be present. A comparison of monthly evaporation values with the monthly rainfall in other parts of Derbyshire (*Table* 8) shows that over a period of years a rainfall in excess of evaporation is normally present through most of the year excepting the months of May, June and July. No station has a rainfall so low (18 inches) as to imply very little or no soil leaching in porous soils; the lowlands in Derbyshire for the most part have a rainfall of more, nearly twice that amount (note the position of the ishoyet for 35 inches in this regard). Clearly, again however, rainfall potential and evaporation figures must be considered closely in relationship to other physiographic factors—notably concerned with variations in landform slope, and soil drainage, and other basic edaphic and soil controls whose distribution in Derbyshire is significant in this regard.

From the standpoint of vegetation—particularly tree growth and afforestation, effective climatic indices will also be much concerned with the factor of wind, for which unfortunately very little precise data is available in Derbyshire. From general estimates elsewhere, however, it may be assumed that in the High Peak average windspeeds at least twice the velocity of prevailing winds in adjacent lowlands are to be expected, and that whilst throughout the year predominant circulation comes from the south-west and west, all points of the compass are well served, with a marked frequency of northerly and north-easterly winds in the spring and early summer. Within the Pennine sector of high relief cols and summits tend always to be exposed and somewhat windy localities, and at times of strong circulation the deeply incised valleys (e.g. the Ashop valley) opening with the direction of the prevailing air flow experience very high winds, whereas valleys otherwise orientated have marked shelter. The sharply scarped plateau surface locally gives rise to interesting and deep local vertical thermal currents, notably along Bradwell and Hucklow Edge above which standing waves develop, much used by and well known to gliders. At times northern Derbyshire is subject to gales of extreme violence, when, under special but not uncommon meteorological conditions lee waves develop progressively east of the Pennines during an overall westerly circulation. The effects in February, 1962, of exposure to gales of this kind, with windspeeds locally exceeding 80 knots, was markedly evident in the severe tree fall that resulted, and which was indeed conspicuous in many localities for its severity on leeward rather than windward facing slopes (see Aanensen 1964, especially Fig. 11 and page 15).

Apart from the greatly varying conditions of windiness an appreciable number of days in each month are associated with the occurrence of calms, all together exceeding $\pm$ 10%, and bringing in their wake important modification to temperature, as has been shown.

From these studies it is clear that within Derbyshire precise definition of climatic sub-regions is hardly possible though clearly there are marked contrasts in many of the characteristics related firstly to the distinctive and somewhat extreme conditions in the highland north-west; secondly to the fringing low plateau and lowlands that form a transverse and transitional belt from south-west to north-east, and finally, thirdly, to the drier, warmer lowland sector in the extreme south and south-east of the county. But whether Derbyshire climates are viewed on macro-, meso-, or micro-scales there is need always to study and assess these closely in relation to the full range of aspects of the physical environment, without which many local differences cannot be appreciated or their significance evaluated.

Acknowledgements

To Miss Margaret Wilkes, M.A., thanks are expressed for assistance in the collection of data and preparation of tables included in this Chapter. My thanks are also due to the Director-General, Meteorological Office, for supplying many of the records included in the tables.

Sources of Climatological Data for Derbyshire

The study of differences of climate in Derbyshire is somewhat handicapped by the paucity of longstanding climatological stations. Though a number have been in existence somewhat sporadically, only at two sites have records been collected for any length of time—viz. at Buxton since 1908 and Belper 1910 (unofficial records for these two places dating back to 1875 and 1880 respectively). For the rest of the county reference can be made only to discontinued or shortlived records. Those for Ashbourne relate only to the years 1943-46, at Wirksworth 1946-61, Chapel-en-le-Frith 1953-66, Shardlow 1952-58 and Chesterfield 1951-66. Records have been maintained at Derby since 1959, Darley Dale 1955 and Morley 1952, but at Chatsworth only since 1961 and at Ashover, 1967. (See Fig. 6.)

Source material of so fragmentary a character makes any assessment of local climatic characteristics and trends somewhat tentative, but for comparative purposes, data for a group of stations have been included in Tables relating only to the years 1961-65 over which period there was overlap in the availability of records, and for which actual rather than weighted or reduced mean figures are given. Though the limitations are clearly evident, some comparisons can be made within the framework of these few years. The rainfall records of Water Boards and other records of research undertaken in the Departments of Geography in the Universities of Sheffield and Nottingham have provided further supplementary data. Some interesting early climatic records have been collected by F. A. Barnes (1966) that have relevance to some parts of Derbyshire.

REFERENCES

Aanersen, C. J. M. ed. (1964). Gales in Yorkshire in February, 1962. Geophysical Memoirs No. 108. H.M.S.O., London.

Barnes, F. A. (1966). 'Weather and Climate' in 'Nottingham and its Region' ed. Edwards, K. C. British Association for the Advancement of Science, Nottingham.

Climatological Atlas of the British Isles (1952). Meteorological Office, London.

Garnett, A. (1939). Diffused Light and Sunlight in Relation to Relief and Settlement in High Latitudes. Scot. Geog. Mag., 55, 271–284.

Garnett, A. (1951). Relief, Latitude and Climate: some local consequences. Indian Geographical Society Silver Jubilee volume, 195, 128–131.

Garnett, A. (1956). 'Climate' in 'Sheffield and its Region' ed. Linton, D. L. British Association for the Advancement of Science, Sheffield.

Investigation of Air Pollution (1967). National Survey Annual Summary Table I for the year ending March, 1967. Ministry of Technology, Warren Spring Laboratory.

Manley, G. (1939). On the occurrence of Snow Cover in Great Britain. Q. J. Roy. Met. Soc., 65, 2–27.

Manley, G. (1949). The Snowline in Britain. Geog. Annaler. 31, 179–193.

Manley, G. (1945). Effective Rate of altitudinal change in Temperate Atlantic Climate Atlantic Climate. Geog. Rev. 35, 408–417.

Pearsall, W. H. (1950). Mountains and Moorlands. London.

Chapter Five

VEGETATION

A. MAGNESIAN LIMESTONE REGION

THE north eastern part of the county is on the Permian Magnesian Limestone outcrop which extends south and east into Nottinghamshire and north through Yorkshire into Co. Durham. The whole formation has been intensively cultivated, because of the well-drained, neutral or slightly alkaline soils which are developed on this rock. Natural vegetation is therefore very scattered and in Derbyshire the best examples are found in steep-sided valleys cut into the limestone, locally known as 'grips'. The best preserved is in Markland Grips where Habitat Studies 1 to 3 have been carried out. One side of Creswell Crags, another similar valley, is also in Derbyshire, although its botanical value is lessened by the presence of a road and sewage works. It is well known archaeologically for the extensive remains of Paleolithic Age which have been excavated from caves in the side of the valley.

Markland Grips contains a mixture of semi-natural and planted woodland. The most natural parts of the woodland are on the crags which line the valley sides and are marked by the presence of *Taxus baccata* and *Tilia platyphyllos*. These woodland areas have been described by Jackson and Sheldon (1949) and Habitat Study 1 was made in one such area. The screes below the crags carry a vegetation dominated by ferns described in Study 2. At the upper end of one of the arms of the valley is a remnant of natural grassland which includes *Carex ericetorum* and *Carex montana*; Habitat Study 3 was made in this area. It should be noted that Markland Grips is strictly private and that free access is not allowed.

Habitat Study 1   *Markland Grips, near Clowne*

SK 508744   Alt. 350 ft.

Recorder: C. B. Waite.   Surveyed: 15-5-66.   Area: Transect 20 yards wide.

Woodland at the edge of the cliff on the east side of the valley. Other trees also present in the woodland are *Fraxinus excelsior, Sorbus aucuparia, Ulmus glabra* and *Tilia platyphyllos*.

Tree Layer

OCCASIONAL: Betula pubescens, Quercus robur, Taxus baccata.

RARE: Acer campestre.

Shrub Layer

OCCASIONAL: Corylus avellana, Crataegus monogyna, Euonymus europaeus, Rosa canina, Rubus fruticosus, Sambucus nigra, Thelycrania sanguinea.

Herb Layer

ABUNDANT: Galium aparine, Hedera helix, Mercuralis perennis, Phyllitis scolopendrium, Urtica dioica.

OCCASIONAL: Arctium minus, Cirsium vulgare, Dryopteris filix-mas, Glechoma hederacea, Geranium robertianum, Sanicula europaea.

Habitat Study 2   *Markland Grips, near Clowne*
SK 508744   Alt. 350 ft.
Recorder: C. B. Waite. Surveyed; 15-5-66. Area: Transect 20 yards wide.
Scree slopes below crags.
ABUNDANT: Dryopteris filix-mas, Phyllitis scolopendrium.
FREQUENT: Ligustrum vulgare.
OCCASIONAL: Betula pubescens, Geranium robertianum, Hedera helix, Ilex aquifolium, Mycelis muralis, Oxalis acetosella, Sanicula europaea.
RARE: Taraxacum officinale.

Habitat Study 3   *Markland Grips, near Clowne*
SK 508744   Alt. 350 ft.   Slope: 30°.   Aspect: N.E.
Recorder: C. B. Waite. Surveyed; 15-5-66. Area: 1 sq. metre.
Grassy slope with small limestone crags.
FREQUENT: Campanula rotundifolia, Carex caryophyllea, Festuca ovina, Festuca rubra, Helianthemum chamaecistus, Poterium sanguisorba.
OCCASIONAL: Brachypodium pinnatum, Pimpinella saxifraga, Plantago lanceolata, Thymus drucei.
RARE: Carex ericetorum, Crataegus monogyna (seedling), Helictotrichon pratense, Lotus corniculatus, Prunella vulgaris, Primula veris.

Most of the other woodlands on the Magnesian limestone are now very largely planted, but some of them contain mature mixed woodland with a characteristically rich shrub layer. One such wood is Scarcliffe Park Wood which, although largely felled and replanted after the last war with *Fraxinus excelsior* and *Picea abies*, does contain an older area of mature trees where felling has not been carried out for at least seventy years.

Habitat Study 4   *Scarcliffe Park Wood, near Bolsover*
SK 509707   Alt. 350 ft.   Slope: Level.
Recorder: C. B. Waite. Surveyed: 16-7-66. Area: 20 yards square.

Sample taken in old mature woodland which contained the following trees and shrubs in addition to those in the sample: *Betula pubescens, Fagus sylvatica, Acer campestre, Quercus robur, Salix alba, Salix fragilis, Taxus baccata, Tilia platyphyllos, Ligustrum vulgare, Frangula alnus, Rhamnus catharticus, Thelycrania sanguinea, Euonymus europaeus, Prunus spinosa, Salix cinerea and Viburnum opulus.*

Tree Layer
Betula pendula (8), Acer pseudoplatanus (4), Ulmus glabra (4), Fraxinus excelsior (3), Sorbus aucuparia (2).

Shrub Layer

Acer pseudoplatanus (young trees) (6), Betula pendula (young trees) (6), Corylus avellana (4), Crataegus monogyna (3), Lonicera periclymenum (2).

Herb Layer

ABUNDANT: Lysimachia nemorum, Mercurialis perennis, Pteridium aquilinum, Sanicula europaea.

FREQUENT: Brachypodium pinnatum, Prunella vulgaris, Rubus fruticosus, Viola riviniana.

OCCASIONAL: Ajuga reptans, Cirsium palustre, Deschampsia caespitosa, Dryopteris filix-mas, Geum urbanum, Holcus lanatus, H. mollis, Hypericum perforatum, H. hirsutum, Lithospermum officinale, Luzula pilosa, Mentha arvensis, Nepeta hederacea, Plantago major, Senecio jacobæa.

RARE: Arctium minus.

Apart from the 'Grips' natural vegetation is very scattered and the best examples are to be found around springs and ponds, where the ground is unsuitable for agriculture, and in disused quarries. The best quarry area is probably Pebley Sand Quarry immediately to the east of Pebley Pond, which has been disused for approximately forty years. This shallow quarry is cut through the limestone into the Basal Sands of the Permian formation. It has a mixture of grassland and scrub on the irregular sides and a striking mosaic of basic flush and acid bog vegetation on the floor, the latter perhaps due to the close proximity of Coal Measure rocks. The next two habitat studies are on the grassland and the wetland vegetation of the floor.

Habitat Study 5   *Pebley Sand Quarry*

SK 490787   Alt. 400 ft.   Slope: Level.

Recorder: C. B. Waite. Surveyed: 31-7-66. Area: 15 x 5 yards.

Dry hillock with grassland near the side of the quarry with scrub beginning to invade.

FREQUENT: Acer campestre, Agrimonia eupatorium, Arrhenatherum elatius, Clinopodium vulgare, Holcus lanatus, Leontodon hispidus, Senecio erucifolius, Tussilago farfara, Zerna erecta.

OCCASIONAL: Briza media, Centaurea nigra, Chrysanthemum leucanthemum, Cirsium palustre, Crataegus monogyna, Dactylis glomerata, Deschampsia caespitosa, Equisetum arvense, Fraxinus excelsior, Heracleum sphondylium, Lathyrus pratensis, Listera ovata, Lotus corniculatus, Plantago lanceolata, Potentilla anserina, Primula vulgaris, Prunella vulgaris, Rumex acetosa, Ulex europaeus.

RARE: Potentilla reptans, Trifolium pratense.

Habitat Study 6   *Pebley Sand Quarry*

SK 490787   Alt. 400 ft.   Slope: Level.

Recorder: C. B. Waite. Surveyed: 31-7-66. Area: 15 yards square.

Area on the wet floor of the quarry including both flush and bog components of the vegetation. Other parts of the mosaic have a completely different composition, often only two or three species being present. Thus some areas have *Eriophorum angustifolium* dominant with abundant *Carex demissa*, others *Ranunculus flammula* dominant with *Carex flacca* abundant, while a third type has *Juncus inflexus* dominant and *Chamaenerion angustifolium* and *Filipendula ulmaria* abundant.

There are scattered trees, mainly *Alnus glutinosa*, with some *Fraxinus excelsior* and *Crataegus monogyna*.

ABUNDANT: Carex flacca, Eriophorum angustifolium, Juncus articulatus, Juncus inflexus.

FREQUENT: Agrostis stolonifera, Briza media, Cirsium palustre, Filipendula ulmaria, Holcus lanatus, Mentha arvensis, Pulicaria dysenterica, Ranunculus flammula, Tussilago farfara.

OCCASIONAL: Angelica sylvestris, Arrhenatherum elatius, Blackstonia perfoliata, Carex demissa, Centaurea nigra, Dactylorhiza fuchsii, Equisetum arvense, Gymnadenia conopsea, Listera ovata, Mentha aquatica, Primula vulgaris, Prunella vulgaris, Ranunculus acris.

RARE: Bellis perennis, Centaurium erythraea, Euphrasia officinalis, Hypericum tetrapterum, Leontodon hispidus, Lotus corniculatus.

The best flush and marsh vegetation is around the margins of ponds and springs which emerge from the limestone, and Studies 7 and 8 illustrate the vegetation to be found in such areas.

Habitat Study 7   *The Walls, Whitwell*

SK 503781   Alt. 400 ft.   Slope: Level.

Recorders: F. W. Adams and C. B. Waite. Surveyed: 3-6-65.

Area: Margin of pond.

The sample is a composite one of the margin of the pond which is approximately 25 yards in diameter. The pond lies in a small wood and is supplied with water from a small stream and nearby springs. The pond itself contains *Myriophyllum spicatum* and the margin is partially shaded by trees and shrubs.

Tree and Shrub Layers

FREQUENT: Fraxinus excelsior, Salix fragilis.

OCCASIONAL: Betula pubescens.

RARE: Corylus avellana, Crataegus monogyna, Populus cf. gileadensis.

Herb Layer

ABUNDANT: Equisetum fluviatile, Filipendula ulmaria, Hippuris vulgaris, Mentha aquatica, Ranunculus sceleratus, Rubus spp., Sparganium ramosum, Typha latifolia.

FREQUENT: Angelica sylvestris, Brachypodium sylvaticum, Circea lutetiana, Deschampsia caespitosa, Epilobium palustre, E. parviflorum, Eupatorium cannabinum, Polygonum amphibium, Ranunculus repens, Rumex conglomeratus, R. sanguineus, Solanum dulcamara, Juncus subnodulosus.

OCCASIONAL: Alisma plantago-aquatica, Caltha palustris, Campanula trachelium, Cirsium palustre, Dactylis glomerata, Galium aparine, Geum,
urbanum, Hypericum tetrapterum, Lamiastrum luteum, Pulicaria dysenterica, Rorippa sp., Stachys sylvatica, Urtica dioica, Zerna ramosa.

RARE: Anagallis tenella, Anthriscus sylvestris, Heracleum sphondylium,
Primula veris, Sanicula europaea, Tamus communis, Taraxacum officinale
Urtica urens, Valeriana dioica.

Habitat Study 8 *Whitwell Wood, near Clowne*

SK 521789. Alt. 450 ft. Slope: 10–15°. Aspect: North.
Recorder: C. B. Waite. Surveyed: 6-7-66. Area: 15 yards square.

Whitwell Wood has been almost entirely felled and replanted in the last
fifty years. The only portion to be left in a natural state is at the northern end
along the stream which is the Derbyshire–Yorkshire boundary. The whole
area is marshy owing to the presence of a number of springs which run into
the main stream and the trees and shrubs are rather scattered. The tree and
shrub species present are *Alnus glutinosa, Betula pubescens, Betula pendula,
Frangula alnus, Fraxinus excelsior, Ligustrum vulgare* and *Salix cinerea.*

Each spring is surrounded by a well developed flush and the study was made
on one of these. The soil is humus-rich, with particles of limestone and has a
pH of 7·5.

ABUNDANT: Carex flacca, Carex hostiana, Carex panicea, Eleocharis quinqueflora, Molinia caerulea.

FREQUENT: Agrostis stolonifera, Anagallis tenella, Epipactis palustris,
Equisetum arvense, Eupatorium cannabinum, Filipendula ulmaria, Gymnadenia conopsea, Juncus articulatus, Mentha aquatica, Parnassia palustris,
Pinguicula vulgaris, Quercus sp. (seedling), Succisa pratensis, Triglochin
palustre.

OCCASIONAL: Angelica sylvestris, Aquilegia vulgaris, Briza media, Cirsium
palustre, Crataegus monogyna (seedling), Eriophorum vaginatum, Festuca
rubra, Galium uliginosum, Juncus inflexus, Leontodon hispidus, Poa
trivialis, Viola palustris.

Apart from these scattered remains of natural vegetation, the main reservoirs
of native plants are probably the hedgerows and Study 9 gives a typical list
from a hedgerow on the Magnesian limestone.

Habitat Study 9 *Lane between Whitwell Church and Wood*

SK 526772 Alt. 450 ft. Slope 5°. Aspect: East.
Recorders: I. Sollitt and M. R. Shaw. Surveyed: 11-9-66.
Area: Hedgerow 5 yards long and 2 yards wide.

A typical hedgerow of the area, kept trimmed and with a fairly thin shrub
layer.

Shrub Layer

Mainly Crataegus monogyna, with a little Corylus avellana, interspersed
with Hedera helix and Tamus communis.

Herb Layer

FREQUENT: Dactylis glomerata, Matricaria matricarioides, Plantago lanceolata, Ranunculus repens, Rubus fruticosus, Urtica dioica.

OCCASIONAL: Galium cruciata, Lamium album, Mercurialis perennis, Plantago
major, Polygonum aviculare, Stellaria graminea, Tamus communis, Trifolium sp., Vicia sepium.

RARE: Heracleum sphondylium, Torilis japonica.

B. EASTERN COAL MEASURES REGION

The majority of the eastern part of the county is situated on Carboniferous
rocks of the Coal Measures. The land-use of this region is typified by a mixture
of agriculture and industry, the latter depending very largely on coalmining,
which is both by traditional methods and open-casting. This, together with
widespread urban development, particularly in the southern part of the
region, has meant that few areas remain with natural or semi-natural vegetation.

The country is hilly with streams in the valleys, but the majority of these
are now, unfortunately, rather polluted. The Coal Measure rocks are predominantly shales, which give rise to poorly drained clayey soils which are
often gleyed, with some sandstones which give rise to more sandy, better
drained soils (see Bridges, 1966, for a detailed description of the soils in the
south of the area). The climax vegetation of the majority of the region is
almost certainly oak woodland, but woodland of any type is now very
scattered and the majority of agricultural land is permanent grassland, often
improved, or temporary ley, with a minority of arable land on the better
drained soils.

Some of the best remaining examples of oak woodland are now within the
boundaries of the city of Sheffield, although formerly in Derbyshire. Scurfield
(1953) has described the vegetation of some of these, including Ladies' Spring
Wood, in which Habitat Study 10 was made. Pigott (1956) also comments on
this wood (here called Totley Wood) in his more general description of woodland in the Sheffield area.

Habitat Study 10   *Ladies' Spring Wood, West Sheffield*

SK 325813   Alt. 500 ft.

Recorder: B. Fearn.   Surveyed: May 1967.

A.   Sample near top of wood. Slope: 22°. Aspect: West.

The upper part of wood has a podsolized soil with a pH of about 3·8 in the
upper layers, which are of mor humus.

Tree Layer

ABUNDANT: Quercus petraea.

FREQUENT: Betula pubescens, Betula pendula.

OCCASIONAL: Ilex aquifolium, Sorbus aucuparia.

Herb Layer

ABUNDANT: Deschampsia flexuosa.

FREQUENT: Pteridium aquilinum.

B. Sample near bottom of wood. Slope: 28°. Aspect North-west.

The lower part of the wood has a brown-earth soil with a pH of approximately 5 in the upper layers, which are of mull humus. The ground vegetation of this part of the wood is dominated by *Holcus mollis* and the relationship between this grass and *Deschampsia flexuosa* and the soils on which they grow has been investigated by Jowett & Scurfield (1949, 1952).

Tree and Shrub Layers

ABUNDANT: Quercus petraea.

FREQUENT: Acer pseudoplatanus, Betula pubescens, Betula pendula, Sambucus nigra.

OCCASIONAL: Alnus glutinosa, Corylus avellana, Fraxinus excelsior, Ulmus glabra.

RARE: Prunus avium, Ribes nigrum.

Herb Layer

ABUNDANT: Endymion non-scriptus, Holcus mollis.

FREQUENT: Lamiastrum luteum, Milium effusum, Oxalis acetosella, Silene dioica.

OCCASIONAL: Deschampsia flexuosa, Pteridium aquilinum, Rubus fruticosus, Stachys sylvatica.

RARE: Anemone nemorosa, Geranium robertianum, Hedera helix, Lonicera periclymenum, Stellaria holostea, Teucrium scorodonia.

A few similar woods with either or both these types of ground vegetation are present elsewhere in the county on the Coal Measures, e.g. at Hardwick Hall and Ault Hucknall. Most of the scattered woodland is, however, planted, but although the trees may not be native to the area the ground vegetation is usually similar as is shown in the next Study.

Habitat Study 11   *Wood N.N.W. of Stubbing Court Pond, near Chesterfield*
SK 360675   Alt.: 500 ft.   Slope: 25°—35°.   Aspect: South-west.
Recorder: I. Sollitt.   Surveyed: 31-7-66.   Area: 1 sq. metre.

Planted wood on a steep, well-drained slope, soil a dark brown loam with a pH of 3·9.

Tree Layer

FREQUENT: Castanea sativa, Fagus sylvatica, Quercus petraea.

OCCASIONAL: Acer pseudoplatanus.

Herb Layer

ABUNDANT: Holcus mollis.

OCCASIONAL: Acer pseudoplatanus (seedlings), Agrostis tenuis, Deschampsia flexuosa, Dryopteris dilatata, Endymion non-scriptus, Quercus (seedlings).

The majority of the grassland as already mentioned has been reseeded, but occasionally permanent pastures, which do not seem to have been improved, except perhaps by liming, are seen. Habitat Study 12 was made in a pasture of this type near Beauchief Abbey.

Habitat Study 12   *Beauchief, Sheffield*

SK 333817.   Alt. 450 ft.   Slope 15°.   Aspect: East.

Recorders: C. B. Waite and M. R. Shaw. Surveyed: 21–6–66. Area: 1 sq. metre.

Two samples were taken, A from near the top and B from halfway down the permanent pasture, which is grazed by horses. The soil is a medium brown loam with pH (in A) of 5·1.

A.

ABUNDANT: Agrostis stolonifera, Luzula campestris.

FREQUENT: Anthoxanthum odoratum, Festuca rubra, Galium saxatile, Hypochaeris radicata.

OCCASIONAL: Festuca pratensis, Rumex acetosa.

RARE: Holcus lanatus, Quercus petraea (seedling).

B.

FREQUENT: Ranunculus acris, R. repens, Trifolium medium, T. repens.

OCCASIONAL: Agrostis stolonifera, Anthoxanthum odoratum, Cirsium palustre, Equisetum arvense, Festuca ovina, Holcus lanatus, Lotus corniculatus, Luzula campestris, Plantago lanceolata, Succisa pratensis.

RARE: Achillea millefolium, Centaurea nigra.

Although in general the vegetation of the Coal Measures tends to be rather species-poor, where flushing takes place through the presence of springs or streams a richer vegetation may develop. The following two Studies illustrate the species composition of such areas.

Habitat Study 13   *Near Povey Farm*

SK 385805   Alt. 300 ft.   Slope 20°.   Aspect: South.

Recorders: C. B. Waite and M. R. Shaw. Surveyed: 29–6–66. Area: 20 by 10 yards.

Flush in meadow grazed by cattle and surrounded by bushes of *Crataegus monogyna*. The meadow faces mixed woodland across a stream lined with *Alnus glutinosa*.

ABUNDANT: Cynosurus cristatus, Ranunculus repens.

FREQUENT: Alopecurus geniculatus, Dactylis glomerata, Deschampsia caespitosa, Carex flacca, Cirsium palustre, Festuca pratensis, Glyceria fluitans, Juncus articulatus, J. effusus var. congestus, Lotus corniculatus, Plantago lanceolata, Poa trivialis, Ranunculus acris.

OCCASIONAL: Ajuga reptans, Anthoxanthum odoratum, Bellis perennis, Briza media, Carex demissa, Carex ovalis, Carex panicea, Galium palustre, Holcus lanatus, Hypericum tetrapterum, Lathyrus pratensis, Leontodon hispidus, Prunella vulgaris, Rumex acetosa, R. sanguineus, Stellaria uliginosa, Trifolium repens, Veronica beccabunga.

RARE: Acer pseudoplatanus (seedling), Achillea millefolium, Centaurea nigra, Crataegus sp. (seedling), Lolium italicum, Quercus sp. (seedling), Succisa pratensis.

R. H. Hall

5. *Ravenstor, Miller's Dale. The crags on this south-facing cliff, which overlooks the River Wye, carry a rich flora including Geranium sanguineum and Silene nutans.*

R. H. Hall

6. BIRD'S-FOOT SEDGE (*Carex ornithopoda*); *Burfoot, Miller's Dale. The first recorded British locality for this species.*

7. *Dovedale, from Bunster Hill, showing the well-developed ashwoods*

8. ALPINE PENNY-CRESS (*Thlaspi alpestre*); *Matlock*

Habitat Study 14   *Cordwell Valley, near Unthank Lane*

SK 309757   Alt. 600 ft.   Aspect: North-east.

Recorder: C. B. Waite.   Surveyed: 5–7–53.   Area: 60 by 15 yards.

The study was made in part of a shallow valley with a small stream running through it. The sample area is bordered on one side by a group of trees; on the opposite side is the steep bank of an arable field. The almost level ground has some boggy parts and a cart-track through it. The trees include the following species: *Acer pseudoplatanus, Corylus avellana, Crataegus monogyna, Fraxinus excelsior, Quercus petraea, Salix cinerea.*

FREQUENT: Agrostis tenuis, Angelica sylvestris, Anthoxanthum odoratum, Arrhenatherum elatius, Bellis perennis, Betonica officinalis, Chamaenerion angustifolium, Cynosurus cristatus, Dactylis glomerata, Deschampsia caespitosa, Digitalis purpurea, Equisetum arvense, Galium saxatile, Hieracium pilosella, Holcus lanatus, Leontodon hispidus, Lolium perenne, Lotus corniculatus, Phleum pratense, Plantago lanceolata, Plantago major, Potentilla erecta, Pteridium aquilinum, Ranunculus acris, R. repens, Rhinanthus minor, Sieglingia decumbens, Stachys sylvatica, Taraxacum officinale, Trifolium pratense, T. repens, Trisetum flavescens, Urtica dioica.

OCCASIONAL: Achillea millefolium, Alchemilla xanthochlora, Athyrium filix-femina, Briza media, Caltha palustris, Centaurea nigra, Cerastium fontanum, Chrysosplenium oppositifolium, Cirsium palustre, Dactylorhiza fuchsii, Dryopteris dilatata, Filipendula ulmaria, Galium cruciata, Geranium robertianum, Heracleum sphondylium, Hieracium spp., Hypochaeris radicata, Lathyrus montana, L. pratensis, Melampyrum pratense, Oxalis acetosa, Poterium polygamum, Rubus fruticosus agg., Rumex acetosa, R. obtusifolius, Stellaria graminea, S. media, Teucrium scorodonia, Trifolium medium, Tussilago farfara, Valeriana officinalis, Vicia cracca.

Although many of the streams and pools in this region have become polluted to such an extent that they no longer have any aquatic or sub-aquatic vegetation, there are still a few, particularly on private estates, which have escaped contamination and Studies 15 to 17 are examples of these. Some stretches of the now disused Chesterfield Canal have also escaped pollution and Habitat Study 18 was made on one of these stretches.

Habitat Study 15   *Third Pond, near Great Pond, Hardwick Hall*

SK 458639   Alt. 370 ft.   Slope: Level.

Recorder: C. B. Waite.   Surveyed: 4–9–66.   Area: 10 by 10 yards.

Study made on a flat area at the head of the pond which had been recently disturbed and would normally be under-water. The soil was a humus-rich clay with a pH of 7·1.

ABUNDANT: Mentha aquatica, Ranunculus sceleratus.

FREQUENT: Apium nodiflorum, Bidens cernua, Epilobium hirsutum, Epilobium montanum, Equisetum fluviatile.

OCCASIONAL: Cardamine flexuosa, Cirsium arvense, Cirsium palustre, Deschampsia caespitosa, Juncus effusus, Juncus inflexus, Myosotis arvensis, Polygonum aviculare, Polygonum persicaria, Rumex sanguineus, Stellaria alsine, Urtica dioica.

E

RARE: Angelica sylvestris, Cirsium vulgare, Galium saxatile, Holcus lanatus, Ranunculus repens, Rumex acetosa, Scutellaria galericulata, Tussilago farfara, Veronica persica.

Habitat Study 16   *Fourth Pond, near Great Pond, Hardwick Hall*

SK 459639   Alt. 400 ft.   Slope: Level.

Recorder: C. B. Waite.   Surveyed: 4-9-66.   Area: 3 by 5 yards.

Study made at the side of the pool in an area which would normally be under-water. The soil is a brown silt with a pH of 7·3.

ABUNDANT: Juncus acutiflorus.

FREQUENT: Alisma plantago-aquatica, Bidens cernua, Equisetum fluviatile, Glyceria fluitans, Mentha aquatica, Potamogeton polygonifolius, Ranunculus sceleratus, Scutellaria galericulata.

OCCASIONAL: Epilobium hirsutum.

Habitat Study 17   *Wingerworth Pond*

SK 378667   Alt.: 400 ft.   Slope: 15°.   Aspect: South-west.

Recorder: I. Sollitt.   Surveyed: 31-7-66.   Area: 1 sq. metre at north-east corner of pool.

This is an artificial pool which is cleared periodically to improve fishing. Trees have been planted on two sides, giving some shelter, but the area where the study was made was not shaded. The soil is a yellow clay with a pH of 5·1.

ABUNDANT: Galium palustre, Juncus effusus.

FREQUENT: Elodea canadensis, Holcus lanatus, Juncus articulatus, Mentha arvensis, Poa annua, Ranunculus repens, Trifolium repens.

OCCASIONAL: Carex ovalis, Eleocharis palustris, Hydrocotyle vulgaris, Plantago major, Ranunculus flammula, Rumex conglomeratus.

RARE: Myriophyllum alterniflorum.

Habitat Study 18   *Section of Chesterfield Canal at Renishaw*

SK 444772   Alt. 200 ft.   Slope: 30°.   Aspect: West.

Recorder: C. B. Waite.   Surveyed: 28-8-66.   Area: 10 by 1 yards.

Stretch of canal side along the edge of a well-grazed permanent pasture. The soil is brown loam with a pH of 6·8.

FREQUENT: Cirsium arvense, Elodea canadensis, Epilobium hirsutum, Juncus inflexus, Lemna minor.

OCCASIONAL: Arrhenatherum elatius, Epilobium montanum, Holcus lanatus, Lycopus europaeus, Potentilla reptans, Scutellaria galericulata, Senecio jacobaea.

In this region of long established industry and widespread urban development there are numerous waste and derelict areas so that there are many man-made habitats. One of the most conspicuous and characteristic of these are disused coal-tips. The following information on their vegetation has been kindly provided by Mr. J. H. Johnson.

The Derbyshire tips vary greatly in size and age; the oldest, up to perhaps 200 years old are the smallest, covering less than a quarter of an acre and not more than twenty or thirty feet high. The recent tips have been constructed by depositing waste from aerial ropeways, and this has led to the production of mounds up to 200 ft. high and half a mile long, covering twenty or more acres. The tips are very largely of shale and coal waste, and many have been set on fire by spontaneous combustion, leaving the shale a characteristic red colour. The vegetation and plants on these tips are very varied so that it is difficult to give a representative sample. Usually, however, the first colonisers are plants with wind-borne seeds including *Chamaenerion angustifolium*, *Betula pendula* and *Betula pubescens*, *Salix* species, *Dactylis glomerata*, *Holcus mollis*, and species of *Rumex* and *Cirsium*. Occasionally species which are uncommon in the area may become dominant on a tip. Thus one tip at Clay Cross was covered with *Reseda luteola* one year, and the following year by *Melilotus officinalis*. After a number of years without tipping, shrub and tree species become dominant. Another ridged tip at Clay Cross on which tipping ceased about thirty years ago may be taken as a representative sample. The sides support a scattered scrub of *Ulex europaeus*, *Cytisus scoparius*, *Rubus fruticosus* and *Crataegus monogyna*, interspersed with patches of *Pteridium aquilinum*, and with numerous saplings of *Acer pseudoplatanus*. Towards the base the main species are waste-land plants including *Chamaenerion angustifolium*, *Artemisia vulgaris*, *Artemisia absinthium*, *Senecio jacobaea*, *Cirsium arvense* and *Cirsium vulgare*. In early Autumn the most conspicuous plant is *Aster novi-belgii*, which has become naturalized.

Many of the tips, after about twenty years without tipping of further material, become covered with a dense growth of *Betula* saplings, while the oldest tips, e.g. at Parkhouse Colliery, near Danesmoor, and near Oakerthorpe Colliery, support fully developed oakwood, although the trees are usually rather short.

There is a wide variety of other artificial habitats in this region, and here it is only possible to give information on a few of these to illustrate the variation in their floristic composition. The next four studies are, therefore, from an arable field, a railway embankment, a hedgebank, and from disturbed ground by a river.

Habitat Study 19   *Loads Head near Wadshelf*

SK 313703   Alt. 950 ft.

Recorder: A. R. Clapham.   Surveyed: August, 1967.   Area: 4 sq. metres.

Weed vegetation near the side of a Barley field.

ABUNDANT: Galeopsis tetrahit, Ranunculus repens, Sonchus arvensis, Spergula arvensis.

FREQUENT: Lathyrus pratensis, Leontodon autumnalis, Stellaria media, Viola arvensis.

OCCASIONAL: Aethusa cynapium, Agropyron repens, Sinapis arvensis, Chrysanthemum leucanthemum, Heracleum sphondylium, Plantago lanceolata, Polygonum persicaria, Vicia hirsuta.

Habitat Study 20 *Beighton*

SK 449825   Alt. 250 ft.

Recorder: C. B. Waite.   Surveyed: 27-7-65.   Area: 50 by 10 yards.

Railway embankment; soil consisting of shale and clay.

FREQUENT: Agrostis stolonifera, Agrostis tenuis, Chrysanthemum vulgare, Deschampsia flexuosa, Holcus mollis, Lotus corniculatus, Tripleurospermum maritimum subsp. inodorum, Senecio squalidus, Senecio sylvaticus, Senecio viscosus, Trifolium dubium, Trifolium medium, Tussilago farfara.

OCCASIONAL: Achillea millefolium, Artemisia absinthium, Artemisia vulgaris, Chrysanthemum leucanthemum, Convolvulus arvensis, Epilobium angusti-folium, Epilobium montanum, Equisetum arvense, Hypericum perforatum, Juncus effusus var. congestus, Knautia arvensis, Lamium album, Lathyrus montanus, Lathyrus pratensis, Silene dioica, Ononis repens, Potentilla reptans, Rumex acetosella, Salix caprea, Sanguisorba officinalis, Cytisus scoparius, Senecio erucifolius, Silene vulgaris, Ulex europaeus.

Habitat Study 21 *Beighton*

SK 448825   Alt. 250 ft.

Recorder: C. B. Waite.   Surveyed: 27-7-65.   Area: 30 by 2 yards.

Hedge bank between lane and barley field. *Acer campestre* is a characteristic species of the hedges in this region.

Shrub Layer

FREQUENT: Crataegus monogyna.

OCCASIONAL: Acer campestre, Corylus avellana, Fraxinus excelsior.

Herb Layer

FREQUENT: Capsella bursa-pastoris, Carduus crispus, Centaurea nigra, Cirsium arvense, Humulus lupulina, Lotus corniculatus, Matricaria matri-carioides, Plantago lanceolata, Plantago major, Urtica dioica.

OCCASIONAL: Agrimonia eupatoria, Agropyron caninum, Agropyron repens, Betonica officinalis, Campanula rotundifolia, Galeopsis tetrahit var. bifida, Galium verum, Linaria vulgaris, Lolium perenne, Mercurialis perennis, Nepeta hederacea, Pimpinella major, Rumex crispus, Scrophularia nodosa, Senecio jacobaea, Solanum dulcamara, Stachys sylvatica, Tamus communis, Torilis japonica.

Habitat Study 22 *Beighton*

SK 454825   Alt. 250 ft.

Recorder: C. B. Waite.   Surveyed: 27-7-65.   Area: 30 by 10 yards.

Disturbed soil by riverbank near railway bridge.

FREQUENT: Senecio squalidus, Senecio viscosus.

OCCASIONAL: Alliaria petiolata, Alnus glutinosa, Alopecurus pratensis, Athyr-ium filix-femina, Brassica rapa, Impatiens glandulifera, Polygonum persicaria, Reseda luteola, Sonchus oleraceus, Stellaria graminea, Thely-crania sanguinea.

RARE: Rumex obtusifolius.

C. MILLSTONE GRIT REGION

The Millstone Grit region of Derbyshire is the southern end of a long, continuous outcrop of this series which runs south from central Yorkshire, where it caps the central part of the Pennines for some fifty miles. In the northern part of Derbyshire it also caps the central anticlinal Pennine ridge, the highest point being on the plateau of Kinder Scout, which reaches 2,088 ft. To the south the outcrop divides to run down each side of the central Carboniferous Limestone region of the county. Much of the outcrop forms high moorland country with impressive escarpments which are known locally as 'edges'. These areas have the largest areas of semi-natural vegetation in the county, although it should be stressed that it has been much altered from the original state both by grazing and burning over hundreds of years, and also through the effects of air pollution since the beginning of the Industrial Revolution. The bog areas in particular have suffered drastic changes in this way.

The moorlands extend southwards for a considerable distance along the high ground, but towards the south, particularly along the eastern extension, the outcrops form progressively lower hills which, to the south of Matlock, are mostly covered with permanent pasture. To the north, agricultural land is confined to the flanks of the hills, seldom rising above about 1,000 ft., and to the valleys.

The Millstone Grit series include both shales and sandstones (see p. 9). The best known and most conspicuous of the latter are the massive, coarse-grained, gritstone bands. The shales have given rise to poorly drained clay soils which are often covered with acid grasslands, while the sandstones form rather sandy, well drained soils which are often podsolized. Bridges (1966) gives a detailed description of these sandstone soils in the south of the region. In the north of the county these soils are only usually formed, however, on steep slopes and just below rock outcrops, the majority of the gentler slopes, particularly at higher altitudes, being covered by a variable thickness of peat.

The whole of the Millstone Grit and Carboniferous Limestone regions together comprise the Peak District. The first description of the vegetation of this area was by Moss in 1913 in his classic 'Vegetation of the Peak District'. This is the only general account of the vegetation of the area apart from the relevent chapters in Tansley's 'The British Islands and their Vegetation' which are, in fact, taken very largely from the work of Moss, and the ecological account of Pearsall in 'Mountains and Moorlands'. Latterly, Eyre (1966) has given a general account of the vegetation of the moorlands on the south-east extension of the Millstone Grit series.

The majority of the lower ground in this region up to 1,000 ft. or perhaps somewhat higher, would under natural conditions be covered by woodland, but today only a few remnants of woods exist, which have not obviously been planted. Some of the best of these are on the west side of Sheffield. They are mainly oakwoods, dominated by *Quercus petraea*, but grading into birchwood at the higher altitudes. Thus Priddock Wood, overlooking Ladybower Reservoir, has a mixture of *Quercus petraea* and *Betula pubescens* in the upper part, with the upper limits at just over 1,000 ft. consisting mainly of scattered birch trees. The presence of scattered trees of *Betula pubescens* on both Ringinglow Moor and Houndkirk Moor, at between 1,300 ft. and 1,350 ft., give some indication of the possible tree limit. One of the best known of these woods is Padley Wood, an oakwood in a steep sided valley cut into the Millstone Grit. The wood has previously been described by Pigott (1956), who considers it to be a relict of much more extensive woodland in the area, owing its survival to the rough and unsuitable nature of the ground.

Habitat Study 23 *Padley Wood*

SK 253795.   Alt. 750 ft.   Slope: 15°.   Aspect: South-east.
Recorder: T. T. Elkington.   Surveyed: 20-9-66.   Area: 4 sq. yds.

The wood is dominated by *Quercus petraea*, with *Betula pubescens*, and *B. pendula* frequent and occasional specimens of *Sorbus aucuparia*. The general ground vegetation (in which the sample was taken) is dominated by *Deschampsia flexuosa* and *Vaccinium myrtillus*, and has a rich bryophyte and lichen flora which is listed by Pigott (1956).

Trees of *Alnus glutinosa* are frequent by the stream and a ground vegetation more characteristic of wet flushes is developed.

ABUNDANT: Deschampsia flexuosa, Pteridium aquilinum, Vaccinium myrtillus.

OCCASIONAL: Agrostis tenuis, Galium saxatile, Luzula pilosa, Oxalis acetosella.

RARE: Blechnum spicant, Dryopteris dilatata.

The vegetation of the Gritstone region, apart from the woodlands, may be divided into three main groups: moor dominated by dwarf shrub species usually developed on peat soils with some drainage; bogs on peat where the water table is near the surface; grasslands usually on mineral soils. Each of these groups may also include flush vegetation where springs emerge or by the sides of streams. It is extremely difficult to give a representative selection of the vegetation types within these groups, particularly for the moors and grasslands, because the species, although few in number, may be found in all combinations forming a complex vegetation mosaic even within a relatively small area. This mosaic is probably both the result of man's influence, particularly by burning, and of the natural variation in soil profile and depth and water drainage. The studies which are given here, although indicating the main floristic composition of the vegetation, do not, therefore, reflect the variations in frequency which may be found.

Three studies were made on the dwarf shrub moor at Stanage Edge. They illustrate some of the variation in floristic composition and frequency and differences in soil profile which may be found on similar sites all over the northern Gritstone region.

Habitat Study 24 *Stanage Edge*

SK 255828   Alt. 1,450 ft.   Slope: Level.
Recorder: T. T. Elkington.   Surveyed: 19-9-66.   Area: 4 sq. yds.

Mixed dwarf shrub moor on a small plateau, interspersed with patches, several yards across, of *Nardus stricta*. Soil consisting of 27 cm. black peat, pH 3·7, resting on gritstone bedrock.

ABUNDANT: Calluna vulgaris, Deschampsia flexuosa, Empetrum nigrum.
FREQUENT: Carex binervis.
OCCASIONAL: Nardus stricta.

Habitat Study 25   *Stanage Edge*
SK 255828   Alt. 1,450 ft.   Slope: 12°.   Aspect: South-west.
Recorder: T. T. Elkington.   Surveyed: 19-9-66.   Area: 4 sq. yds.
   Moor on an even sloping hillside below the edge. Soil profile consisting of
1·5 cm raw litter, 10-12 cm black peat, pH 3·7, 4 cm chocolate-coloured
weathered gritstone resting on bedrock.
ABUNDANT: Calluna vulgaris, Deschampsia flexuosa, Empetrum nigrum,
   Vaccinium myrtillus.
OCCASIONAL: Agrostis tenuis.
RARE: Anthoxanthum odoratum, Galium saxatile.

Habitat Study 26   *Stanage Edge*
SK 381248   Alt. 1,350 ft.   Slope: 10°.   Aspect: South-west.
Recorder: T. T. Elkington.   Surveyed: 19-9-66.   Area: 4 sq. yds.
   Calluna moor on evenly sloping hillside. Soil profile consisting of 2 cm.
Calluna litter, 8-10 cm. black peat, pH 3·2, and 5·7 cm brown sandy soil
resting on gritstone bedrock.
ABUNDANT: Calluna vulgaris.
OCCASIONAL: Deschampsia flexuosa.
RARE: Vaccinium myrtillus.

   In some areas *Vaccinium vitis-idaea* may also be a component of the moor
vegetation as in Study 27.

Habitat Study 27   *Slippery Stones, Upper Derwent*
SK 168955   Alt. 1,000 ft.   Slope: 22°   Aspect: South-east Valley.
Recorder: Conservancy Unit.   Surveyed: 29-5-68.   Area: 4 sq. metres.
   Moor vegetation on hillside; soil, peat with a pH of 3·5.
ABUNDANT: Deschampsia flexuosa, Vaccinium myrtillus.
FREQUENT: Festuca ovina, Vaccinium vitis-idaea.
RARE: Nardus stricta.

   The bog areas of the Derbyshire Gritstone region are similar to those found
over the whole of the Southern Pennines, and which have been described by
Pearsall (1950). They are characterised by a paucity of *Sphagnum*, of which
the underlying peat is largely made, and a general dominance of *Eriophorum
vaginatum*.
   The best known bog area in Derbyshire is undoubtedly Ringinglow
Bog, to the west of Sheffield, where both present vegetation and past history
has been studied by Conway (1947, 1949). The history of the area, as shown
by pollen analysis, has also been compared with other sites in the South
Pennines by Conway (1954). It seems most likely that the almost complete
absence of *Sphagnum*, except for *S. recurvum*, which is present mainly in

drainage channels, is due to atmospheric pollution since the beginning of the Industrial Revolution, and that the dominance of *Eriophorum vaginatum* is a comparatively recent feature. (See Conway (1949) and Tallis (1964)).

Study 27 was made in the type of vegetation which seems most widespread. A detailed vegetation description of the whole bog has been given by Conway (1949).

Habitat Study 28   *Ringinglow Bog*

SK 250833   Alt. 1,400 ft.   Slope: Level.

Recorder: T. T. Elkington.   Surveyed: 19–9–66.   Area: 4 sq. yds.

Widespread vegetation on bog surface which has scattered pools one to two feet across. The substratum is peat with a pH of 3·4. In wetter areas *Eriophorum angustifolium* becomes dominant.

FREQUENT: Calluna vulgaris, Deschampsia flexuosa, Eriophorum vaginatum.

OCCASIONAL: Tetraphis pellucida.

RARE: Eriophorum angustifolium.

These species recur constantly in similar areas over the whole of the northern Gritstone region. The peat is dissected by erosion channels in many places and a slightly different vegetation is often found on the redistributed peat. *Empetrum nigrum* is a common species in this habitat, and at higher altitudes *Rubus chamaemorus* is also found along the edge of these channels.

Habitat Study 29   *Grinds Brook, Edale*

SK 116870   Alt. 1,575 ft.   Slope: 11–17°.   Aspect: North-east.

Recorder: Conservancy Unit.   Surveyed: 21–5–65.   Area: 4 sq. metres.

Vegetation colonizing peat (pH. 3·3) over six feet deep in the top of a drainage gully.

FREQUENT: Deschampsia flexuosa, Empetrum nigrum, Eriophorum vaginatun, Vaccinium myrtillus.

RARE: Carex sp., Festuca ovina, Nardus stricta.

The grasslands of the Millstone Grit region are mainly developed on mineral soils. On the grit bands themselves grassland is particularly characteristic of a zone just below the edges (Study 30) and also along stream sides where mineral soil is exposed (Study 31).

Habitat Study 30   *Stanage Edge*

SK 248832   Alt. 1,350 ft.   Slope: 25°.   Aspect: West-south-west.

Recorder: T. T. Elkington.   Surveyed: 19–9–66.   Area: 4 sq. yds.

Grassland on steeply sloping hillside immediately below the crags, ground irregular with projecting rocks. Soil 32 cm deep of secondary origin consisting of a mixture of peat and sand, pH 3·9.

ABUNDANT: Festuca ovina.

FREQUENT: Galium saxatile.

OCCASIONAL: Agrostis tenuis, Anthoxanthum odoratum, Calluna vulgaris, Rumex acetosella.

RARE: Deschampsia flexuosa, Vaccinium myrtillus.

Habitat Study 31 *Burbage Brook*

SK 261829 Alt. 1,300 ft. Slope: 41°. Aspect: West-north-west.

Recorder: Conservancy Unit. Surveyed: 2–8–68. Area: 4 sq. metres.

Steeply sloping grassland by stream.

ABUNDANT: Deschampsia flexuosa, Holcus mollis.

FREQUENT: Agrostis canina.

OCCASIONAL: Luzula multiflora, Nardus stricta.

RARE: Anthoxanthum odoratum, Galium saxatile, Potentilla erecta, Vaccinium myrtillus.

Grasslands may also be developed on more level areas, where there has been no accumulation of peat. The grassland may be dominated by *Deschampsia flexuosa* without or with *Festuca ovina* as in Study 32, or by *Nardus stricta*.

Habitat Study 32 *Moscar Moor*

SK 223867 Alt. 1,150 ft. Slope: 11°. Aspect: West.

Recorder: Conservancy Unit. Surveyed: 11–8–65. Area: 4 sq. metres.

Acid sheep-grazed grassland. Total soil depth 17 inches, the upper 6 inches of raw greasy humus, the lower part of dark grey-brown sandy loam with pH 3·2 (at 10 inches depth).

ABUNDANT: Deschampsia flexuosa, Festuca ovina.

FREQUENT: Nardus stricta.

RARE: Empetrum nigrum, Vaccinium myrtillus.

Similar grasslands are found on the shales and fine-grained sandstones, although they often tend to be more species-rich. Two studies have been included, one (Study 33) on a sloping site with relatively free drainage, and the other (Study 34) on a level site with underlying shale, where the drainage is impeded.

Habitat Study 33 *Grinds Brook, Edale*

SK 116870 Alt. 1,400 ft. Slope: 13–18°. Aspect: North-east.

Recorder: Conservancy Unit. Surveyed: 21–5–65. Area 4 sq. metres.

Acid grassland dominated by *Deschampsia flexuosa* and *Nardus stricta*. Soil depth 19 inches (surface pH 3·6) developed on fine-banded sandstone.

ABUNDANT: Deschampsia flexuosa, Galium saxatile, Nardus stricta.

FREQUENT: Anthoxanthum odoratum, Festuca ovina.

RARE: Agrostis tenuis, Carex pilulifera, Luzula campestris, Vaccinium myrtillus.

Habitat Study 34 *Mam Tor*

SK 130833. Alt. 1,200 ft. Slope: Level. General Aspect: East.

Recorder: Conservancy Unit. Surveyed: 25–5–65. Area: 4 sq. metres.

Grassland on shale rock. Drainage is impeded with the soil profile, composed mainly of clayey soil, 19 inches deep, with pH 5·0.

The water table is at 9 inches depth and roots are not found deeper than this.

FREQUENT: Agrostis tenuis, Deschampsia caespitosa, Festuca rubra, Holcus lanatus, Ranunculus repens, Rumex acetosa.

OCCASIONAL: Anthoxanthum odoratum, Poa pratensis.

RARE: Cirsium palustre, Festuca ovina, Juncus effusus, Viola riviniana.

Flushes are frequent throughout the region by streams and springs, the latter being common at the bases of sandstone bands, where they rest on underlying impermeable shales. The vegetation of the flushes is variable and presumably depends to a great extent on the mineral content of the water. Two of the commonest types are dominated by *Sphagnum* and *Juncus effusus* or by *Molinia caerulea*, and examples of these are given in Studies 35 and 36. Others are much more species-rich, and examples of these are given in Studies 37 and 38.

Habitat Study 35   *Stanage Edge*
SK 244833.   Alt. 1,150 ft.   Slope: 15°.   Aspect: West-south-west.
Recorder: A. R. Clapham.   Surveyed: August, 1967.

Flush with (i) a central area dominated by *Sphagnum recurvum* and without *Juncus effusus* and (ii) a surrounding zone dominated by *Juncus effusus*.

(i)   Agrostis   stolonifera   Anthoxanthum   odoratum   Carex   nigra, Deschampsia flexuosa, Erica tetralix, Eriophorum angustifolium, Polytrichum commune, Potentilla erecta, Sphagnum palustre, Sphagnum recurvum.

(ii)   Agrostis   stolonifera,   Agrostis   tenuis,   Deschampsia   flexuosa, Juncus effusus, Molinia caerulea, Polytrichum commune, Sphagnum recurvum.

Habitat Study 36   *Stanage Edge*
SK 244833   Alt. 1,150 ft.   Slope: Level.
Recorder: A. R. Clapham.   Surveyed: August, 1967.

Flush at bottom of hillside dominated by *Molinia caerulea*.

ABUNDANT: Molinia caerulea.

OCCASIONAL: Deschampsia flexuosa, Nardus stricta.

RARE: Vaccinium myrtillus.

Habitat Study 37   *White Edge*
SK 259773.   Alt. 950 ft.   Slope: 10–15°.   Aspect: West.
Recorder: C. B. Waite.   Surveyed: 17-7-66.   Area: Triangular area, 50 yards long.

Flush draining into a small stream beneath a gritstone escarpment and surrounded by moorland. The whole area is dominated by *Crepis paludosa* and *Dactylorhiza maculata* ssp. *ericetorum* when they are in flower.

ABUNDANT: Crepis paludosa, Dactylorhiza maculata ssp. ericetorum, Equisetum fluviatile Holcus mollis, Juncus acutiflorus.

FREQUENT: Agrostis tenuis, Angelica sylvestris, Anthoxanthum odoratum, Cirsium palustre, Filipendula ulmaria, Lotus uliginosus, Luzula multiflora (var. compacta), Potentilla erecta, Ranunculus acris, Succisa pratensis, Viola palustris.

OCCASIONAL: Athyrium filix-femina, Calluna vulgaris, Carex echinata, Deschampsia caespitosa, Deschampsia flexuosa, Dryopteris carthusiana, Epilobium palustre, Erica tetralix, Festuca ovina, Galium palustre, Hypericum tetrapterum, Lychnis flos-cuculi, Molinia caerulea, Plantago lanceolata, Pteridium aquilinum, Rumex acetosa, Stellaria graminea.

RARE: Blechnum spicant, Eriophorum angustifolium, Holcus lanatus, Vaccinium vitis-idaea.

Habitat Study 38 *Salter Sitch, near Owler Bar*

SK 290778 Alt. 1,100 ft. Slope: Level.

Recorder: C. B. Waite. Surveyed: 3–7–66. Area: 25 by 12 yds. approx.

Flat marshy area bordering a stream. Soil black humus overlying clay.

ABUNDANT: Equisetum fluviatile, Hydrocotyle vulgaris.

FREQUENT: Carex nigra, Deschampsia flexuosa, Festuca rubra, Glyceria fluitans, Juncus subuliflorus, Juncus effusus, Poa trivialis, Viola palustris.

OCCASIONAL: Achillea ptarmica, Anthoxanthum odoratum, Briza media, Carex echinata, Cirsium palustre, Deschampsia caespitosa, Epilobium palustre, Festuca pratensis, Galium palustre, Galium uliginosum, Holcus lanatus, Ophioglossum vulgatum, Triglochin palustre.

There are no natural ponds or lakes of any size on the Millstone Grit Series, and the best locality for sub-aquatic and aquatic vegetation is probably Combs Reservoir, near Chapel-en-le-Frith, which is on the boundary of this Series and Coal Measures. Two studies are included from this area, one made in a marsh area and the other on bare mud exposed at a time of severe drought.

Habitat Study 39 *Combs Reservoir, near Chapel-en-le-Frith*

SK 043792 Alt. 750 ft. Slope: Level.

Recorder: R. H. Hall. Surveyed: 25–8–46.

Marshy ground at head of Reservoir. The area is liable to occasional inundation, when the level of the reservoir is high.

FREQUENT: Achillea ptarmica, Bidens tripartita, Eleocharis palustris, Polygonum amphibium, Polygonum hydropiper.

OCCASIONAL: Epilobium obscurum, Hypericum tetrapterum, Mentha × verticillata, Myosotis palustris, Senecio aquaticus.

Habitat Study 40 *Combs Reservoir, near Chapel-en-le-Frith*

SK 037795. Alt. 750 ft.

Recorder: R. H. Hall. Surveyed: Aug. and Sept., 1949.

The summer and autumn periods of 1947 and 1949 were extremely dry, and large areas of shore mud were uncovered for four months up to the beginning of November in each year. The flora of flowering plants and bryophytes of this mud was of considerable ecological interest and showed marked zonation from the water's edge to the high water mark. *Limosella aquatica, Peplis portula,* and *Littorella uniflora* occurred in the open association of freshly uncovered mud near to the water's edge. Higher up the shore seedlings of *Bidens tripartita* and *Gnaphalium uliginosum* were plentiful, in association with *Callitriche stagnalis, Ranunculus flammula* and *Polygonum aviculare.*

On October 22nd, 1949, *Limosella aquatica*—not previously seen at Combs —was present in vast quantities and formed pure communities in a belt approximately ten yards wide from the margin of the water. Higher up the shore *Limosella* was associated with the species already recorded.

The bryophyte flora of the site was equally striking, consisting primarily of ephemeral species. Of exceptional interest was the occurrence in great quantity of *Physcomitrium sphaericum,* a rare annual, first recorded for the locality in 1894 and not seen again until 1947.

D. CARBONIFEROUS LIMESTONE REGION

The central and western parts of the county are occupied by a large continuous outcrop of Carboniferous Limestone covering an area of approximately 180 square miles from Buxton and Castleton in the west and north to Matlock and Brassington in the east and south, and extending west into Staffordshire. The only other rocks to outcrop in this area are a series of basaltic volcanic rocks which are known locally as 'toadstone'. They are mainly covered by grassland which is generally more calcifuge in character than that of the surrounding limestone. There are also small inliers of Carboniferous Limestone both to the east (at Crich and Ashover) and to the south (at Ticknall and Calke Abbey) of the main outcrop.

This limestone region is basically a plateau, which is lower to the south, at about 900 ft.—1,100 ft. and higher and more rolling to the north, with the hills reaching up to 1,400 ft.—1,500 ft. The plateau is dissected by numerous valleys or 'dales', many of them deeply incised into the limestone. A few still have rivers or streams flowing in them, but many are dry, both because the main valleys are cut so deeply, and also because of the drainage channels known as 'soughs' cut by miners in the eighteenth and nineteenth centuries in an attempt to drain the lead mines.

The major part of the limestone plateau is covered by permanent grassland, which is either used for grazing or for hay. Woodland is confined to a few plantations and shelter belts round farms, the natural woodland cover, which is thought to have been the climax vegetation for most of the area having been destroyed many centuries ago, probably before Norman times, partially at least to provide fuel for smelting lead. There is little arable land, mainly due to the relatively high altitude of the plateau area.

The soils of the plateau are mainly brown-earths, which Pigott (1962) has shown to have been formed partly from detrital limestone material and partly from loess, the two components having been intimately mixed under the periglacial conditions which the area was subject to when it was near the edge of the ice-sheet during the last (Weichsel) glaciation. The deeper soils have now become leached and partially podsolized and carry a calcifuge vegetation.

The dales, because of their steep slopes, have not been utilized to the same extent as the plateau for agriculture, and are the main reservoirs of natural and semi-natural vegetation of the limestone region. Even here, some sheep and cattle grazing takes place, and this, together with past exploitation of the woodland and varying rabbit pressure, has led to the formation of the mixture of woodland, scrub and grassland which is characteristic of the dales today.

The most natural vegetation in the dales is almost certainly that of the cliffs and crags, and many of the uncommon British plants present in Derbyshire are confined to this habitat. Both herb and woodland communities are present, and the floristic composition of both is greatly influenced by aspect which affects the microclimate to a considerable degree as has been shown by Pigott (1958) and Jarvis (1963). Two examples are given of herbaceous communities, one (Study 41) from a north-facing cliff and the other (Study 42) from a south-facing one.

Habitat Study 41 *Winnats Pass*

SK 135825   Alt. 1,250 ft.   Slope: 43°—62°.   Aspect: North-east.
Recorder: Conservancy Unit. Surveyed: 14-7-65. Area: 1 sq. metre.

Saxifraga hypnoides dominant as hanging mats on small Carboniferous Limestone crags interspaced with terraces covered with tall herbs, the whole cliff moist because of flushing.

ABUNDANT: Festuca rubra, Geranium robertianum, Saxifraga hypnoides.

FREQUENT: Arrhenatherum elatius, Epilobium montanum, Rumex acetosa, Scabiosa columbaria, Silene dioica, Trisetum flavescens, Valeriana officinalis.

OCCASIONAL: Cerastium fontanum, Geranium lucidum, Holcus lanatus, Mercurialis perennis, Poa pratensis, Polemonium caeruleum, Taraxacum officinale.

RARE: Galium sterneri, Veronica chamaedrys.

Habitat Study 42 *Miller's Dale*

SK 155732   Alt. 650 ft.   Slope: 40°.   Aspect: South-south-east.
Recorder: Conservancy Unit. Surveyed: 11-9-68. Area: 1 sq. metre.

ABUNDANT: Centaurea scabiosa, Festuca ovina, Helianthemum chamaecistus.

FREQUENT: Anthoxanthum odoratum, Campanula rotundifolia, Centaurea nigra, Dactylis glomerata, Geranium sanguineum, Helictotrichon pratense, Helictotrichon pubescens, Hieracium pilosella, Koeleria cristata, Lotus corniculatus, Poterium sanguisorba, Plantago lanceolata, Silene nutans, Trisetum flavescens.

OCCASIONAL: Arrhenatherum elatius, Carex flacca, Festuca arundinacea.

RARE: Briza media, Cerastium fontanum, Galium sterneri, Hieracium sp., Leontodon hispidus, Linum catharticum, Medicago lupulina, Pimpinella saxifraga, Scabiosa columbaria, Senecio jacobaea, Solidago virgaurea, Viola hirta.

Undisturbed woodland in the dales area is probably confined to crags and cliffs, and there are several tree species, e.g. *Taxus baccata* and *Sorbus rupicola*, which are restricted to these situations on the limestone. Study 43 gives the trees and shrubs from one cliff of this type in Cressbrook Dale.

Habitat Study 43   *Cressbrook Dale*
SK 174736   Alt. 900 ft.   Slope: Cliff.   Aspect: West.
Recorder: A. R. Clapham. Surveyed 18-8-66. Area: c. 25 yards of cliff.
Cliff just south of Ravensdale cottages.

Tree Layer
FREQUENT: Fraxinus excelsior.
OCCASIONAL: Sorbus aucuparia, Sorbus rupicola, Taxus baccata.
RARE: Ulmus glabra.

Shrub Layer
ABUNDANT: Crataegus monogyna, Hedera helix.
FREQUENT: Ligustrum vulgare.
OCCASIONAL: Corylus avellana.
RARE: Rhamnus catharticus, Thelycrania sanguinea.

The majority of the woods of the dales have probably been exploited at some time in the past, particularly for fuel for lead smelting, and their species composition today is probably a consequence of this disturbance. They are now largely dominated by Ash (*Fraxinus excelsior*) and as such are some of the best known ashwoods in Britain. Although some of the woods contain only Ash in the tree layer, many have variable proportions of Elm (*Ulmus glabra*) and Sycamore (*Acer pseudoplatanus*).

The shrub layer is often well developed and includes *Acer campestre*, *Corylus avellana*, *Crataegus monogyna*, *Euonymus europeaus*, *Prunus padus*, *Rhamnus catharticus*, *Thelycrania sanguinea* and *Viburnum opulus*. The ground vegetation is quite variable, but may often be divided into two types, that on the dale sides where the slope is relatively steep and the soil shallow with scree near the surface, and the vegetation near the dale bottoms where the soil is deeper and moister. Studies 44 and 45 have been made in these two situations respectively. A detailed study of the woodland in Monk's Dale, with a description of the possible succession, has been given by Scurfield (1959).

Habitat Study 44   *Cressbrook Dale*
SK 173742.   Alt. 900 ft.   Slope: 19°.   Aspect: West-north-west.
Recorder: Conservancy Unit. Surveyed: 25-5-67. Area: 1 sq. metre.

Fraxinus woodland on scree with a shrub layer of *Corylus avellana* and *Crataegus monogyna* giving moderate shade.
ABUNDANT: Melica uniflora, Mercurialis perennis.
FREQUENT: Allium ursinum, Hedera helix.
OCCASIONAL: Anemone nemorosa, Arum maculatum, Galium aparine, Rubus
   saxatilis.

Habitat Study 45 *Cressbrook Dale*

SK 173742   Alt. 850 ft.   Slope: 27°.   Aspect: East.

Recorder: Conservancy Unit. Surveyed: 25–5–67. Area: 1 sq. metre.

Woodland of *Fraxinus excelsior* and *Acer pseudoplatanus* near dale bottom on a clay soil.

ABUNDANT: Deschampsia caespitosa, Oxalis acetosella.

FREQUENT: Anemone nemorosa, Melica uniflora, Mercurialis perennis.

OCCASIONAL: Allium ursinum, Festuca rubra, Geranium robertianum, Glechoma hederacea, Heracleum sphondylium, Sanicula europaea, Torilis japonica.

RARE: Brachypodium sylvaticum, Geum × intermedium, Myosotis sylvatica, Pimpinella major, Prunus spinosa (seedling), Taraxacum officinale.

Other common members of the ground flora are *Endymion non-scriptus* and *Galium odoratum*, while *Primula vulgaris* is also to be found occasionally.

Around the edges of the woodland and in clearings is a vegetation intermediate between that of the woodland and grassland, often with abundant *Convallaria majalis*. Study 46 is an example of this community.

Habitat Study 46 *Via Gellia, near Matlock*

SK 275570   Alt. 600 ft.   Slope: 32°.   Aspect: North-north-west.

Recorder: Conservancy Unit. Surveyed: 23–7–65. Area: 1 sq. metre.

Open scree grassland with soil of pH 7·8 at 10 cm depth.

ABUNDANT: Convallaria majalis, Melica nutans, Mercurialis perennis.

FREQUENT: Arrhenatherum elatius, Brachypodium sylvaticum, Carex flacca, Festuca ovina, Teucrium scorodonia.

OCCASIONAL: Festuca rubra, Helianthemum chamaecistus, Tussilago farfara.

RARE: Campanula rotundifolia, Scrophularia nodosa.

The non-wooded parts of the dales contain a complex mixture of communities. The most open vegetation is found on the screes which are common on the steeper slopes and in the early stages of colonization *Arrhenatherum elatius* and *Geranium robertianum* may be the only species present, as in Study 47.

Habitat Study 47 *Lathkill Dale*

SK 175655   Alt. 650 ft.   Slope: 34°.   Aspect: South.

Recorder: Conservancy Unit. Surveyed: 20–7–66. Area: 1 sq. metre.

Scattered vegetation on unstabilized scree with almost no soil.

OCCASIONAL: Arrhenatherum elatius, Geranium robertianum.

As the vegetational cover increases, so does the number of species and eventually the scree becomes stabilized by the development of calcareous grassland. An intermediate stage in this succession, where there are still large areas of bare rock, but a considerable diversity of species, is shown by Study 48.

Habitat Study 48   *Biggin Dale*

SK 146583   Alt. 750 ft.   Slope: 31°.   Aspect: West-north-west.

Recorder: Conservancy Unit.   Surveyed: 4–7–68.   Area: 1 sq. metre.

ABUNDANT: Arrhenatherum elatius, Festuca rubra, Viola riviniana.

FREQUENT: Briza media, Carex caryophyllea, Carex flacca, Festuca ovina, Galium sterneri, Geranium robertianum, Hieracium pilosella, Thymus drucei.

OCCASIONAL: Anthoxanthum odoratum, Campanula rotundifolia, Fragaria vesca, Helictotrichon pratense, Leontodon hispidus, Plantago lanceolata, Senecio jacobaea, Trifolium repens.

RARE: Poterium sanguisorba, Scabiosa columbaria.

The floristic composition of the grasslands of the dale slopes is very variable and is dependent on a number of factors. Firstly, there is variation in the species composition of the grassland between dales unrelated to physical factors, which seems to be due to each dale or group of dales having been isolated from one another for a considerable time. Secondly, both the species composition and diversity is greatly affected by grazing. Grazed grassland generally has a greater abundance of *Festuca ovina*, less *Arrhenatherum elatius* and *Helictotrichon pratense*, more annuals and a greater diversity of species in terms of number per unit area than ungrazed grassland and these differences are well shown in Studies 49 and 50. Burning is another important factor, which has recently been considered in detail by Lloyd (1968). He points out that the greater amount of litter present in ungrazed grassland means that this tends to be subject to more accidental fires than grazed grassland. Furthermore, the effect of burning is to reinforce the species differences between grazed and ungrazed grassland in that *Festuca ovina* is sensitive to and decreases in frequency after burning, while *Helictotrichon pratense* and *Arrhenatherum elatius*, when present, increase in frequency.

Habitat Study 49   *Cressbrook Dale*

SK 173746   Alt. 900 ft.   Slope: 37°.   Aspect: South-east.

Recorder: Conservancy Unit.   Surveyed: 30–6–65.   Area: 1 sq. metre.

Open grassland on shallow rendzina; soil surface pH of 6·7.

ABUNDANT: Anthyllis vulneraria, Carex caryophyllea, Festuca ovina, Helictotrichon pratense, Hieracium pilosella, Leontodon hispidus, Linum catharticum, Poterium sanguisorba, Thymus drucei.

FREQUENT: Centaurea nigra, Galium sterneri, Helianthemum chamaecistus, Koeleria cristata, Pimpinella saxifraga, Plantago lanceolata, Scabiosa columbaria.

OCCASIONAL: Campanula rotundifolia, Euphrasia officinalis, Hieracium sp., Polygala vulgaris, Primula veris, Sieglingia decumbens, Succisa pratensis.

RARE: Anthoxanthum odoratum, Cerastium fontanum, Crataegus monogyna (seedlings), Fraxinus excelsior (seedlings), Gymnadenia conopsea, Minuartia verna, Plantago media, Silene nutans.

9. *Aerial view of the upper part of Lathkill Dale looking north-west; showing the dale incised in the limestone plateau*

10. MARSH ANDROMEDA (*Andromeda polifolia*); *Axe Edge, Buxton*

11. *Head of Grinds Brook, Kinder Scout showing a gritstone edge and block scree below.*

12. SPRING CINQUEFOIL (*Potentilla tabernaemontani*); *Back Dale.*

Habitat Study 50 *Lathkill Dale*

SK 183657 Alt. 650 ft. Slope: 32°. Aspect: South-west.

Recorder: Conservancy Unit. Surveyed: 5-7-66. Area: 1 sq. metre.

Ungrazed, previously burnt, daleside grassland; surface soil pH 6·3.

ABUNDANT: Arrhenatherum elatius, Festuca rubra, Poa pratensis, Poterium sanguisorba.

FREQUENT: Helictotrichon pratense, Koeleria cristata.

OCCASIONAL: Anthoxanthum odoratum, Carex flacca, Festuca ovina, Galium sterneri, Linum catharticum, Trisetum flavescens.

RARE: Campanula rotundifolia, Crataegus monogyna (seedlings), Heracleum sphondylium, Senecio jacobaea, Sonchus asper, Taraxacum officinale.

Another important physiographic factor is that of aspect and, as in the case of cliffs and crags mentioned earlier, this has a very considerable effect on the species composition of the grassland. This is discussed by Grime and Blythe (1969) in relation to their work in the Winnats Pass, and Studies 51 and 52 made here illustrate these differences.

Habitat Study 51 *Winnats Pass*

SK 135825 Alt. 1,250 ft. Slope: 41°. Aspect: North.

Recorder: Conservancy Unit. Surveyed: 19-7-65. Area: 1 sq. metre.

Grassland on steep, north-facing slope. Soil a shallow rendzina with surface pH of 6·4.

ABUNDANT: Arrhenatherum elatius, Festuca rubra.

FREQUENT: Helictotrichon pubescens, Heracleum sphondylium, Holcus lanatus, Mercurialis perennis, Poa pratensis, Poa trivialis, Urtica dioica.

OCCASIONAL: Anthriscus sylvestris, Campanula rotundifolia, Cerastium fontanum, Epilobium montanum, Galium sterneri, Rumex acetosa, Silene dioica, Trisetum flavescens.

RARE: Cirsium vulgare, Galium aparine, Galium cruciata, Mycelis muralis, Oxalis acetosella, Polemonium caeruleum, Scabiosa columbaria, Taraxacum officinale.

Habitat Study 52 *Winnats Pass*

SK 135825 Alt. 1,250 ft. Slope: 50°. Aspect: South.

Recorder: Conservancy Unit. Surveyed: 19-7-65. Area: 1 sq. metre.

Grassland on steep, south-facing slope. Soil a shallow rendzina with surface pH of 7·3.

ABUNDANT: Arrhenatherum elatius, Festuca ovina, Festuca rubra.

FREQUENT: Campanula rotundifolia, Dactylis glomerata, Draba muralis, Koeleria cristata, Myosotis arvensis, Poa pratensis, Scabiosa columbaria, Trisetum flavescens, Veronica agrestis.

OCCASIONAL: Aphanes arvensis, Arabidopsis thaliana, Arenaria serpyllifolia, Bellis perennis, Cerastium fontanum, Galium sterneri, Hieracium pilosella, Lotus corniculatus, Medicago lupulina, Thymus drucei.

RARE: Carduus nutans, Geranium molle, Pimpinella saxifraga, Rumex acetosa, Sedum acre, Senecio vulgaris, Taraxacum officinale.

The grassland vegetation of the south-facing slopes is similar to that of south-facing crags (see Study 42) and grades into it as the slope increases. Within such communities a very distinctive vegetation dominated by annual species is often developed on small terraces and rock-outcrops. These areas have a shallow soil, which is often a protorendzina, almost completely composed of black humus. Habitat Study 53 illustrates the species composition of this vegetation. The adaptations of the main species present to this habitat and its characteristics are discussed in detail by Ratcliffe (1961).

Habitat Study 53   *Monsal Head*

SK 184716   Alt. 750 ft.   Slope: 5°.   Aspect: South-south-west.

Recorder: Conservancy Unit. Surveyed: 6-6-67. Area: 1 sq. metre.

Open grassland on terrace with abundant annuals; soil surface pH 7·1.

ABUNDANT: Arenaria serpyllifolia, Festuca ovina, Medicago lupulina, Poa pratensis.

FREQUENT: Aphanes arvensis, Arrhenatherum elatius, Cerastium semidecandrum, Dactylis glomerata, Erophila verna, Festuca rubra, Myosotis ramosissima, Plantago lanceolata, Saxifraga tridactylites, Sedum acre.

OCCASIONAL: Catapodium rigidum, Galium verum, Geranium molle, Koeleria cristata, Sagina apetala, Trifolium dubium, Veronica arvensis.

RARE: Bellis perennis, Lotus corniculatus, Poa annua.

In contrast to the south-facing slopes, the grassland vegetation of the north-facing slopes has affinities both with that of crags and cliffs of similar aspect (see Study 41), and also with that found in flushed seepage areas, which may be found on slopes of any aspect, although perhaps more commonly on those with a northern one. The vegetation of such areas is commonly dominated by tall herbs. A number of uncommon Derbyshire plants are either restricted or most common in them. These include *Polemonium caeruleum*, the habitat of which is described in detail by Pigott (1958), *Trollius europaeus* and *Cirsium heterophyllum*. There is considerable variation in this vegetation type, and Pigott (l.c.) has pointed out that sites which have *Polemonium caeruleum* as a component species only rarely have *Deschampsia caespitosa*, *Trollius europaeus* and *Cirsium heterophyllum*. Habitat Study 54 was made on a north-facing slope of this type with *Polemonium caeruleum*.

Habitat Study 54   *Chee Dale*

SK 117726   Alt. 750 ft.   Slope: 41°.   Aspect: North-north-east.

Recorder: Conservancy Unit. Surveyed: 26-9-68. Area: 1 sq. metre.

Ungrazed slope; soil surface pH 7·3.

ABUNDANT: Arrhenatherum elatius, Festuca rubra, Geum rivale, Pimpinella major.

FREQUENT: Mercurialis perennis, Polemonium caeruleum, Viola riviniana.

OCCASIONAL: Angelica sylvestris, Dactylis glomerata, Deschampsia caespitosa, Gymnadenia conopsea, Heracleum sphondylium, Holcus lanatus, Hypericum hirsutum, Succisa pratensis.

RARE: Helictotrichon pubescens, Poa pratensis, Solidago virgaurea.

Sometimes, because of grazing, a species-rich turf has developed in seepage areas. Small herbs, e.g. *Parnassia palustris* and *Succisa pratensis* are conspicuous. Study 55 is an example of this type of vegetation.

Habitat Study 55 *Topley Pike*

SK 113726   Alt. 900 ft.   Slope: 47°.   Aspect: East-north-east.

Recorder: Conservancy Unit.   Surveyed: 13–9–68.   Area: 1 sq. metre.

Steep, moist, lightly grazed slope; soil surface pH 6·0.

ABUNDANT: Agrostis canina, Festuca ovina, Potentilla erecta, Succisa pratensis.

FREQUENT: Anthoxanthum odoratum, Carex caryophyllea, Carex flacca, Carex panicea, Deschampsia flexuosa, Hieracium sp., Lotus corniculatus, Viola riviniana.

OCCASIONAL: Agrostis tenuis, Angelica sylvestris, Arrhenatherum elatius, Briza media, Calluna vulgaris, Campanula rotundifolia, Cirsium palustre, Galium sterneri, Festuca rubra, Geum rivale, Helictotrichon pubescens, Holcus lanatus, Hypericum pulchrum, Leontodon hispidus, Parnassia palustris, Plantago lanceolata, Poa pratensis, Scabiosa columbaria, Dactylis glomerata, Helictotrichon pratense, Linum catharticum, Potentilla sterilis, Poterium sanguisorba, Senecio jacobaea, Solidago virgaurea.

Seepage areas on steep slopes may also be colonized by calcifuges, even where leaching is minimal, and the same slope in Chee Dale on which Study 54 was made also has *Calluna vulgaris* and *Vaccinium myrtillus* growing on it.

Towards the upper edges of the dales, where the slopes are more gradual, the soil depth is generally deeper, the pH lower, and a profile similar to that of a brown earth is developed. The relationships of this zone with that of the dale slopes and the plateau above were first considered in detail by Balme (1953), who interpreted the sequence as a catena with the increase in soil depth and associated leaching dependent on the progressive reduction in slope. Pigott (1962), however, showed that the plateau soils have a considerable loessic component and that the soil found along the brows of the dales is a mixture derived from the *in situ* soil and plateau soil which has slipped down. Furthermore, in areas where the dale sides are not steep, plateau soil may be found covering parts of the slope as in Coombs Dale, and such areas are often indicated by the presence of *Ulex europaeus*. The relationship between the vegetation of such areas of plateau soil and the adjoining calcareous grassland on rendzina soils derived wholly from the limestone has been discussed by Grime (1963). The grassland found along the brows on these soils of lower pH is particularly characterized by the presence of *Agrostis tenuis* (see Balme l.c.), and Study 56 illustrates the floristic composition of this vegetation. Two of the most common associate species are *Lathyrus montanus* and *Viola lutea*, and when the latter is·in flower the community often becomes conspicuous through its presence.

Habitat Study 56   *Wardlow Hay Cop, Cressbrook Dale*

SK 177744   Alt. 1,000 ft.   Slope: 20°—25°.   Aspect: North-east.

Recorder: Conservancy Unit.   Surveyed: 8–7–65.   Area: 1 sq. metre.

Soil a leached brown-earth with a surface pH of 5·2.

ABUNDANT: Agrostis tenuis, Campanula rotundifolia, Festuca ovina, Festuca rubra, Galium saxatile, Luzula campestris.

FREQUENT: Anthoxanthum odoratum, Briza media, Carex caryophyllea, Helictotrichon pratense, Koeleria cristata, Viola riviniana.

OCCASIONAL: Achillea millefolium, Anemone nemorosa, Carex flacca, Carex panicea, Deschampsia flexuosa, Helictotrichon pubescens, Potentilla erecta, Trifolium repens, Viola lutea.

RARE: Sieglingia decumbens, Succisa pratensis.

Other common species found on these soils of intermediate pH status include *Ulex europaeus, Hypericum pulchrum, Potentilla erecta* and *Conopodium majus*.

On the plateau itself leaching of the deeper soils has taken place and in some areas the profile has become podsolized with accumulation of peat and a limestone heath vegetation has developed. The degree of drainage is variable and Study 57 was made in an area with only slightly impeded drainage, while Study 58 was made in one with considerable impedance to drainage.

Habitat Study 57   *Wardlow Hay Cop*

SK 178743   Alt. 1,100 ft.   Slope: 15°—20°.   Aspect: North-north-east.

Recorder: Conservancy Unit.   Surveyed: 7–7–65.   Area: 1 sq. metre.

ABUNDANT: Agrostis canina, Deschampsia flexuosa, Festuca ovina, Galium saxatile, Luzula campestris, Nardus stricta, Vaccinium myrtillus.

FREQUENT: Anthoxanthum odoratum.

OCCASIONAL: Agrostis tenuis, Anemone nemorosa, Deschampsia caespitosa, Helictotrichon pratense.

RARE: Cerastium fontanum.

Habitat Study 58   *Bradwell Moor*

SK 148798 Alt. 1,350 ft.   Slope: 4°.   Aspect: East-south-east.

Recorder: Conservancy Unit.   Surveyed: 30–5–68.   Area: 1 sq. metre.

Heather moor, surface pH of soil 3·8.

ABUNDANT: Calluna vulgaris, Deschampsia flexuosa, Trichophorum caespitosum, Vaccinium myrtillus.

OCCASIONAL: Carex nigra, Empetrum nigrum, Erica tetralix, Luzula campestris.

Only a minority of the dales now contain permanent rivers, but many others contain springs which result in temporary streams of varying length according to the season and rainfall. The margins of the rivers and springs are often occupied by marsh vegetation, and one example of this is included here.

Habitat Study 59 *Monsal Dale*

SK 172705   Alt. 450 ft.   Slope: 5°.   Aspect: North-east.

Recorders: I. Sollitt and M. R. Shaw. Surveyed: 7–8–66. Area: 20 yards by 7 yards.

Marsh developed round a spring in a field by the River Wye. Area always wet, but water flow to river may not be permanent.

ABUNDANT: Cardamine pratensis, Carex flacca, Holcus lanatus, Juncus articulatus.

FREQUENT: Blysmus compressus, Pinguicula vulgaris.

OCCASIONAL: Apium nodiflorum, Caltha palustris, Carex nigra, Cirsium palustre, Cynosurus cristatus, Epilobium palustre, Equisetum arvense, Filipendula ulmaria, Juncus effusus, Juncus inflexus, Lotus uliginosum, Prunella vulgaris, Potentilla anserina, Succisa pratensis.

RARE: Briza media, Carex hirta, Eriphorum angustifolium, Hypericum tetrapterum, Lychnis flos-cuculi, Myosotis scorpioides, Ranunculus acris, Triglochin palustris, Tussilago farfara.

Sub-aquatic vegetation is present both along river and stream margins and also around artificial lakes and pools, including some of the many dew ponds. An indication of the variation in this vegetation is given in the next three Habitat Studies.

Habitat Study 60 *Monk's Dale*

SK 137740   Alt. 700 ft.   Slope: 3°.   Aspect: South.

Recorder: Conservancy Unit. Surveyed: 20–8–68. Area: 1 sq. metre.

Stream course in bottom of dale; surface soil pH 7·9.

ABUNDANT: Apium nodiflorum.

FREQUENT: Phalaris arundinacea, Potentilla anserina, Nasturtium microphyllum.

OCCASIONAL: Agrostis stolonifera, Equisetum palustre, Mentha × verticillata, Veronica beccabunga.

RARE: Deschampsia caespitosa, Epilobium parviflorum, Trifolium repens.

Habitat Study 61 *Via Gellia*

SK 289572   Alt. 300 ft.   Slope: 1°.   Aspect: South-south-west.

Recorder: Conservancy Unit. Surveyed: 24–9–68. Area: 1 sq. metre.

Marsh vegetation developed in silted-up disused mill pond.

ABUNDANT: Cardamine flexuosa, Epilobium hirsutum, Phalaris arundinacea.

FREQUENT: Filipendula ulmaria.

RARE: Angelica sylvestris, Cirsium palustre, Fraxinus excelsior (seedling), Urtica dioica.

Other common species to be found dominating river and pond vegetation include *Filipendula ulmaria, Solanum dulcamara* and *Typha latifolia.*

Habitat Study 62   *Longstone Moor*

SK 195747   Alt. 1,100 ft.   Slope: Level.

Recorder: Conservancy Unit.   Surveyed: 1-8-67.   Area: 1 sq. metre.

Margin of artificial pool associated with old mineral workings; surrounding vegetation limestone heath.   Surface soil pH of pool margin 5·4.

ABUNDANT: Glyceria fluitans, Lemna minor.

FREQUENT: Juncus articulatus.

OCCASIONAL: Cardamine pratensis, Eleocharis palustris.

RARE: Ranunculus flammula.

Artificial habitats are not important on the limestone, the two commonest being those associated with quarrying and mineral working. Disused quarries in the limestone have very similar vegetation to the corresponding natural habitats, particularly that of crags and screes, the degree of development depending on the length of time for which the quarry has been disused. The most distinctive artificial habitats of the limestone are the spoil heaps associated with mineral working, mainly for lead and fluorspar.   These are colonised by an open vegetation, the species composition of which depends particularly on the abundance of rock containing lead ore. Where the amounts are high the only species found are *Festuca ovina* and *Minuartia verna*, but where there is a greater admixture of limestone a more species-rich vegetation develops, which often contains *Cochlearia pyrenaica* in the Castleton area and *Thlaspi alpestre* in the Matlock area. Studies are included from both these areas.

Habitat Study 63   *Gratton Dale, near Pikehall*

SK 192595   Alt. 900 ft.   Slope: 30°.   Aspect: North-west.

Recorder: K. M. Hollick.   Surveyed: 3-6-66.   Area: 1 sq. metre.
Spoil heap of old lead working.

ABUNDANT: Campanula rotundifolia, Festuca ovina, Minuartia verna.

FREQUENT: Achillea millefolium, Galium sterneri, Plantago lanceolata, Ranunculus repens, Thlaspi alpestre, Viola lutea.

OCCASIONAL: Cerastium fontanum, Rumex acetosella, Thymus drucei, Veronica chamaedrys.

RARE: Carex caryophyllea, Euphrasia nemorosa.

Habitat Study 64   *Dirtlow Rake, near Castleton*

SK 154822   Alt. 1,050 ft.   Slope: 25°.   Aspect: South.

Recorder: T. T. Elkington.   Surveyed: 20-9-66.   Area: 4 sq. metres.

Spoil heaps from lead vein, material mainly limestone and fluorspar; pH of soil 7·6.

OCCASIONAL: Arabis hirsuta, Arrhenatherum elatius, Campanula rotundifolia, Cochlearia pyrenaica, Dactylis glomerata, Euphrasia sp., Festuca ovina, Galium verum, Hieracium pilosella, Leontodon hispidus, Linum catharticum, Lotus corniculatus, Minuartia verna, Sedum acre, Thymus drucei.

Hedges are not common in the limestone region, their place being taken by dry-stone walls. Some of these have an interesting flora and two studies made on north and south-facing embankment walls in the Via Gellia are included here. The north-facing walls often have an abundance of ferns, *Polypodium vulgare* often being found with or instead of those listed in Study 65.

Habitat Study 65 *Via Gellia*

SK 285574   Alt. 450 ft.   Slope: 91°.   Aspect: North-north-west.
Recorder: Conservancy Unit.   Surveyed: 24-9-68.   Area: 1 sq. metre.
    Embankment wall of limestone facing road.
FREQUENT: Cystopteris fragilis.
OCCASIONAL: Asplenium ruta-muraria, Dryopteris filix-mas, Epilobium montanum, Sagina procumbens.
RARE: Mycelis muralis, Poa trivialis, Taraxacum officinale.

Habitat Study 66 *Via Gellia*

SK 285574   Alt. 450 ft.   Slope: 75°.   Aspect: South.
Recorder: Conservancy Unit.   Surveyed: 24-9-68.   Area: 1 sq. metre.
    Embankment wall of limestone facing road.
ABUNDANT: Cymbalaria muralis.
FREQUENT: Arrhenatherum elatius.
OCCASIONAL: Geranium robertianum, Poa pratensis.
RARE: Hieracium sp., Plantago major, Taraxacum officinale.

    As has been mentioned in the introductory paragraphs, the only other rock types to outcrop in the limestone area are the basaltic lavas known as 'toadstone'. These are mainly covered by an interesting series of grasslands which are intermediate in character between the calcareous grasslands of the dale slopes and the calcifuge grasslands developed, for example, on the Millstone Grit series. It is not surprising, therefore, that they have a number of similiarities with the grasslands developed on the leached brown-earth soils of the dale margins and have a number of the same characteristic species. These include *Agrostis tenuis*, *Potentilla erecta*, *Betonica officinalis*, *Galium verum* and *Lathyrus montanus*. There is considerable variation in floristic composition in the toadstone grasslands, depending partly on the degree of mineral enrichment from the surrounding limestone, and also on whether the grassland is grazed or not. Two studies are included here from toadstone outcrops, of which the first is grazed and the second ungrazed and has by comparison a greater abundance of tall herb species than the grazed sample.

Habitat Study 67 *Knotlow, Monk's Dale*

SK 134736   Alt. 1,020 ft.   Slope: 20°—24°.   Aspect: North-east.
Recorder: Conservancy Unit.   Surveyed: 4-6-65.   Area: 1 sq. metre.
    Rough, well-drained, cow-grazed pasture on toadstone. Mineral soil with surface pH 5·1.
ABUNDANT: Agrostis tenuis, Conopodium majus, Festuca rubra, Holcus lanatus.

FREQUENT: Anthoxanthum odoratum, Betonica officinalis, Festuca ovina, Galium saxatile, Rumex acetosa.

OCCASIONAL: Cerastium fontanum, Dactylis glomerata, Helictotrichon pratense, Lotus corniculatus, Luzula campestris, Plantago lanceolata, Poa pratensis, Potentilla erecta, Trifolium repens, Veronica chamaedrys.

RARE: Crataegus monogyna (seedling), Galium verum.

Habitat Study 68   *Miller's Dale (Blackwell Dale)*

SK 134731   Alt. 850 ft.   Slope: 44°.   Aspect: North-west.

Recorder: Conservancy Unit. Surveyed: 15-7-68. Area: 1 sq. metre.

Grassland on toadstone which had been burnt in the Spring, but not grazed; surface soil pH 4·9.

ABUNDANT: Agrostis tenuis, Festuca rubra, Potentilla erecta.

FREQUENT: Achillea millefolium, Angelica sylvestris, Anthoxanthum odoratum, Arrhenatherum elatius, Centaurea nigra, Chamaenerion angustifolium, Dactylis glomerata, Deschampsia caespitosa, Hieracium sp., Holcus lanatus, Lotus uliginosus, Plantago lanceolata, Poa pratensis, Vaccinium myrtillus, Viola riviniana.

OCCASIONAL: Agrostis canina, Campanula rotundifolia, Carex caryophyllea, Epilobium montanum, Galium verum, Lathyrus montanus, Luzula campestris, Rubus idaeus, Rumex acetosa, Sieglingia decumbens.

E. SOUTHERN LOWLAND REGION

The southern part of the county, south of a line from Ashbourne in the west to Sandiacre in the east, is a lowland area, more or less bisected by the River Trent and its alluvial plain. The region mainly occupies part of a large outcrop of Triassic deposits which extends into the surrounding counties, although in the extreme south it includes part of a Coal Measures outlier and exposures of shales of the Millstone Grit series are also present. As well as alluvial deposits, there are considerable amounts of glacial drift in places.

The Triassic deposits are of two main types. The Bunter beds are mainly composed of unconsolidated sandstone and pebble beds which give rise to light, free-draining, sandy soils. These are not good arable soils because of low inherent fertility and low available water capacity. Nevertheless they are extensively cultivated, about half the area being devoted to arable crops, the rest being either ley or permanent grassland. The Keuper series, which occupies the majority of the area, includes both sandstones and clay bands, and gives rise to much heavier loam and clay soils, which are largely covered by permanent grassland.

The Trent valley has a wide flood plain, the surface deposits of which are silts and clays. The soils developed on these are heavy, poorly drained and often gleyed, and many parts are subject to flooding. The majority of the area is therefore used for permanent pasture or long-term ley. These superficial deposits are underlain by extensive gravel beds. Large-scale extraction of the gravel is taking place with the subsequent production of lakes, some of which develop an extensive marginal vegetation. Some of the lakes are now being filled in with ash from the large electrical power stations which have been built along the Trent to utilize the river water for cooling purposes.

This land-use pattern, together with urban development, particularly in the Derby area, has meant that semi-natural vegetation is very limited in extent. Woodland in particular is very restricted and is more or less confined to areas unsuitable for agriculture, except for a few places where timber is grown commercially, e.g. Shirley Park. Two woodland studies are included here, one from a Bunter sandstone rock outcrop and the other from a wet valley bottom.

Habitat Study 69   *Anchor Church, near Repton*

SK 340272   Alt. 150 ft.   Slope: 70°.   Aspect: North.

Recorders: E. Wain and K. M. Hollick. Surveyed: 11-6-66. Area: 1 sq. metre.

Bunter sandstone cliff with semi-natural woodland dominated by *Ulmus glabra* with *Quercus robur* subdominant. There are also planted trees of *Acer pseudoplatanus*, *Fagus sylvatica* and *Tilia × vulgaris*.

Thin soil on steep slopes with surface pH of 5·8. Two samples of ground vegetation recorded.

(i)

ABUNDANT: Mycelis muralis, Poa nemoralis.

FREQUENT: Chamaenerion angustifolium, Silene dioica.

OCCASIONAL: Holcus lanatus.

RARE: Digitalis purpurea, Taraxacum officinale.

(ii)

FREQUENT: Chrysanthemum parthenium, Galium aparine, Hedera helix, Mercurialis perennis.

OCCASIONAL: Festuca gigantea, Lamiastrium luteum, Silene dioica.

RARE: Digitalis purpurea, Dryopteris filix-mas, Myosotis arvensis.

Habitat Study 70   *Tinkers' Inn, Clifton*

SK 178445   Alt. 500 ft.   Slope: Slight.   Aspect: North.

Recorder: K. M. Hollick. Surveyed: 18-10-68. Area: 1 sq. metre.

Carr woodland, dominated by *Alnus glutinosa*, in a small valley on Keuper deposits. Surface soil pH 6·8.

Herb Layer

ABUNDANT: Equisetum arvense, Valeriana dioica.

FREQUENT: Angelica sylvestris, Carex paniculata, Galium palustre, Juncus effusus, Lychnis flos-cuculi, Phleum pratense, Prunella vulgaris, Ranunculus repens, Rubus fruticosus, Rumex acetosa, Salix cinerea, Valeriana officinalis.

RARE: Cirsium palustre, Crataegus monogyna, Rubus idaeus, Taraxacum officinale.

Other fragments of carr woodland of this type exist, some with *Quercus robur*, e.g. Cubley Coppice, and presumably oak-alder woodlands occupied many of the poorly drained areas in this region in the past. One of these is at Hulland Moss (Study 71) a valley bog which has been partly drained with development of heath, but has still a number of wetland species. Marsh vegetation survives in various places including Radbourne Rough, along the Bradley and Wyaston brooks, and at Tinker's Inn, some areas being dominated by *Carex paniculata*.

Habitat Study 71   *Hulland Moss*

SK 250460 Alt. 600 ft. Slope and Aspect: Almost level, draining to South. Recorder: K. M. Hollick. Surveyed: 10–10–66. Area: 1 sq. metre.

Drained bog area generally dominated by *Agrostis tenuis* and *Molinia caerulea* with bushes of *Rubus fruticosus*, *Ulex europaeus* and *Ulex gallii*. Area being colonized by *Betula pendula* with trees of *Alnus glutinosa* and *Quercus robur* also frequent. Soil humus-rich, with a surface pH of 4·7. Area at junction of Bunter deposits and shales of the Millstone Grit series.

Two samples recorded:

(i)

ABUNDANT: Agrostis tenuis.

FREQUENT: Carex panicea, Hydrocotyle vulgaris, Juncus acutiflorus, Potentilla erecta, Sphagnum sp., Polytrichum sp.

OCCASIONAL: Carex echinata, Holcus mollis, Juncus articulatus, Molinia caerulea, Viola palustris.

RARE: Quercus robur (seedling).

(ii)

ABUNDANT: Narthecium ossifragum.

FREQUENT: Agrostis tenuis, Molinia caerulea.

OCCASIONAL: Holcus mollis, Nardus stricta, Potentilla erecta, Rumex acetosa, Succisa pratensis.

RARE: Angelica sylvestris, 'Anthoxanthum 'odoratum, Cardamine pratensis, Juncus acutiflorus, Rubus fruticosus, Ulex europaeus, Viola palustris.

Much of the region where not arable is covered with meadow and pasture, and only small areas retain a semi-natural appearance. The Bunter sandstones weather to form an acid, sandy soil and grassland on such outcrops usually has abundant *Agrostis tenuis* and *Deschampsia flexuosa*. Where the soil has become podsolized, heather moor may develop as at Carver's Rocks, where Study 72 was made.

Habitat Study 72   *Carver's Rocks*

SK 332228   Alt. 440 ft.   Slope: Slight.   Aspect: South-west. Recorder: A. L. Thorpe. Surveyed: 22–6–66. Area: 4 yards square.

Grassland on brown sandy soil, surface pH 3·8.

ABUNDANT: Agrostis tenuis.

FREQUENT: Deschampsia flexuosa, Rumex acetosella.

OCCASIONAL: Calluna vulgaris, Pteridium aquilinum.

RARE: Cerastium fontanum, Polygonum aviculare, Quercus (seedling), Rubus fruticosus, Sorbus aucuparia (seedling), Teucrium scorodonia.

When the grassland is used as pasture and manuring takes place, forage grasses, e.g. *Anthoxanthum odoratum*, *Alopecurus pratensis* and *Dactylis glomerata* usually become abundant, as in Study 73. Similar grassland is present on the Keuper, although areas of rough grassland only occasionally grazed often have poor drainage, and species such as *Cirsium palustre* and *Juncus effusus* may then be present as in Study 74.

Habitat Study 73 *Near Ashbourne*
SK 171463. Alt. 500 ft. Slope: Slight. Aspect: North-west.
Recorder: K. M. Hollick. Surveyed: 27-5-65. Area: 25 sq. metres.

Meadow on Bunter deposits, ploughed once in the 1940's, cut for hay, followed by some grazing by cows. Occasionally manured.

Soil pH at 4 inches depth 5·8.

ABUNDANT: Alopecurus pratensis, Anthoxanthum odoratum, Dactylis glomerata, Ranunculus acris.

FREQUENT: Bellis perennis, Heracleum sphondylium, Holcus lanatus, Lolium perenne, Plantago lanceolata, Poa pratensis, Rumex acetosa.

OCCASIONAL: Cerastium fontanum, Ranunculus bulbosus, Sanguisorba officinalis, Taraxacum officinale.

RARE: Achillea millefolium.

Habitat Study 74 *Tinker's Inn, Clifton*
SK 178445 Alt. 500 ft. Slope: Slight. Aspect: South.
Recorder: K. M. Hollick. Surveyed: 18-10-68. Area: 1 sq. metre.

Rough pasture on Keuper.

ABUNDANT: Cynosurus cristatus.

FREQUENT: Agrostis tenuis, Bellis perennis, Hypochaeris radicata, Juncus effusus, Leontodon autumnalis, Lotus corniculatus, Plantago lanceolata, Prunella vulgaris, Ranunculus repens.

RARE: Cirsium arvense, Cirsium palustre, Taraxacum officinale.

The alluvial deposits of the river flood plains in the region, particularly those of the River Trent, are mainly covered by pasture, although latterly some of these have been ploughed. Formerly many of the pastures had a rich flora, but continued agricultural improvement has considerably reduced their number. An example of one of these is given in Study 75.

Habitat Study 75 *Near Repton*
SK 303273 Alt. 140 ft. Slope: Level.
Recorder: E. Wain. Surveyed: 31-5-65.

Pasture on edge of old flood plain. Attempt at drainage made in past and remaining hollows are waterlogged in winter. Surface soil pH from 'hummock' 6·4 and from 'hollow' 5·0; some evidence of gleying. *Alopecurus pratensis* and *Ranunculus acris* co-dominant.

ABUNDANT: Alopecurus pratensis, Festuca rubra, Holcus lanatus, Ranunculus acris, Rumex acetosa, Rumex acetosella.

FREQUENT: Anthoxanthum odoratum, Cardamine pratensis, Cerastium fontanum.

OCCASIONAL: Achillea millefolium, Bellis perennis, Centaurea nigra, Conopodium majus, Cynosurus cristatus, Plantago lanceolata, Salix fragilis, Scrophularia nodosa, Senecio jacobaea, Trifolium pratense.

RARE: Taraxacum officinale.

The region contains a number of rivers, including the Trent and the lower reaches of the Dove and Derwent. In addition there are several canals and numerous ponds and lakes, including those resulting from gravel extraction. It is not surprising, therefore, that the best developed sub-aquatic and aquatic vegetation in the county is in this region. Ponds, particularly on the heavy Keuper soils, often have a rich marginal vegetation, and Study 76 is an example from one of the small ponds which occur in the Darley Moor–Yeaveley area.

Habitat Study 76   *Yeaveley*

SK 181393   Alt. 475 ft.   Slope: Level.

Recorder: K. M. Hollick.   Surveyed: 8-8-66.   Area: 9 sq. metres.

Small pond in pasture on Keuper series; partly shaded by nearby trees. Pond mud with pH 5·0.

ABUNDANT: Glyceria fluitans, Sparganium fluitans.

FREQUENT: Agrostis stolonifera, Juncus articulatus, Lemna minor, Potentilla palustris.

OCCASIONAL: Bidens cernua, Carex pseudocyperus, Juncus effusus, Ranunculus flammula.

RARE: Galium palustre, Polygonum persicaria.

At Morley, in the east of the county, an interesting sub-aquatic and aquatic vegetation is associated with pools which have resulted from the extraction of clay for brick-making:

Habitat Study 77   *Morley*

SK 389418   Alt. 440 ft.   Slope: Level.

Recorder: Derby N.H.S.   Surveyed: June, 1966.   Area: margin of pool.

Marginal vegetation of old clay-pit. Site on shales of the Millstone Grit series with overlying boulder clay. Soil pH 3·6.

ABUNDANT: Sphagnum sp., Typha latifolia.

FREQUENT: Callitriche stagnalis, Equisetum palustre, Juncus effusus, Potamogeton natans, Sparganium erectum.

OCCASIONAL: Alisma plantago-aquatica, Cardamine hirsuta, Cirsium palustris, Epilobium palustre, Galium palustre, Glyceria fluitans, Hydrocotyle vulgaris, Lotus uliginosus, Myosotis scorpioides.

RARE: Hottonia palustris.

There is a wide variety of aquatic and sub-aquatic habitats in the Trent Valley area, and three samples from these are included here; from a meadow ditch, a canal bank, and the margin of a gravel pit.

Habitat Study 78 *Walton-on-Trent*

SK 215190   Alt. 160 ft.   Slope: Level.

Recorder: E. Wain. Surveyed: 23-6-65.

Banks of ditch passing through meadows by River Trent. Soil a thick layer of clay over gravel.

ABUNDANT: Ranunculus sceleratus.

FREQUENT: Filipendula ulmaria, Juncus subuliflorus, Rosa canina, Solanum dulcamara.

OCCASIONAL: Bryonia dioica, Callitriche sp. (in water), Crataegus laevigata, Epilobium hirsutum, Galium palustre, Juncus effusus, Myosotis palustris, Phleum pratensis, Prunella vulgaris, Rumex acetosa.

RARE: Apium nodiflorum, Carex acutiformis, Carex otrubae, Cirsium vulgare, Galium verum, Iris pseudacorus, Lychnis flos-cuculi, Potentilla reptans, Nasturtium officinale, Sanguisorba officinalis, Sonchus asper, Vicia cracca.

Habitat Study 79 *Trent and Mersey Canal, near Shardlow*

SK 450306   Alt. 100 ft.

Recorder: R. H. Hall. Surveyed: 19-6-65. Area: 5 by 3 yards.

Stretch of south bank of canal, where there is a belt of marsh vegetation between the tow-path and the canal.

FREQUENT: Epilobium hirsutum, Filipendula ulmaria, Lythrum salicaria, Mentha aquatica, Nuphar lutea, Sparganium neglectum, Thalictrum flavum.

OCCASIONAL: Achillea millefolium, Achillea ptarmica, Acorus calamus, Alisma plantago-aquatica, Butomus umbellatus, Carex acuta, Carex acutiformis, Carex otrubae, Impatiens glandulifera, Iris pseudacorus, Juncus effusus, Lycopus europaeus, Phalaris arundinacea, Rorippa islandica, Rumex crispus, Scrophularia nodosa, Solanum dulcamara, Stachys palustris, Chrysanthemum vulgare, Valeriana officinalis, Vicia cracca.

RARE: Galium uliginosum, Rumex hydrolapathum.

Habitat Study 80 *Near Trent Station*

SK 488317   Alt. 90 ft.

Recorder: R. H. Hall. Surveyed: 7-11-65.

One of many pits resulting from the extraction of gravel in connection with the railway system. Soil on pool-margin, clay with pH 6·5.

Vegetation within a 6—8 foot belt along the pool-margin, water depth 0—24 inches.

FREQUENT: Chara hispida, Glyceria maxima, Nuphar lutea, Potamogeton natans, Schoenoplectus lacustris.

OCCASIONAL: Apium nodiflorum, Elodea canadensis, Glyceria fluitans, Juncus articulatus, Myosotis scorpioides.

Vegetation growing on banks of pool.

FREQUENT: Juncus inflexus, Poa pratensis.

OCCASIONAL: Apium nodiflorum, Ranunculus repens.

Gravel is also worked from the glacial drift about Hulland, the spoil-heaps developing a rich ephemeral flora of which Study 81 is an example.

Habitat Study 81 *Hulland Gravel Pit*

SK 263454   Alt. 650 ft.   Slope: Level, with hummocks.
Recorder: K. M. Hollick.   Surveyed: 9–7–68.

Surface of waste material, chiefly sand, which had not been tipped on for a number of years.

ABUNDANT: Agrostis tenuis, Dactylorhiza fuchsii, Holcus mollis, Lotus corniculatus, Trifolium hybridum, Trifolium repens.

FREQUENT: Equisetum arvense, Hypochaeris radicata, Silene alba.

OCCASIONAL: Aira caryophyllea, Aira praecox, Silene cucubalus, Vicia angustifolia, Vicia tetrasperma, Viola tricolor.

In this region hedgerows form an important reservoir for wild life, particularly for woodland species, and two examples are given here to illustrate their floristic composition.

Habitat Study 82 *Ashbourne Green*

SK 187477   Alt. 625 ft.   Slope: Level.
Recorder: K. M. Hollick.   Surveyed: 5–6–66.   Area: 10 yard length of hedge.

Length of well tended hedgerow with a wet ditch on the north-west side. Soil a sticky loam with a pH of 4·8, below the well-developed humus layer. On shales of the Millstone Grit series.

Shrub Layer

ABUNDANT: Crataegus monogyna.

FREQUENT: Acer campestre, Rubus fruticosus agg, Ulmus procera.

OCCASIONAL: Fraxinus excelsior, Prunus spinosa, Rosa canina agg, Sambucus nigra, Ulmus glabra.

Herb Layer

ABUNDANT: Mercurialis perennis.

FREQUENT: Adoxa moschatellina, Geranium robertianum, Poa pratensis, Ranunculus ficaria, Stellaria holostea, Urtica dioica, Vicia sepium.

OCCASIONAL: Arrhenatherum elatius, Agropyron repens, Arum maculatum, Chaerophyllum temulentum, Dactylis glomerata, Galium aparine, Ranunculus auricomus, Rumex sanguineus, Zerna ramosa.

Habitat Study 83 *Mapleton Road, Ashbourne*

SK 177468   Alt. 475 ft.   Slope: 10°.   Aspect: North-east.

Recorder: K. M. Hollick. Surveyed: 5-6-66. Area: 10 yard length of hedge.

Well tended hedgerow with sheep-grazed pasture on one side. Humus-rich soil with pH 6·8 developed on Bunter sandstone.

Shrub Layer

ABUNDANT: Crataegus monogyna.

FREQUENT: Corylus avellana, Rubus fruticosus agg, Thelycrania sanguinea.

OCCASIONAL: Acer pseudoplatanus, Sambucus nigra, Ulmus glabra.

Herb Layer

ABUNDANT: Arrhenatherum elatius, Festuca ovina.

FREQUENT: Anthriscus sylvestris, Dryopteris filix-mas, Geranium robertianum, Geum urbanum, Lamium album, Stachys sylvatica, Urtica dioica.

OCCASIONAL: Alliaria petiolata, Dryopteris dilatata, Epilobium montanum, Festuca gigantea, Hedera helix, Holcus lanatus, Lathyrus pratensis, Silene dioica, Veronica chamaedrys, Zerna ramosa.

RARE: Fragaria vesca.

Finally, where selective herbicides have not been used, an interesting annual flora is to be found in arable fields, and Study 84 was made in a barley field with such a flora.

Habitat Study 84 *Hulland*

SK 250464   Alt. 625 ft.   Slope: Level.

Recorder: K. M. Hollick. Surveyed: 16-8-65. Area: 1 sq. metre.

Weeds of a barley field.

ABUNDANT: Matricaria matricarioides, Polygonum aviculare, Spergula arvensis, Veronica persica, Viola agrestis.

FREQUENT: Chrysanthemum segetum, Myosotis arvensis, Polygonum convolvulus, Polygonum persicaria, Sinapis arvensis.

RARE: Stellaria media.

Other species present in the same field were:

FREQUENT: Convolvulus arvensis, Fumaria officinalis, Vicia angustifolia, Vicia hirsuta.

RARE: Aethusa cynapium, Geranium molle, Papaver rhoeas, Sonchus arvensis.

Acknowledgements

I wish to thank all those who have made the habitat studies which are included in this chapter and especially the Nature Conservancy Unit at Sheffield University who in particular have provided many of the limestone studies. I also wish to thank Miss K. M. Hollick and Dr. J. P. Grime for their helpful discussion of the vegetation of the southern lowland and limestone regions respectively.

REFERENCES

Balme, O. E. (1953). Edaphic and Vegetational Zoning on the Carboniferous Limestone of the Derbyshire Dales. J. Ecol., 41, 331–344.

Bridges, E. M. (1966). The Soils and Land Use of the District North of Derby. Memoir of the Soil Survey of Great Britain. Harpenden.

Conway, V. M. (1947). Ringinglow Bog, near Sheffield. I. Historical. J. Ecol., 34, 149–181.

Conway, V. M. (1949). Ringinglow Bog, near Sheffield. II. The present surface. J. Ecol., 37, 148–170.

Conway, V. M. (1954). Stratigraphy and Pollen Analysis of Southern Pennine Blanket Peats. J. Ecol., 42, 117–147.

Eyre, S. R. (1966). 'The Vegetation of a South Pennine Upland' in 'Geography as Human Ecology'. ed. Eyre, S. R. & Jones, G. R. J. London.

Grime, J. P. (1963). An Ecological Investigation at a junction between two plant communities in Coombsdale on the Derbyshire Limestone. J. Ecol., 51, 391–402.

Grime, J. P., and Blythe, G. M. (1969). Relationships between snails and vegettaion at the Winnats Pass. J. Ecol., 57, 43–66.

Jackson, G., and Sheldon, J. (1949). The Vegetation of Magnesian Limestone Cliffs at Markland Grips, near Sheffield. J. Ecol., 37, 38–50.

Jarvis, M. S. (1963). A comparison between the water relations of species with contrasting types of geographical distribution in the British Isles, in 'The Water Relations of Plants', ed. Rutter, A. J., and Whitehead, F. H., Oxford.

Jowett, G. H., and Scurfield, G. (1949). A statistical investigation into the distribution of *Holcus mollis* L. and *Deschampsia flexuosa* (L.) Trin. J. Ecol., 37, 68–81.

Jowett, G. H., and Scurfield, G. (1952). Statistical investigations into the Success of *Holcus mollis* L. and *Deschampsia flexuosa* (L.) Trin. J. Ecol., 40, 393–404.

Lloyd, P. S. (1968). The ecological significance of fire in limestone grassland communities of the Derbyshire Dales. J. Ecol., 56, 811–826.

Moss, C. E. (1913). Vegetation of the Peak District. Cambridge.

Pearsall, W. H. (1950). Mountains and Moorlands. London.

Pigott, C. D. (1956). 'Vegetation' in 'Sheffield and its Region', ed. Linton, D. L. British Association for the Advancement of Science. Sheffield.

Pigott, C. D. (1958). Biological Flora of the British Isles: *Polemonium caeruleum* L. J. Ecol., 46, 507–525.

Pigott, C. D. (1962). Soil formation and development on the Carboniferous Limestone of Derbyshire. I. Parent Materials. J. Ecol., 50, 145–156.

Ratcliffe, D. (1961). Adaptation to habitat in a group of annual plants. J. Ecol., 49, 187–203.

Scurfield, G. (1953). Ecological Observations in Southern Pennine Woodlands. J. Ecol., 41, 1–12.

Scurfield, G. (1959). The Ashwoods of the Derbyshire Carboniferous Limestone: Monk's Dale. J. Ecol., 47, 357–369.

Tallis, J. H. (1964). Studies on Southern Pennine Peats. III. The Behaviour of *Sphagnum*. J. Ecol., 52, 345–353.

Tansley, A. G. (1939). The British Islands and their Vegetation. Cambridge.

13. *Aerial view of Kinder Scout with light snow-cover, looking east and showing Kinder Downfall and the plateau with very dissected peat-bog*

R. H. Hall

14. HYBRID BILBERRY (*Vaccinium* x *intermedium*); *Poorlot's Quarries, Tansley*

15. *Aerial view of part of the Trent valley near Repton, looking east*

R. H. Hall

16. *Aquatic vegetation in the Trent Valley*

Chapter Six

GEOGRAPHICAL DISTRIBUTION

It has never been the intention for the Flora to include a large number of distribution maps of species within the county. Records have been collected so as to ensure the inclusion of at least one record for each species from each 10-kilometre grid square in which it grows, but lists of species have not been systematically made from grid squares of any smaller size. Nevertheless it has been thought useful to include a short section of maps both to illustrate the types of species distribution within the county and the varying distribution patterns of Derbyshire plants within Britain.

The Derbyshire maps have been constructed by making a dot for all localities for which a four figure grid reference is available. In the case of comparatively rare species e.g. *Silene nutans, Polemonium caeruleum* the dots give a nearly accurate picture of all localities. In the case of common species e.g. *Vaccinium vitis-idaea* this is not true and the dots, although showing more or less accurately the area and limits of distribution, should not be taken to represent all localities within the county. The maps of British distribution are based on those in the *Atlas of the British Flora* (reproduced by permission of the Botanical Society of the British Isles and Thomas Nelson & Sons Ltd.) as revised by the Biological Records Centre, Nature Conservancy, Monks Wood, Huntingdon. Each dot represents at least one record in a 10-kilometre square of the National Grid.

The types of distribution within Derbyshire, which are illustrated, are essentially controlled and limited by combinations of geological formation, topography and climate. The boundary between the highland and lowland zones of Britain crosses the county from south-west to north-east and therefore the pattern of distribution within the county for a species often closely reflects the species' distribution in Great Britain as a whole. Many species present in the county are limited by the occurrence of suitable habitats, thus many aquatic and sub-aquatic species are much commoner in the south of the county than in the north.

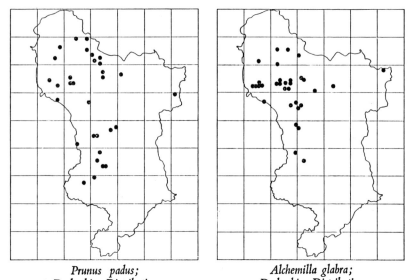

Prunus padus;
Derbyshire Distribution

Alchemilla glabra;
Derbyshire Distribution

Bryonia dioica;
Derbyshire Distribution

Coronopus squamatus;
Derbyshire Distribution

Prunus padus and Alchemilla glabra have predominantly northern and western distributions in the county and are not restricted by geological formation. They similarly both have a general northern and western distribution in Britain, although P. padus is also present in East Anglia.

Bryonia dioica and Coronopus squamatus have predominantly southern distributions in the county and this pattern is reflected in Britain as a whole where they are most common in the south and east of the country.

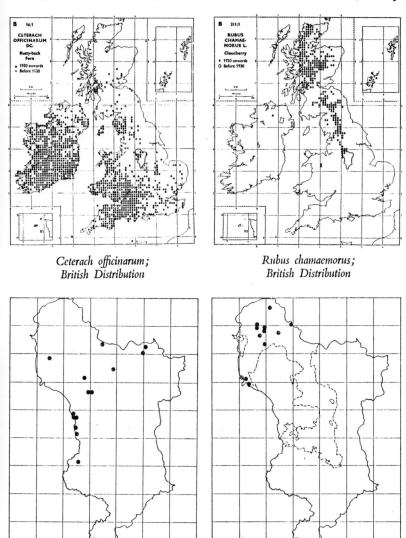

Ceterach officinarum;
British Distribution

Rubus chamaemorus;
British Distribution

Ceterach officinarum;
Derbyshire Distribution

Rubus chamaemorus;
Derbyshire Distribution

Ceterach officinarum has a northern and western distribution in Derbyshire, not limited by geological formation, and is most common in the south and west of Britain, the Derbyshire localities forming a somewhat isolated group towards the eastern boundary.

Rubus chamaemorus is restricted to relatively high altitudes on the Millstone Grit in the north of the county, which is at the southern limit of this north boreal species in Britain.

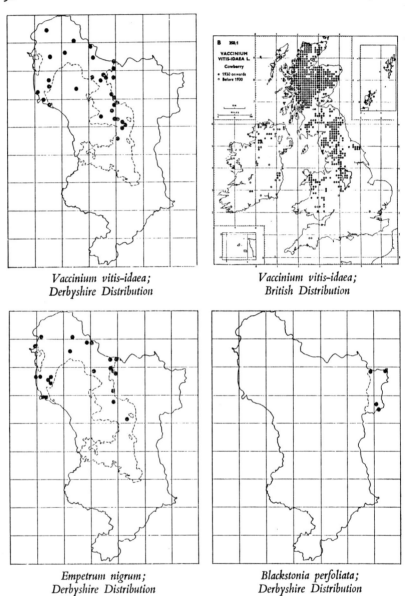

Vaccinium vitis-idaea;
Derbyshire Distribution

Vaccinium vitis-idaea;
British Distribution

Empetrum nigrum;
Derbyshire Distribution

Blackstonia perfoliata;
Derbyshire Distribution

Vaccinium vitis-idaea and *Empetrum nigrum* are both almost entirely confined to moorlands on rocks of the Millstone Grit Series in Derbyshire. Both are predominantly northern species with very similar British distributions, Derbyshire being near their south-east limits.

Blackstonia perfoliata is one of the few Derbyshire plants to be confined to the Magnesian Limestone outcrop. Other such species include *Carex ericetorum* and *Carex montana*.

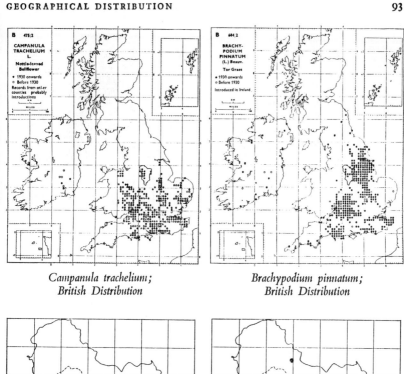

Campanula trachelium;
British Distribution

Brachypodium pinnatum;
British Distribution

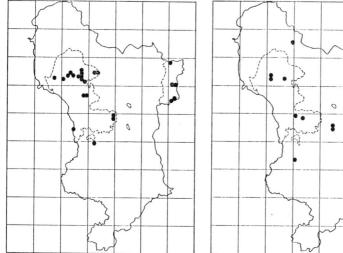

Campanula trachelium;
Derbyshire Distribution

Brachypodium pinnatum;
Derbyshire Distribution

Campanula trachelium and Brachypodium pinnatum, like a number of other species, e.g. Zerna erecta, are restricted to limestone in Derbyshire, but are present both on the Carboniferous and Magnesian Limestones. Both are southern species in Britain, but Brachypodium pinnatum has a much more eastern tendency which is reflected in its much greater abundance on the Magnesian Limestone.

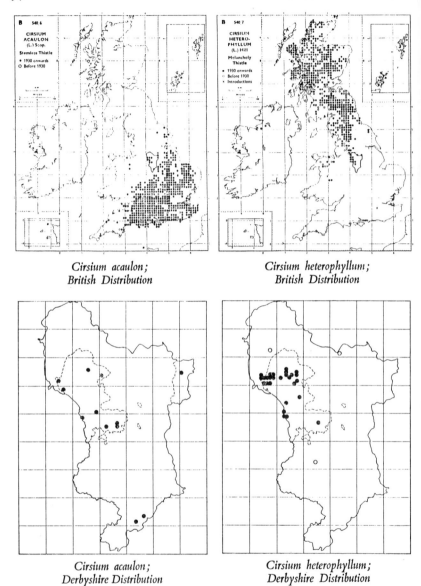

Cirsium acaulon;
British Distribution

Cirsium heterophyllum;
British Distribution

Cirsium acaulon;
Derbyshire Distribution

Cirsium heterophyllum;
Derbyshire Distribution

Cirsium acaulon and *C. heterophyllum* are two thistles largely confined to limestone in Derbyshire. In the map of *C. heterophyllum* open circles indicate long-established plants not on limestone, which may, however, be introductions. The maps of British distribution show that both are near their geographical limits in the county, *C. acaulon* being a southern species and *C. heterophyllum* being a northern one. *C. acaulon* is only found on south to south-west facing slopes in Derbyshire indicating how close it is to its northern climatic limit.

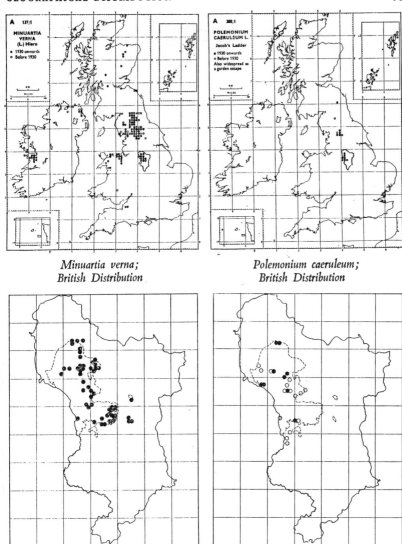

Minuartia verna;
British Distribution

Polemonium caeruleum;
British Distribution

Minuartia verna;
Derbyshire Distribution

Polemonium caeruleum;
Derbyshire Distribution

Minuartia verna is an often abundant plant of the Carboniferous Limestone of Derbyshire and is present on the two inliers to the east. It is one of several species which have their centre of distribution in northern England, with Derbyshire forming one of the main areas.

In the Derbyshire map of *Polemonium caeruleum* black circles indicate undoubted native localities (pers. comm. C. D. Pigott), which are all on Carboniferous Limestone. Open circles represent other localities, not examined by him at some of which the plants may be garden escapes. This rare species is confined to northern England with the dales of Derbyshire and Staffordshire forming one of its main areas.

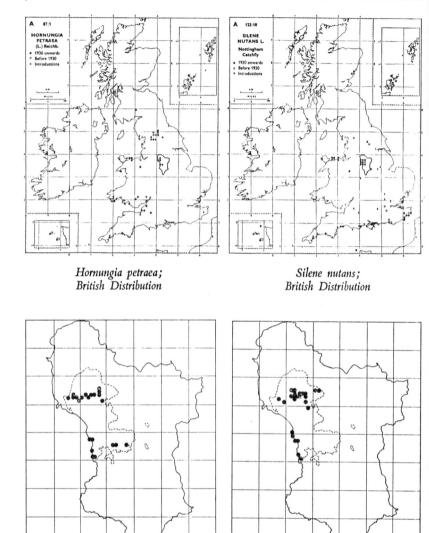

Hornungia petraea;
British Distribution

Silene nutans;
British Distribution

Hornungia petraea;
Derbyshire Distribution

Silene nutans;
Derbyshire Distribution

Hornungia petraea and *Silene nutans* are both confined to the Carboniferous Limestone in Derbyshire and are particularly centred on the valley systems of the Wye and Dove. In Britain they are both disjunct species with localities in widely scattered areas.

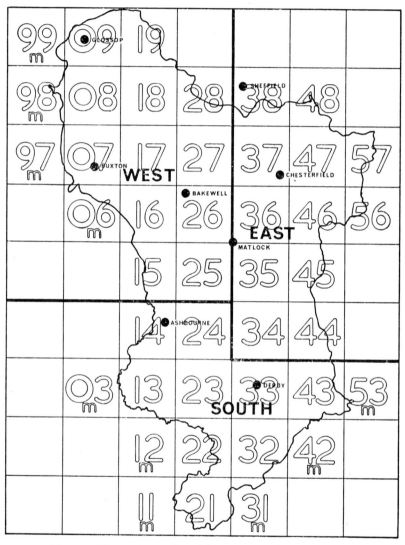

Fig. 10.   Flora regions and 10 kilometre National Grid references.
'Minor' squares indicated by M (see page 99).

THE FLORA OF DERBYSHIRE

EXPLANATORY NOTES

NOMENCLATURE. In general nomenclature and arrangement of the Flora follows the *Excursion Flora of the British Isles* by A. R. Clapham, T. G. Tutin and E. F. Warburg, 2nd edn., Cambridge. In a few cases changes have been made from these names in order to bring the nomenclature of this Flora into line with current usage, particularly as reflected by the *Flora Europaea*. Synonyms are given where the names differ from those used by Linton or in the *Excursion Flora of the British Isles*.

REGIONS AND MAP REFERENCES. The county has been divided into three regions; West, East and South, for citing localities. Their limits are indicated in fig. 10, which also includes the 10-kilometre squares of the National Grid and their references as used in the Flora. The whole county lies within the 100-kilometre square SK (43), except for a small portion in the extreme north-west (in 10-kilometre squares 99, 98 and 97), which lies in the 100-kilometre square SJ (33). For convenience, the 10-kilometre squares which cover Derbyshire have been divided into two groups, 'major' squares which either lie completely within the county boundary or have a substantial county area within them, and 'minor' squares which only include a small area of the county. 'Minor' squares are marked on fig. 10 by an M. The county boundary has been taken as that in existence at the present day, except in the Sheffield area where the boundary in being in 1903 (the date of publication of Linton's Flora) has been used.

CITATION OF LOCALITIES. For common and generally abundant species distribution is described by listing occurrence in 10-kilometre squares or alternatively the phrase 'all squares except . . . .' is often used and in this context only refers to major squares (see above). For nearly all other species all available localities have been cited together with one-kilometre (four figure) National Grid references. In the case of more locally common species, it has been the policy to secure at least one locality for each 10-kilometre grid-square, and it must be recognised that for such species no attempt has been made to record every locality within the county.

ORIGINAL RECORDS. The master set of records from which this Flora has been compiled is housed in the County Borough of Derby Museum and Art Gallery, and may be seen by appointment with the Curator.

PTERIDOPHYTA

LYCOPODIACEAE

LYCOPODIUM L.

L. selago L.   *Fir Clubmoss*

Native. Formerly a rare plant of moorland on the Millstone Grit, collected by Jonathan Salt (*c.* 1800) on 'high moors about six miles S.W. of Sheffield'. Now known only from an area of quarries and tip-heaps of sands used for making refractory bricks and as isolated plants in the peaty soil of a limestone ledge and at the foot of a limestone outcrop; all these localities are on acid substrata in the Carboniferous Limestone area. Very rare.

W. Limestone ledge above Brook Bottom, Tideswell 1477, Monk's Dale 1375, Harborough Brick-works 2355.

Linton: Charlesworth Coombs 09, Kinder Scout 08, above Edale Chapel 18, Diamond Hill and near Cat & Fiddle 07, East Moor 27.

L. inundatum L.   *Marsh Clubmoss*

Native. Formerly a very rare plant of bogs on the Millstone Grit. No recent records.

Linton: Kinder Scout 08, Chinley Hill 08, Tansley Common 36.

L. clavatum L.   *Stag's-horn Moss*

Native. Moors and heaths; tolerating shade. Formerly not infrequent but now found in very few of its old localities, though occasional as a colonist of old gritstone quarries and, with *L. alpinum,* of quarries and tip-heaps of acid refractory sands on the limestone plateau.

W. Between Hathersage and Grindleford 2379, under Froggatt Edge 2476, Padley Wood 2579, Stanton Moor, two small patches 2463, Harthill Moor 2262, Stancliffe Old Quarry 2663, Harborough Brick-works 2355.

E. Between Whatstandwell and Wirksworth 3154.

S. Shirley Wood, a small patch 2042.

Linton: Above Stirrup Wood 99, Charlesworth Coombs 09, Back Tor, Edale 18, Bole Hill, Hathersage 28, Back Dale, Harpur Hill 07, three-quarters of a mile S.E. of Wirksworth 25, Edlaston Coppy 14, Breadsall Moor 33.

L. alpinum L.   *Alpine Clubmoss*

Native. Formerly a very rare plant of 'stony moors and heaths', collected by Jonathan Salt on 'high moors about five miles S.W. of Sheffield'. There are no recent records from these localities, but *L. alpinum* is now an occasional colonist of quarries and tip-heaps of acid refractory sands at various points on the Carboniferous Limestone plateau.

It is of great interest that three species of *Lycopodium* persist in the county and that they all behave as colonists of newly available and suitably moist and acid habitats.

W. Friden 1662, Minninglow Gravel-pit 2057, Harborough Brick-works 2355.

Linton: Nr. Derwent 19, five miles S.W. of Sheffield 38, Abney Moor 17, near Eyam 27.

SELAGINELLACEAE

SELAGINELLA Beauv.

S. selaginoides (L.) Link   *Lesser Clubmoss*

Native. Recorded for Kinder Scout in 1805 but there have been no recent records.

Linton: Kinder Scout 08.

EQUISETACEAE

EQUISETUM L.

E. fluviatile L. (*E. limosum* Sm.)   *Water Horsetail*

Native. In shallow water at the edge of reservoirs, ponds and ditches; less often in marshes and fens. Local but widespread.

W. Watford Bridge 0086, Chapel Reservoirs 0479, Hogshaw Houses 0073, Totley Moss 2279, Calver New Bridge 2475, White Edge 2675, Salter Sitch 2877, Dove Head 0368, Dale Head 0469, Wye Valley, nr. Ashford 1769, Minninglow 2057.

E. Wingerworth Great Pond 3667, Hardwick 4564, Ambergate 3452, Alfreton Brook 4056.

S. Recorded for all squares except 43, 32.

Linton: Edale 18, Heeley 38, Killamarsh 48, Pebley Pond 47, Long Eaton 43.

E. palustre L.   *Marsh Horsetail*

Native. Marshes, fens, bogs and damp places. Common.

Recorded for all squares except:

> **W.** 09, 19
> **E.** 38, 45, 34
> **S.** 33

Linton: 'Common . . . Generally distributed though scarcely plentiful anywhere'.

E. sylvaticum L.   *Wood Horsetail*

Native. Damp woods on acid soils; streamsides and flushes on moorland. Locally common.

W. Warhurst Fold 9993, nr. Glossop 09, Alport Dale 1491, nr. Whaley Bridge 9980 & 0080, Burbage Edge 0373, Longshaw 2578, Glutton Bridge 0866, nr. Bakewell 26.

E. Cordwell 3176, Holmesfield 3277, Hate Wood, Ashover 3463, Holloway 35.

S. Hough Park 2445, Hulland Carr 2645, Hulland Ward 2747.

Linton: Hurst Brook, Bamford 18, Bradwell 18, nr. Totley 28, Hathersage Booths 28, Great Hucklow 17, Froggatt 27, Black Rocks 25, Killamarsh 48, Walton Wood 47, Horsley Car 34, Kirk Langley 23, Calke 32, Bretby 22.

E. arvense L.   *Field Horsetail*

Native. Fields, waysides, banks etc. Common.

Recorded for all squares.

Linton: 'Common'.

E. telmateia Ehrh. (*E. maximum* Lam.)   *Great Horsetail*

Native. Damp shady banks and streamsides. Local.

W. Birchill Bank 2271, Chatsworth 2670, Bakewell 2367, Wensley 2661, Stanton in Peak 2363, Thorpe 1650, nr. Wirksworth 2653, Hopton 2552, Via Gellia 2857.

E. Ford nr. Eckington 4081, Wadshelf 3271, Lea Bridge, Glapwell Wood 4766, Cromford 3156.

S. Ashbourne Green 1949, Hulland Carr 2645, nr. Hulland Hall 2446, Bradley Brook 2244, Alkmonton 2037.

Linton: Hague Bar, New Mills 98, Chinley Station 08, Hucknall 46, Wyaston Grove 14.

OPHIOGLOSSACEAE

BOTRYCHIUM Swartz

B. lunaria (L.) Swartz   *Moonwort*

Native. Short grassland and heath and on old mineral spoil-heaps in the limestone dales. Widespread but in small quantity.

W. Conies Dale 1280, Ringinglow 2983, Bamford 2083, Hathersage 2381, Grin Plantation 0571 and other places in square 07, Tideswell Dale 1573, Miller's Dale 1573, High Rake 1678, Coombs Dale 2274, Sir William Hill 2178, Youlgreave Moor 1964, Thorpe 1650, Carsington Pastures 2553, and elsewhere in square 25.

S. Hough Park, Hulland 2445, between Bradley Wood and Ladyhole 2045, Osmaston 2043.

Linton: Charlesworth Coombs 98, Brampton 37, Calke 32, Swarkestone 32, Foremark Hall 32.

OPHIOGLOSSUM L.

O. vulgatum L.   *Adder's Tongue*

Native. Damp meadows and pastures. Frequent and widespread.

W. Roadside, Houndkirk Moor 2881, Stanley Moor Reservoir 0471, Cowdale 0871, Priest's Way, Deepdale 0970, Cressbrook Dale 1773, Standwood 2670, Brierlow Bar 1869, Beeley Hill Top 2768, Biggin Dale 1458, nr. Brassington 2154.

E. Hazelbarrow Farm 3781, Ford nr. Eckington 4081, Horsleygate 3177, Whitwell Wood 5279, Ashover 36, Pleasley Vale 5265, Alfreton Brook 4056.

S. Ashbourne Green 1847, Fenny Bentley 2247, Collycroft 1644, Atlow 2247, Gun Hill 2445, Calke 3622, Ticknall 3523.

Linton: Mellor 98, Chinley 08, Edale 18, Morley 34, Cubley Common 13, Radbourne 23.

OSMUNDACEAE

OSMUNDA L.

O. regalis L.   *Royal Fern*

Native. No recent records from the few former localities; now extinct 'through the operation of fern-sellers' or because 'carried off to cottagers' gardens'.

Linton: Castleton 18, Darley Dale 26, Breward's Car, Mugginton 24, Osmaston **Ponds** 24.

HYPOLEPIDACEAE

PTERIDIUM Scop.

P. aquilinum (L.) Kuhn (*Pteris aquilina* L.)   *Bracken*
Native. Woods, heaths, moors, hedge-banks; on the gritstone moors probably
marking areas of deeper and unpodsolized soil in former woodland. Abundant
everywhere except on shallow limestone, strongly podsolized or waterlogged
soils or at altitudes above 1,750 ft.
Recorded for all squares.
Linton: 'Common, generally distributed'.

HYMENOPHYLLACEAE

HYMENOPHYLLUM Sm.

H. wilsonii Hooker   *Wilson's Filmy Fern*
Native. Rocks, tree-trunks, etc., in a moist atmosphere. Very rare.
W.   Cressbrook Dale 1773.
Linton: Not recorded.

THELYPTERIDACEAE

THELYPTERIS Schmidel

T. limbosperma (All.) H. P. Fuchs (*Lastraea oreopteris* (Ehrh.) Bory)
                                                             Mountain Fern
Native. Woods, mountain pastures, steep banks above streams, screes; avoids
limestone. Common, especially on the gritstone.
W.   Warhurst Fold 9993, nr. Charlesworth 0194, Alport Dale 1390-1, Chew
Wood 9982, recorded for 08, Win Hill 1984, also Jaggers Clough and
Derwent 18, Turncliff 0470, also Goyt Valley and Burbage Edge 07, Longshaw
2578, also Bar Brook, Salter Sitch, Padley and Hollins Bank 27, Bakewell
2268.
S.   Hulland Dumble 2446, Cubley 1539.
Linton: Nr. Dore 28, Wirksworth 25, Scarcliffe 57, Dethick 35, Horsley 34,
Robin Wood and Repton Shrubs 32.

T. phegopteris (L.) Slosson (*Phegopteris polypodioides* Fée)   *Beech Fern*
Native. Damp woods and shady rocks, especially on the high gritstone moors.
Rare.
W.   Lynch Clough 1594, Alport Dale 1491, recorded for 08, Grindsbrook
1286, Jaggers Clough 1487, Mill Clough 1889, recorded for 07, Priestcliffe
Lees 1472.
E.   Shining Cliff Woods, Ambergate 3352.
Linton: Ernocroft and Stirrup Wood 99, Mellor 98, Hathersage 28, Abney
Clough 27, nr. Hartington 16, nr. Windley 34.

ASPLENIACEAE

PHYLLITIS Hill

P. scolopendrium (L.) Newman   (*Scolopendrium vulgare* Symons)
Hart's-tongue Fern
Native. Rocky woods and hedge-banks, shady rocks and walls, old wells.
Frequent, especially on limestone.

W. Recorded for 18 and 07, Cressbrook Dale 1773, also Chee Dale and
Monk's Dale in 17, Lathkill Dale 1865, recorded for 26, Dovedale 1452,
rocks about Brassington 2154-5.

E. Unstone, one plant in an old brick-kiln 3777, limestone rocks, Markland
Grips 5074, Stubbin Court 3667, Pleasley Vale 5265, recorded for 35.

S. Railway bridge, Fenny Bentley 1748, Snelston Lime-pits 1541, and old
walls, Ashbourne 1746, bridge nr. Offcote Grange (now destroyed) 2047,
Etwall 2632, Ticknall 3523.

Linton: Stirrup Wood 99, Mellor & Ludworth 98, Whaley Bridge towards
Buxton 08, Mickleover, Littleover and Breadsall 33, Stapenhill 22.

ASPLENIUM L.

A. adiantum-nigrum L.   *Black Spleenwort*
Native. Rocky woods, shady walls and banks. Rare.

W. Jaggers Clough 1686, railway bridge, Edale Valley 1285, river bridge,
Hathersage 2380, Bamford 2083, Under Harpur Hill 0571, Abney 1979,
Robin Hood, Baslow, 2872, Coombs Dale 2274, Lathkill Dale 1765, Alport
2264, Dove Holes, Dovedale 1453, Via Gellia 2265, Cromford 2959.

S. A few plants on Doveridge Church 1134.

Linton: Horsley and Breadsall Moor 34, Between Yeaveley and Rodsley 14,
Brailsford 24, Calke, Repton and Anchor Church 32.

A. obovatum Viv.   *Lanceolate Spleenwort*
Probably native. Rock outcrops. Very rare.

W. Little Hucklow in 17.

Linton: Not recorded.

A. trichomanes L.   *Maidenhair Spleenwort*
Native. Rocks and walls. Frequent, especially on the limestone. Our plant,
whether on limestone or gritstone, appears to be ssp. *quadrivalens* D. E. Meyer,
with long narrow rhizome-scales and oblong pinnae.

W. Recorded for all squares except 09, 19, 28.

E. Recorded for 47, Markland Grips 5074, Pleasley Vale 5165, Holloway
3256.

S. Wyaston 1842, Osmaston 2042, recorded for 22 and 32.

Linton: Mellor 98, Quarndon 34, Derby and Breadsall 33, Dale 43.

A. viride Huds.   *Green Spleenwort*
Native. Damp limestone crevices. Local and rather rare.
W. Dirtlow Rake 1481, Cave Dale 1582, Conies Dale 1380, Hartle Dale 1680, Ashwood Dale and Woo Dale 0972, Back Dale 0870, Brook Bottom, Tideswell 1477, Miller's Dale 1573, Monk's Dale 1374, High Edge 0669, Parsley Hay 1463, Gratton Dale 2060.
Linton: Dovedale 15, Coxbench Wood 34.

A. ruta-muraria L.   *Wall-Rue*
Native. Walls and rocks. Frequent and widespread.
W. Recorded for all squares except 09, 19.
E. Beauchief 3282, Brampton 3371, Markland Grips 5074, Creswell Crags 5274, Stubbing 3567, Scarcliffe 46, Pleasley Vale 5165, Crich 3553, Morley 3940.
S. Ashbourne churchyard 1746, Birchwood Quarry 1541, Brailsford Hall 2640, Doveridge 1134, Etwall church 2361, Repton 3027, recorded for 21 and 31.
Linton: River Etherow 09, Mellor 98, Duffield and Morley 34, Derby and Breadsall 33.

CETERACH DC.

C. officinarum DC.   *Rusty-back Fern*
Native. Limestone rocks and walls. Very local and rare.
See also map (page 91).
W. Coombs 0379, Monsal Dale Station 1771, Ramsley Reservoir 2874, Lathkill Dale 1966, 2066, Dovedale 1453, Wolfscote Dale 1457, recorded for 25.
E. Ford nr. Eckington 3980, 4082.
S. Snelston copper mines 1541.
Linton: Newton 22.

ATHYRIACEAE

ATHYRIUM Roth

A. filix-femina (L.) Roth   *Lady Fern*
Native. Shady places on acid soils, by streams etc. Frequent.
W. Recorded for all squares except 17, 09.
E. Beauchief 3381, Killamarsh 4682, Holmesfield 3277, Whitwell Wood 5278, Wingerworth Gt. Pond 3667, Griff Wood, Ault Hucknall 4665, Pleasley Vale 5165, nr. Ambergate 3552, Alfreton Brook 4056.
S. Offcote 1847, Spinnyford Brook 2444, Bentley Carr 1838, Barton Blount 2234, Dale Abbey 4338, Grange Wood, Lullington 2714.
Linton: Longstone Edge 17/27, Kedleston 34, Breadsall Moor 33, Drakelow 22, Calke and Repton 32.

CYSTOPTERIS Bernh.

C. fragilis (L.) Bernh.   *Bladder Fern*

Native. Rocky woods especially on limestone; mortared walls. Locally frequent.

W.   Recorded for all squares except 09, 08, 28.

E.   Whitwell Wood 5278, Ravensnest 3461, Pleasley Vale 5265.

Linton: Ashover 36, between Ellastone and Ashbourne 14.

ASPIDIACEAE

DRYOPTERIS Adanson

D. filix-mas (L.) Schott (*Lastraea filix-mas* C. Presl)   *Male Fern*

Native. Woods, hedge-banks, scrub. Common and widely distributed. Recorded for all squares except **E.** 44.

Linton: 'Generally distributed and mostly abundant'.

D. abbreviata (DC.) Newman (*Lastraea filix-mas* var. *abbreviata* Bab.)

Possibly native. On screes in upland areas. Linton's single record goes back to 1855 and there have been no more recent finds. Should be looked for.

Linton: 'In the Peak, Derbyshire, *Johnson and Sowerby, Ferns of Gt. Brit., 1855*'.

D. borreri Newman   *Borrer's Male Fern*

Native. Much less common than *D. filix-mas* and usually on acid soils.

W.   Jaggers Clough, Golden Clough and Ashopton in 18, Bakewell 2268.

S.   Woodcock Dumble 1849, Edlaston Coppice 1743.

Linton: Chinley, south of Hayfield 08, Combes Moss 07.

D. carthusiana (Vill.) H. P. Fuchs (*Lastraea spinulosa* C. Presl)
                                                                Narrow Buckler-fern

Native. Damp and wet woods, marshes and wet heaths. Frequent but local.

W.   Nr. Hayfield 0387, Longshaw 2579, White Edge 2677, Leash Fen 2874, Tissington 1751.

E.   Eckington Moss 4080.

S.   Nr. Clifton 1744, Edlaston Coppice 1743.

Linton: Axe Edge 07, Grindsbrook 18, Shatton 18, Ashwood Dale and Miller's Dale 07/17, Cratcliffe Wood 26, Pebley Pond 47, Ashover 36, Walton Wood 36, Shirley Wood 24, Hulland Moss 24, Repton Shrubs 32.

D. dilatata (Hoffm.) A. Gray (*Lastraea dilatata* (Hoffm.) C. Presl)
                                                                Broad Buckler-fern

Native. Woods, hedge-banks, moorlands, chiefly on acid substrata. Common and widespread.

Recorded for all squares except **E. 34, 44.**

S. Ashbourne 1746, Hulland Carr 2645, Bentley Carr 1838, Dale Abbey 4338, Robin Wood 3525.

Linton: 'Generally distributed and abundant'.

POLYSTICHUM Roth

P. setiferum (Forskål) Woynar (*P. angulare* (Willd.) C. Presl) *Soft Shield-fern*
Native. Woods, hedgebanks etc. Rare and perhaps decreasing.
W. In wooded area above Chee Dale 1273, Thornbridge Hall 1971.

Linton: Hartle Dale 18, Fernilee 07, Stand Wood, Chatsworth 27, Lovers' Walk, Matlock 25, Nether Loads and Old Brampton 37, Edlaston Wood and top of Quilow Hill 14, Breadsall 33.

P. aculeatum (L.) Roth   (*P. lobatum* (Hudson) Chevall.)   *Hard Shield-fern*
Native. Woods, rocks and hedge-banks, chiefly on limestone. Locally frequent.
W. Alport Dale below waterfalls 19, Pig Tor Dale 0872, also Chee Tor and Cressbrook Dale in 17, Middleton Dale 2175, Coombs Dale 2274, recorded for 16, Conksbury Bridge 2165, Thorpe 1550, also Dovedale 15, Via Gellia 2556.
E. Nether Loads 3177, Grasscroft Wood 3476, Markland Grips 5074, Whitwell Wood 5278, Pleasley Vale 5265.
S. Ashbourne Green 1947, also Fenny Bentley, Wyaston, Roston, Norbury, Snelston and Mapleton in 14, Offcote 2048, Boylestone 1835, Cubley 1737.
Linton: Morley 34, Trusley 23.

GYMNOCARPIUM Newman

G. dryopteris (L.) Newman (*Phegopteris dryopteris* (L.) Fée)   *Oak Fern*
Native. Mountain woods, shaded rocks and stream-banks. Rare.
W. Alport Castles 1491, Lynch Clough 1594, Jaggers Clough 1487, Blackden Clough 1288, Fairbrook Clough 1089, Ladybower (one plant) 2086, recorded for 07, Bretton Clough 2078, Umberley Brook 2672.
E. Beighton 48.
Linton: Ernocroft and Stirrup Wood 99, Charlesworth Coombs 09, Chinley 98, nr. Rowsley 26, Matlock 25, Pleasley 56, Cubley 13.

G. robertianum (Hoffm.) Newman (*Phegopteris calcarea* (Sm.) Fée)
Limestone Polypody
Native. Limestone screes and rocks.   Occasional; confined to the Carboniferous Limestone.
W. Deep Dale 0971, Back Dale 0870, Diamond Hill 0570, Ashwood Dale 0872, Monk's Dale 1373, Miller's Dale 1672, Chee Dale 1273, Priestcliffe Lees 1473, Ravensdale and Cressbrook Dale 1774, Topley Pike and east end of Deep Dale 1072, Dam Dale 1178, Stoney Middleton 2174, Coombs Dale 2274, recorded for 16, Via Gellia 2858.
Linton: 'Rather rare'.

BLECHNACEAE

BLECHNUM L.

B. spicant (L.) Roth   *Hard Fern*

Native. Woods, heaths, moors, rocks, moist banks; avoiding limestone.
Common on the gritstone and shale-grits, rather rare elsewhere.

W. Chew Wood 9992, nr. Glossop 09, Alport Castles 1490, Hayfield 0487,
Derwent Moors 1889, Goyt's Moss 0177, Padley Wood 2579, Minninglow
2057.

S. Sturston 1946, Biggin 2547, recorded for 32.

Linton: Moors nr. Dore 38, Walton Wood 37, Breadsall Moor 34, Burnaston
23.

POLYPODIACEAE

POLYPODIUM L.

P. vulgare L.   *Common Polypody*

Native. Banks, rocks, old trees, walls. Frequent and widespread. The
common form in the county is ssp. *vulgare*, with the pinnae of equal length
except for a few shorter ones near the base and apex and with the young
sori circular in outline. Ssp. *prionodes* Rothm. (*P. interjectum* Shivas), with the
more or less acute pinnae longest near or below the middle of the ovate to
ovate-lanceolate leaf and with the young sori oval, has been recorded from a
few localities.

W. Recorded for all squares except 09, 19.

E. Milltown 3561, recorded for 56, 35, 34.

S. Quilow Hill, Edlaston 1943, Offcote 1947, nr. Hulland 2547, recorded
for 13.

Linton: 'General through the county'.

SPERMATOPHYTA
GYMNOSPERMAE

PINACEAE

PINUS L.

P. sylvestris L.   *Scots Pine*

Introduced. Extensively planted and frequently self-sown.

Recorded for all squares except:

> **W.** 09, 19.
> **E.** 34, 44.
> **S.** 33, 21.

Linton: Mackworth 33.

ABIES Mill.

A. alba Mill.   *Silver Fir*
Introduced.   Occasionally planted.
W.   Lathkill Dale 1966.
Linton: Not recorded.

CUPRESSACEAE

JUNIPERUS L.

J. communis L.   *Juniper*
Formerly native. Recorded for limestone slopes and cliffs in Ashwood Dale
in 1884, but not seen in recent years.
W.   Grin Wood, Buxton 0572.
Linton: Ashwood Dale 07.

TAXACEAE

TAXUS L.

T. baccata L.   *Yew*
Native. Woods and cliffs on limestone; probably planted elsewhere. Local
and by no means frequent in the limestone dales, but in greater abundance on
cliffs of Magnesian Limestone.
W.   Recorded for 08 and 18, Ashwood Dale 0772, Cressbrook Dale 1773,
recorded for 16, Cratcliffe 2262, Dovedale 1452, Kniveton Churchyard 2050.
E.   Beauchief 3381, Millthorpe 3176, Pebley Pond 4878, Markland Grips
5074, Stubbin 3567, Ault Hucknall 4665, Pleasley 5164, recorded for 35.
S.   Ashbourne Churchyard 1746, woods at Bradley 2245, Shirley Wood
2041, Sudbury 1532, Recorded for 23, 33, 43 and 32, Caldwell 2517.
Linton: Middleton 27, Morley 34, Smalley 44.

ANGIOSPERMAE
DICOTYLEDONES

RANUNCULACEAE

CALTHA L.

C. palustris L.   *Marsh Marigold*
Native. Wet meadows, alder woods and osier beds; moorland rills and flushes.
Locally abundant. The moorland form has much smaller leaves and flowers and
is often prostrate and rooting at the nodes (*C. minor* Miller).
Recorded for all squares.
Linton: 'Common, except on limestone'.

TROLLIUS L.

T. europaeus L.   *Globe-flower*

Native. Moist pastures, scrub and open woodland on the Carboniferous Limestone. Local.

W. Grin Wood, Buxton 0572, Dale Road, Buxton 0672, Lovers' Leap 0772, Cowdale 0871, Cunning Dale 0773, recorded in 10 localities in the dales of the Wye Valley 17, Gratton Dale 1959, Rocks, Cromford 2957, south side of Via Gellia 2656, Brassington Rocks (extinct?) 2154.

E. Shining Cliff Woods, Ambergate 3451.
Linton: 'Rare'.

HELLEBORUS L.

H. foetidus L.   *Stinking Hellebore*

Doubtfully native. Woods, plantations and scree, especially on limestone. Rare. Probably always a relic of former cultivation, but readily reproducing by seed.

W. Brough nr. Hope 1882, Between Baslow and Stony Middleton (planted?) 27, Hartington Dale 1360, Bakewell 2269, Hopton Wood 2656, Bonsall 2758, Via Gelia 2556, rock nr. Cromford 2957.

E. Totley, one plant 3080.
Linton: Codnor Castle walls 44, Dethick 35, Mickleover 33.

H. viridis L.   *Green Hellebore*

Probably native, but often a relic of former cultivation. Woods, hedges and pastures, chiefly on limestone. Rare.

W. Grin Plantation, Buxton 0571, Cunning Dale 0872, Staden (planted?) 0772, Wormhill 1273, Smerrill Grange 1961, Bradford Dale 2063, Conksbury Bridge, Lathkill Dale 2165, bank nr. Tissington Church 1752, Washbrook, Tissington 1651, nr. Lea Hall, Tissington 1952, river bank, Thorpe Rough 1550, Whim Farm, Wolfscote Hill 1357, nr. the Pig of Lead, Via Gellia 2857.

S. Roston 1240. One plant in derelict shrubbery, Drakelow, probably now destroyed, 2321.

Linton: Woodhouse, Bolsover 47.

ERANTHIS Salisb.

E. hyemalis (L.) Salisb.   *Winter Aconite*

Introduced and naturalized in parks, plantations and shrubberies. Local.

W. Great Longstone 1971, nr. Haddon Hall 26, Newhaven 1659, nr. Tissington Hall 1752.

E. Whirlow Brook Park 3083.

S. Lullington Hall 2414.

Linton: Haddon Hall 26, Repton 32, Coton Hall 21, Stapenhill Vicarage Grounds 22.

ACONITUM L.

A. napellus L.   *Monkshood*

Introduced. Woods, hedgebanks and shaded streamsides. An occasional garden escape.

Most cultivated forms differ from the native *A. anglicum* Stapf, though this has been introduced in some parts of the country. There are no certain records for Derbyshire, all established escapes being of continental origin.

W. Nr. Taddington 1470, railway nr. Friden 1660, Stanton Woodhouse 2564, Birchover 2362, Tissington 1851, nr. Bradbourne Mill 2052.

E. Totley Bank Wood 3281, Pleasley Vale 5165, Alfreton Brook 4056.

S. Bradley Wood 2046.

Linton: Wormhill 17, Drakelow Brook 22.

CONSOLIDA (DC.) S. F. Gray

C. ambigua (L.) P.W. Ball and Heywood (*Delphinium ajacis* Reichenb.)
                                                                   Larkspur

Introduced. Cornfields. A very rare casual; no recent records.

Linton: Willington 32, Repton 32.

ANEMONE L.

A. nemorosa L.   *Wood Anemone*

Native. Woods, hedgebanks and shaded roadsides on all except strongly podsolized soils; also in limestone pastures and scrub on the site of former woodland. Common and widespread.

Recorded for all squares.

Linton: 'Common. General throughout the county'.

CLEMATIS L.

C. vitalba L.   *Traveller's Joy, Old Man's Beard*

Native or introduced. Hedges, scrub and wood-margins, especially on limestone. Rare and probably always introduced except at Ticknall, where it is abundant.

W. Monsal Dale 17, Beeley 2667.

E. Hardwick Hall 4663, Pleasley Vale 5265.

S. Mugginton 2842, Alvaston 3933, Dale Abbey 4338, Ticknall 3523.

Linton: Matlock Bath 25, Heights of Abraham 25, Wingfield Manor 35, Shottle Gate 34.

RANUNCULUS L.

R. acris L.   *Meadow Buttercup*

Native. Damp meadows and pastures on calcareous or mildly acid soils; waysides. Common and widespread.

Recorded for all squares.

Linton: 'Common'.

R. repens L.   *Creeping Buttercup*

Native. Wet meadows, pastures and woods; roadsides, gravel-heaps, cultivated ground. Common and widespread.

Recorded for all squares.

Linton: 'Common'.

R. bulbosus L.   *Bulbous Buttercup*

Native. Dry meadows and pastures; grassy slopes and waysides. Common and widespread, especially on calcareous or base-rich acid soils.

Recorded for all squares except:

W. 19.

Linton: 'Common'.

R. arvensis L.   *Corn Crowfoot*

Probably native. Cornfields and disturbed ground. Formerly not uncommon but now rarely seen.

W.   Nr. Diamond Hill, Buxton 0570, Bradford Dale 2164.

E.   Cartledge nr. Holmesfield 3277, Newbold 37, Cutthorpe 37, nr. Markland Grips 5074, Whitwell Wood 5278, nr. Ashover Hay 3560, recorded for 56, Cornfields, Stanley 4140.

S.   Clifton 1645, fowl-run, Yeldersley Lane 2243, Littleover 3334, Cornfields, Darley Abbey 4338, Long Eaton 4933, Thrumpton 5031, recorded for 22 and 21.

Linton: Nr. Stirrup Wood 99, North Wingfield 46, Shirebrook 56, Matlock 25/35, Marston Montgomery 13, Hilton, Boylestone and Church Broughton 23, Repton, Askew Hill and Ticknall 32.

R. sardous Crantz

Probably native. Cultivated and waste ground. Rare and local; no recent records.

Linton: Nr. Derby 33, Brizlincote, Repton and Calke 32, Burton 22.

R. parviflorus L.   *Small-flowered Buttercup*

Probably native. Dry grassy banks, waysides and waste ground. Very rare; no recent records.

Linton: Canal bank nr. Borrowash 43, Repton 32, Burton 22.

R. auricomus L.   *Goldilocks*

Native. Woods, hedgebanks. Frequent and widespread. The form usually met with in the county has more or less well-developed petals.

Recorded for all squares except:

W.   09, 19, 08, 18.
E.   35, 45.

Linton: Dethick 35.

R. lingua L.   *Greater Spearwort*

Native. Ponds, ditches, canals. Very rare.

E.   Ditch on Holy Moor 3368, pond, Wingerworth 3867, Whatstandwell 3354.

S.   Dale Abbey 4338, Melbourne Pool 3824, Ingleby 3427.

Linton: Cottage Pond, Chatsworth 27, Matlock 35, South Normanton 45, 'Melbourne Pool, formerly', Ingleby and Knowl Hills 32.

R. flammula L. ssp. flammula   *Lesser Spearwort*

Native. Wet places; moorland rills and flushes. Common.
Recorded for all squares except:

S. 22.

Linton: 'Frequent. Generally distributed throughout the county'.

Very variable in size and habit. A robust erect form, with strongly serrate leaves, named var. *serratus* DC. by J. Freyn (1898), is reported by Linton from nr. Rowarth 08. A similar form is recorded from Cutthorpe 3473. A notably narrow-leaved form is found nr. Axe Edge 0371.

Prostrate forms, rooting at the nodes, were named var. *radicans* Nolte by Linton and records were given for Rowarth Brook, below Rowarth, Brackenfield Green, Wingerworth Great Pond, and Linton Heath. It does not seem possible to draw a line between these extreme habit-forms and other creeping forms.

R. sceleratus L.   *Celery-leaved Crowfoot*

Native. In muddy ditches and ponds and by slow streams. Local.

W.   Combs Lake, Chapel 0379, Water-cum-Jolly Dale 1672, Bar Brook, Baslow 2774, Queen Mary's Bower, Chatsworth Park 2570, Beresford Dale 1259.

E.   Recorded for all squares except 35, 34, 38.

S.   Recorded for all squares except 14.

Linton: Between Norton and Heeley 38.

R. hederaceus L.   *Ivy-leaved Crowfoot*

Native. Wet places, mud at the margin of streams and ponds. Local.

W.   Derwent Edge 1987, Hope 1585, Cave Dale 1482, Bretton Clough nr. Hathersage 2179, also Bubnell, Baslow and Emperor Stream, Chatsworth in 27, sandy pit nr. Harborough Rocks 2454.

E.   Moorhall, Cordwell 3175, also Donkey Racecourse nr. Over Newbold, and Freebirch in 37, Nether Loads 3269, Stanley 4240.

S.   Small ponds about Ashbourne 1647, etc., Ladyhole 2144, also Bradley, Brailsford, Hulland Moss, Mercaston and Mugginton in 24, Findern 3130, Dale Abbey 4338, Bretby 3124, Weston-on-Trent 4027.

Linton: Dinting 09, Snake Inn 19, Lydgate and Eccles Pike 08, Dore Moor and Bamford 28, Buxton and Combs Moss 07, Wardlow Mires and Great Hucklow 17, nr. Birchover and Bradford 26, Sutton Scarsdale 46, Brackenfield Green 35, nr. Sudbury 13, Hatton and Hilton 23, nr. Burton 22.

R. omiophyllus Ten. (*R. lenormandi* F.W. Schultz)

Lenormand's Water Crowfoot

Native. Non-calcareous streams, muddy places and moorland flushes. Locally frequent.

W. Charlesworth Combs 0191, Alport Castles 1491, Alport Dale 1390/1, recorded for 08, Mam Gap, Edale 1283/4, also Win Hill, Ashopton and Derwent Edge in 18, Hathersage 2383, also Ringinglow and Longshaw in 28, Swann's Canal, Buxton 0470, also nr. Watford Wood 0374, Axe Edge 0270, Leash Fen 2874, also Far End and Robin Inn, Baslow, Haywood and Grindleford in 27, Brand Side 0368, Washgate 0567, Carsington 2453.

E. Nr. Woodthorpe Hall, Holmesfield 3178, Calow 4170, Wingerworth 3867, Sutton Scarsdale 4469.

S. Mercaston Brook 2742, Mugginton 2843.

Linton: Mickleover 33, Between Calke and Melbourne 32.

R. fluitans Lam.   *Long-leaved Water Crowfoot*

Native. Rapidly flowing streams and rivers. Frequent.

C. D. K. Cook writes 'R. *fluitans* is normally confined to rivers with pebble-covered bottoms. In the Rivers Dove, Wye and Derwent . . . (it) is almost confined to the non-limestone areas . . . in the same rivers in limestone areas it is replaced by *R. penicillatus*'.

W. By the Wye, Bakewell 26, Dovedale 15.

E. Duffield 3344.

S. Recorded for all squares except 14, 22, 21. In the River Dove *R. fluitans* succeeds *R. penicillatus* downstream from Abbotsholme 1038, the overlap of the two species being less than a quarter of a mile.

Linton: River Derwent, Whatstandwell 35, South Normanton 45, Burton, Egginton and Willington 22.

Some of Linton's records may refer to *R. penicillatus*, q.v.

R. fluitans var. *bachii* (Wirtgen) Wirtgen, with leaves resembling those of *R. fluitans* and flowers those of *R. trichophyllus,* and always sterile, has been recorded from Castleton 1482, Hope 1783 and Alport 2264.

R. circinatus Sibth.   *Stiff-leaved Water Crowfoot*

Native. Slow streams, canals, large ditches, ponds, gravel-pits; usually in fairly deep water. Locally frequent.

W. Hathersage 2381.

E. Canal, Killamarsh 48, Barlborough Park 4778, Hardwick Ponds 4563, Canal, Cromford to Ambergate 3155, 3354, 3452, Kirk Hallam 4540.

S. Ashbourne Hall Pond 1946, Spinnyford Brook and Brailsford Mill-stream 2443, Wyaston Lake 2041, Shirley Brook 2139, Stanton-by-Dale 4637, Canal nr. Swarkestone 3729.

Linton: Matlock Bath 25, Whaley Mill nr. Bolsover 57.

R. trichophyllus Chaix (incl. *R. drouetii* F. Schultz)

Short-leaved Water Crowfoot

Native. Ponds, ditches and slow streams; persisting as terrestrial plants in pools that dry out in summer. Occasional.

Forms with the mature achenes quite glabrous were formerly named *R. drouetii* F. Schultz, but there is no clear-cut distinction and all records for both *R. trichophyllus* and *R. drouetii* have been put together in the lists below.

W. Dirtlow Rake, Castleton 1582, Monk's Dale 1373/4, pond and stream nr. Cressbrook Mill 1772, Lathkill Dale 1765, small pond in Manor Farm, Brassington 2355.

E. Linacre 3272.

S. Sudbury 1631, pond at Malcomsley, Cubley 1537, Spondon Park 3935, backwater west of Sawley 4631, Swarkestone Ponds 3628.

Linton: Between Great Rocks Dale and Fairfield Common 07, Nether Langwith 57, River Wye, Ashford 16, Ault Hucknall 46, Via Gellia 25, N.E. of Fritchley 35, Mickleover 33, Brizlincote, Bretby and brook above Egginton 22, Ticknall and Twyford Brook 32.

All British forms are placed in ssp. *trichophyllus.*

R. aquatilis L. (*R. heterophyllus* Weber)

Native. Ponds and streams. Rare.

W. Peak Forest 1179.

S. Pools, Ticknall Quarries 3524.

Linton: Morley 34, Mickleover, Mackworth and Litchurch 33.

C. D. K. Cook reports a sterile hybrid, probably of *fluitans* and *aquatilis,* from the River Wye in Monsal Dale.

R. peltatus Schrank

Common Water Crowfoot

Native. Ponds, gravel-pits, ditches, canals, slow streams. Frequent; the most commonly encountered of the Batrachian Ranunculi with dissected leaves.

W. Stream nr. Combs Lake 0379, nr. the Saw Mill, Chatsworth 2772, Arbor Low 1662, Minninglow Gravel-pit 2057, Brassington 2355.

E. Beauchief Abbey 3381, also Bradway and Norton in 38, Canal, Killa-, marsh 48, nr. Barlow 3474, Linacre Reservoirs 3372, nr. Barlborough 47, Press Reservoirs, Wingerworth 3565.

S. Small ponds nr. Ashbourne 1948, etc., also Roston and Snelston in 14, Shirley 2242, Yeaveley 1839, Kirk Langley 2838, nr. Trent Station 4932.

Linton: Ashwood Dale 07, Millers Dale 17, West Hallam 44, Mickleover and old bed of Derwent, Litchurch 33, Repton, Twyford, Ticknall and Calke 32.

R. penicillatus (Dumort.) Bab. (*R. pseudofluitans* (Syme) Baker & Foggitt)
Native. Limestone streams. Locally frequent.

Derbyshire records are chiefly of var. *calcareus* (R. W. Butcher) C. D. K. Cook, abundant in many of the rivers of the limestone dales.

W. River Wye throughout 17, 1473, etc., Calver Millstream 2475, Ashford-in-the-Water 1769, Lathkill Dale 1765, 2066, also River Derwent at Rowsley, and Darley Dale, River Wye, Bakewell, and Bradford Dale in 26, River Dove from Hartington 1258, to Thorpe 1450, Havenhill Dale, Bradbourne 2052.

E. River Derwent about Matlock 3059, and Cromford 3056, and down to Whatstandwell 3354.

S. River Dove from above Mapleton 1647, to below Norbury 1242, Reservoir, Snelston 1543, Fenny Bentley Brook 1646, etc., Henmore Brook 1746, etc., River Dove down to Abbotsholme where *R. fluitans* succeeds.

Linton: River Rother 48, 37, 47, Doe Lea nr. Renishaw 47. Some of Linton's *R. fluitans* records may refer to this species.

C. D. K. Cook reports var. *vertumnus* C. D. K. Cook, differing from var. *calcareus* in the shorter leaves with more numerous and widely divergent segments, from the south of the county.

R. baudotii Godron   *Brackish Water-Crowfoot*
The old record for this species is almost certainly erroneous.

Linton: Monk's Dale, Wormhill 17.

R. ficaria L.   *Lesser Celandine*
Native. Woods, pastures and meadows, grassy banks, stream-sides. Abundant and widespread. Ssp. *ficaria*, lacking axillary bulbils and yielding many well-developed achenes per flower; and ssp. *bulbifer* Lawalrée, bearing bulbils in the leaf-axils but developing very few good achenes, are both common in the county.

Recorded for all squares except 09.

Linton: 'Common and general'.

ADONIS L.

A. annua L. (*A. autumnalis* L.)   *Pheasant's Eye*
Introduced. Cornfields, occasional and extremely scarce.

S.   Little Eaton 33.

Linton: Spital nr. Chesterfield 37.

MYOSURUS L.
M. minimus L.   *Mousetail*
Native. Arable land. Formerly in a few localities in the south of the county, but there are no recent records.

Linton: Darley Dale 26, nr. Derby 33, Drakelow 21, Stapenhill 22.

AQUILEGIA L.
A. vulgaris L.   *Columbine*
Native. Woods, thickets, open scrub, moist pastures. Occasional, particularly on limestone; probably often an escape from cultivation.

W. Abney 1980, Lovers' Leap, Buxton 0872, Woo Dale 0972, Miller's Dale 1373, also Monsal Dale 1771, Cressbrook Dale 1773, Fin Cop 1770, Dimmin Dale 1670, Ashford, Deep Dale, Taddington and Monk's Dale in 17, Stony Middleton Dale 2275, Deep Dale nr. Ashford 1669, Rowsley 2666, Via Gellia 2556, also Matlock Bath and Winster in 25.

E. Pratthall above Linacre 3272, Highlightly Farm, Cordwell 3276, Markland Grips 5074, Whitwell Wood 5278, Creswell 5374, Scarcliffe Park Wood 5170, Glapwell 4766, Pleasley Vale 5165.

Linton: Willington and Drakelow 22, Calke and Foremark 32.

THALICTRUM L.
T. flavum L.   *Common Meadow Rue*
Native. Marshy meadows, fens, stream-sides, osier-beds. Local.

W. Ashwood Dale (depauperate) 0972, Gratton Dale 1959.

E. Eckington 4480, Pleasley Park 5165.

S. Roadside, Fenny Bentley (garden escape?) 1849, Spondon 4035, Shardlow 4330, Wilne—Old Sawley 4530, Repton 2927, 3027, 3228, Anchor Church 3327, Swarkestone 3629, Kings Newton 3827, Walton-on-Trent 2018.

Linton: Castleton 18, Monk's Dale 17, Renishaw 47, Alfreton Brook 45, Banks of River Derwent nr. Duffield Church 34, Breadsall 33, Thrumpton 53.

T. minus L. ssp. minus (*T. collinum* Wallr.)   *Lesser Meadow Rue*
Native. Partly stabilized scree-slopes and in rock-crevices in the limestone dales; restricted to protorendzina soil but indifferent to aspect. Local.

W. The Winnats 1382, Peveril Castle and Cave Dale 1482, Dirtlow Rake 1581, Pin Dale 1582, Monsal Dale 1771, Monk's Dale 1374, Miller's Dale 1573, Priestcliffe 1473, Water-cum-Jolly Dale 1672, Cressbrook Dale 1773, Middleton Dale 2175, Via Gellia 2756/7, Heights of Abraham 2958.

Linton: 'Rare'.

BERBERIDACEAE
BERBERIS L.
B. vulgaris L.   *Barberry*
Doubtfully native. Hedges and wood-margins; perhaps usually planted. Local.

W. Shatton 2082, Hathersage 28, Wardlow 1874, Froggatt 2476, Longshaw 2578, Chatsworth 2771, Edensor 2569, Parwich 1953.

E. Cordwell 3176, Markland Grips 5074, Ashover 3463, Stony Houghton 4966, Stanley 4141.

S. Ashbourne Green 1847, Dale Abbey 4338.

Linton: Millhouses 38, Morley 34, Mackworth 33, Calke 32, Bretby 22.

MAHONIA Nutt.

M. aquifolium (Pursh) Nutt.   *Oregon Grape*
Introduced. Commonly planted and naturalized in a few places. Rare.
W. On an island in the River Lathkill 2066.
E. Creswell Crags 5274.
Linton: Not recorded.

NYMPHAEACEAE

NYMPHAEA L.

N. alba L. (*Castalia speciosa* Salisb.)   *White Water-lily*
Native. Lakes, pools and rivers; extending into water poorer in mineral
nutrient than does the yellow water-lily. Occasional; introduced in several
localities.
E. Wingerworth Great Pond 3767, recorded for 57, ponds nr. Hardwick
Hall 4563.
S. Bretby Ponds 3023, Calke Park 3623, Weston-on-Trent 4027.
Linton: Norton 38, Ambergate and Wigwell 35, West Hallam 43, Trent nr.
Burton 22, Drakelow and Walton 21.

NUPHAR Sm.

N. lutea (L.) Sm. (*Nymphaea lutea* L.)   *Yellow Water-lily*
Native. Lakes, pools, rivers. Frequent.
W. Ponds, Chatsworth 2670.
E. Somersall Pond 3570, Renishaw Lake 4478, recorded for 57, Winger-
worth Great Pond 3667, Morley 3940, West Hallam 4341, by the Derwent,
Belper 3548.
S. Recorded for all squares except 14, 13.
Linton: Hassop 27, Hardwick 46, Shipley Gate 44.

CERATOPHYLLACEAE

CERATOPHYLLUM L.

C. demersum L.   *Hornwort*
Native. Ponds, ditches and streams. Rare.
E. The Moss, Eckington 4080, canal, Ambergate 3551.
S. Pond, Brailsford Hall 2640, nr. Sudbury Station 1630, canal, Derby 3636,
Melbourne Pool 3821.
Linton: Sutton Scarsdale 46, Radbourne 23, Drakelow 21.

C. submersum L.   *Hornwort*
Native. Ponds and ditches. Very rare.
E. Canal nr. Ambergate 3452.
Linton: Not recorded.

PAPAVERACEAE

PAPAVER L.

P. rhoeas L. *Field Poppy*

Native. Arable land, waysides and waste places over a wide range of soil-types. Occasional, and chiefly in the south of the county.

Recorded for all squares except:

> **W.** 09, 19, 18, 28, 26, 25.
> **E.** 34.

Linton: Little Eaton 34.

P. dubium L. (*P. lamottei* Boreau) *Long-headed Poppy*

Native. Arable land, waste ground, gravel pits, walls, roadsides, railway banks etc., commonly on light dry sandy or gravelly soils where it may partly or completely replace *P. rhoeas*. Locally frequent.

W. Litton 1675, Baslow 2572, Alport 2264, Middleton Cross 2855.

E. Ford 4081, Over Newbold 3573, Pebley Pond, Barlborough 4878, Elmton 57, Stubbin Court 3667, Scarcliffe 46, Pleasley Vale 5265, Little Eaton 3641, Duffield 3443, West Hallam 4240.

S. Recorded for all squares except 23.

Linton: Church Broughton 23.

P. lecoqii Lamotte *Babington's Poppy*

Native. Quarries, road-margins, field-borders; less commonly in arable fields. Occasional, and perhaps decreasing.

W. Ashford 2068.

E. Pleasley, on wool waste 56.

S. Persistent weed, Ashbourne 1746, rubbish tip, Hanging Bridge 1546, Markeaton 3237, Cat & Fiddle Lane, Dale Abbey 4339, Grange Wood, Lullington 2714.

Linton: Bamford Station 28, Elmton 57, nr. Matlock 25/35, Dalebrook nr. Sudbury 13.

P. hybridum L. *Round Prickly-headed Poppy*

Introduced. A rare casual of arable fields and waste places.

S. By a pig-sty, Hanging Bridge 1545.

Linton: Not recorded.

P. argemone L. *Long Prickly-headed Poppy*

Probably native. Sand-quarries; waste places, waysides and arable land on light sandy soils. Occasional in the south of the county.

E. Boythorpe 3869, Pleasley, on wool waste 56.

S.   Clifton Goods Yard 1644, Ashbourne 1746, Dale Abbey 4338.

Linton: Shirland 45, Hulland 24, Breadsall 33, Burton 22, Ticknall, Calke and Repton 32, Cauldwell, Coton and Linton 21.

P. somniferum L.   *Opium Poppy*

Introduced. Railway banks, waysides, waste places; a relic of cultivation or garden escape. Very rare.

W.   Eaton 0936, on tipped soil, Longstone Edge nr. Calver 2273.

S.   A frequent casual, Ashbourne 1746, persistent on the tip by Clifton Station 1644, nr. Breadsall 3639, Grange Wood, Lullington 2714.

Linton: Nr. Bakewell 26, Cromford and Matlock Bridge 25, Drakelow and Stapenhill 22.

MECONOPSIS Vig.

M. cambrica (L.) Vig.   *Welsh Poppy*

Introduced. Shady rocky places and hedge-banks, as an escape from gardens.

W.   Hay Dale 1772, Ashford 2069, Matlock Bath 2958.

E.   Lumsdale 3160.

Linton: Stapenhill 22.

CHELIDONIUM L.

C. majus L.   *Greater Celandine*

Doubtfully native. Banks, hedgerows, walls, waste places; chiefly near buildings. Frequent.

Recorded for all squares except:

> **W.**   09, 08, 28, 07.
> **E.**   48, 47.

Linton: The Hague, Renishaw 47, Oxcroft, Bolsover 47.

FUMARIACEAE

CORYDALIS Medic.

C. claviculata (L.) DC.   (*Neckeria claviculata* (L.) N. E. Brown)
                                                     White Climbing Fumitory

Native. Woods on rocky slopes, cliff-foot scree and shaded stream-sides on acid substrata and specially characteristic of the Millstone Grit. Locally frequent.

W.   Nr. Shireoaks 0738, Stanage Edge 2483, Yarncliffe, Longshaw 2579, Froggatt Edge 2576, Chatsworth Woods 2671, Baslow Edge 2578, Harthill Moor 2263, Robin Hood's Stride 2262, Bonsall 2758.

E.   Ecclesall Woods 3282, Ringinglow 3083, Crowhole 3375, Monk Wood 3576, Barlow 3474, Brockwell Lane 3672, nr. Stretton 3760, Hay Wood, Walton 3368, Lea Bridge 3156, Morley 3940, Smalley 4441, Dale Abbey 4338.

S.   Edlaston Coppice and The Holts Wood, Clifton 1743, Hulland Dumble 2446, Shirley Wood 2041, Mugginton 2843.

C. lutea (L.) DC. (*Neckeria lutea* (L.) Scop.)   *Yellow   Fumitory*
Introduced. Naturalized on old walls near houses. Occasional.
W.   Wormhill 1274, Eyam 2176, Lathkill Dale 1966, Youlgreave 2164, Parwich 1854, Kniveton 2050, Slaly 2757.
E.   Ashover 3561, Windley 3045, Milford 3544.
S.   Sandybrook 1748, Shirley 2141, Marston Montgomery 1337, Repton 3027, Walton Hall 2117.
Linton: Walton 37, Crich 35.

FUMARIA L.

F. capreolata L. (*F. pallidiflora* Jord.)   *Ramping Fumitory*
Native. Cultivated and waste ground. Very rare.
W.   Bonsall 2757.
Linton: Lumsdale 36, Breadsall 33, Repton 32.
The British form is ssp. *babingtonii* (Pugsley) P. D. Sell.

F. muralis Sonder ex Koch ssp. **boraei** (Jord.) Pugsley   *Boreau's Fumitory*
Probably native. Cultivated and waste ground and old walls. Very rare.
E.   Nr. Dore & Totley Station 3181, Spital, Chesterfield 3871.
Linton: Not recorded.

F. officinalis L.   *Common Fumitory*
Native. Cultivated ground, especially on lighter soils. Frequent, especially in the south of the county.
Recorded for all squares except:

>   **W.**   09, 19, 08, 18, 16.
>   **E.**   45, 34.
>   **S.**   13, 23.

Linton: Shirland 45, Church Broughton 13.

CRUCIFERAE

BRASSICA L.

B. oleracea L.   *Cabbage*
Introduced. An escape from cultivation. Rare.
W.   Tideswell Dale 1573.
Linton: Not recorded.

B. napus L.   *Rape, Cole, Swedish Turnip, Swede*
Probably introduced. Field-borders, waysides, banks of ditches or streams, waste places; an escape from cultivation. Rare. The plant most commonly seen as an escape is ssp. *oleifera* DC., rape, cole or coleseed, with non-tuberous tap-root, which is often cultivated for fodder.

W.   Chelmorton 1170, Rowsley 2465.

E.   Donkey Race Course Chesterfield 3572.

S.   Eaton 1136, Dale Abbey 4338.

Linton: Breadsall 34, Borrowash 43, Askew Hill, Repton 32, Calke 32.

B. rapa L.   *Turnip, Navew*
Probably introduced. Stream-banks and arable and waste ground. Locally abundant.

Wild plants belong chiefly to ssp. *sylvestris* (L.) Janchen, with a non-tuberous tap-root. This may be either biennial with a well-defined basal rosette, the form found on river-banks (var. *sylvestris* H. C.Wats.); or annual and lacking the basal rosette, the weed of arable land and waste places (var. *briggsii* H. C. Wats.).

W.   Nr. Middleton 1762, Ball Cross 2269, Fenny Bentley Brook 1850, Bradbourne 2052.

E.   Recorded for all squares except 35, 45, 34.

S.   Recorded for all squares except 33, 22, 32.

B. nigra (L.) Koch (*B. sinapioides* Roth)   *Black Mustard*
Probably native. Stream-banks, arable land, waysides and waste places. Not common.

W.   Bakewell 26.

S.   Clifton 1644, Kniveton 2049.

Linton: River Derwent, Matlock Bridge 26, Pinxton 45, Copse Hill, Osmaston 24, Breadsall 34, Hollington 23, Repton 32.

ERUCASTRUM C. Presl

E. gallicum (Willd.) O. E. Schultz
Introduced. A rare casual of waste places.

S.   Ashbourne 1746.

Linton: Not recorded.

SINAPIS L.

S. arvensis L. (*Brassica sinapistrum* Boiss.)   *Charlock, Wild Mustard*
Probably native. A weed of arable land. Very abundant.

Recorded for all squares except 19.

Linton: 'A common weed'.

S. alba L. (*Brassica alba* (L.) Rabenh.) *White Mustard*
Introduced. A weed of arable and waste land. Rare.
W. Tideswell Dale 1573, Nether Padley 2578.
E. Little Eaton 3641, Stanley 4141.
S. Ashbourne 1746, Clifton Tip 1644.
Linton: Kelstedge 36, Matlock 35, Breadsall and Mickleover 33, Ockbrook 43, Repton 32, Cauldwell 21.

DIPLOTAXIS DC.

D. muralis (L.) DC. *Wall Rocket, Stinkweed*
Introduced. Waste ground and walls. A rare casual.
W. Chapel-en-le-Frith 0681.
E. Matlock 3060.
Linton: Castleton 18, Willington 22, Stapenhill 22.

ERUCA Miller

E. sativa Miller
Introduced. A rare casual.
W. Newhaven 1660.
S. Clifton Tip 1644, Yeldersley 2043, Walton 2318.
Linton: Not recorded.

RAPHANUS L.

R. raphanistrum L. *Wild Radish, White Charlock, Runch*
Doubtfully native. A weed of arable land, especially on non-calcareous soils. Local.
Both white- and yellow-flowered forms are found in Derbyshire.
Recorded for all squares except:

    W. 09, 19, 08.
    E. 57, 45.
    S. 23.

Linton: Charlesworth 09, Pinxton 45.

CONRINGIA Adanson

C. orientalis (L.) Dumort. *Hare's-ear Cabbage*
Introduced. A rare casual of waste and cultivated ground.
S. Ashbourne (with parsley seed) 1746, Clifton (in a fowl-run) 1545.
Linton: Not recorded.

LEPIDIUM L.

L. sativum L.   *Garden Cress*
Introduced. A rare casual from gardens.
S.   Occasional at Ashbourne as a garden throw-out 1746.
Linton: Not recorded.

L. campestre (L.) R.Br.   *Pepperwort*
Native. A weed of arable land and waste places. Local, and apparently
decreasing.
W.   Between Bonsall and Matlock Bath 2858.
E.   Morley 3941, Dale 4338, Stanton-by-Dale 4638, Mapperley 4343.
Linton: Between Yeldersley and Brailsford 24, Breadsall 33, Drakelow 22,
Calke and Repton 32, Cauldwell 21.

L. heterophyllum Bentham (*L. hirtum* Sm., in part)   *Smith's Cress*
Native. A weed of arable land, waysides and waste places. Very rare.
S.   By Windmill Lane, Ashbourne (site now destroyed) 1847, Mackworth
3137.
Linton: Charity Farm, Wyaston 14.

L. ruderale L.   *Narrow-leaved Pepperwort*
Native. Waste places and waysides. Very rare.
W.   Nr. Great Hucklow 1778, Coombs Dale 2374.
E.   Brockwell, garden weed 3672.
S.   Arable field, Ashbourne 1746, Clifton Goods Yard 1644, Derby 3635,
Coton-in-the-Elms 2514.
Linton: Not recorded.

CORONOPUS Haller

C. squamatus (Forskål) Ascherson (*C. ruellii* All.)   *Swine-cress, Wart-cress*
Native. Trampled places by roads and paths, gateways, etc. Occasional, and
mainly in the south of the county. See also map (page 90.)
W.   Litton Mill 1573.
E.   Duffield 3442, Dale Abbey 3943, Stanley 4041.
S.   Ashbourne Green 1847, Clifton 1645, Brailsford 2441, Offcote 2048,
Radbourne 2836, Littleover 3234, Walton 2318.
Linton: Alfreton/Shirland 45, Ockbrook 43, Trent Lock 43, Repton 32,
Stapenhill 22.

C. didymus (L.) Sm.   *Lesser Swine-cress*
Introduced. A very rare casual of gardens and waste places.
S.   Hanging Bridge 1545.
Linton: 'Garden weed at Hillside, Ockbrook' 43.

CARDARIA Desv.

C. draba (L.) Desv. (*Lepidium draba* L.)   *Hoary Pepperwort*
Introduced. A troublesome weed of arable land. There is only one record in Linton but the plant is now spreading rapidly in the south of the county.

W. Miller's Dale Railway Station 1473, 1373, Great Longstone railway embankment 1971, Parsley Hay railway track 1463.

E. Brockwell Tip 3771, Calthorpe Hall 3473, Babington Hospital, Belper 3447.

S. Ashbourne Goods Yard 1746, Darley Abbey 3538, Trent Ballast Tip 4831, Netherseal 2813.

Linton: 'In some plenty at Stanton' 43.

ISATIS L.

I. tinctoria L.   *Woad*
Probably introduced. There is one old record, but no plant has been seen for many years.

Linton: Willington 22.

IBERIS L.

I. amara L.   *Bitter Candytuft*
Introduced. In short turf of fields and banks. An occasional casual.

W. Nr. Tideswell 1575.

E. Ford Bridge, Allestree 3540.

Linton: Monsal Dale 17, Cressbrook Dale 17, Yeldersley 14.

THLASPI L.

T. arvense L.   *Field Penny-cress*
Doubtfully native. A weed of arable and waste land. Local and chiefly in the south of the county.

W. Diamond Hill 0570, Great Rocks Dale 1074, Glutton Bridge 0866, River Wye, Bakewell 2268.

E. Brockwell 3772, Bramley 4666, Pleasley Vale 5265, Breadsall 3640, between Little Eaton and Duffield 3542, Mapperley 4343.

S. Ashbourne 1746, 1944, Clifton 1644, Allenton 3732, Stanhope Arms 4238, nr. Trent 4931, Staunton Harold 3721, Swarkestone 3628.

Linton: Brizlincote, Bretby and Stapenhill. 22.

T. alpestre L.   *Alpine Penny-cress*
Native. Spoil-heaps of old lead-workings and less commonly on rocks and walls or in woods on the Carboniferous Limestone. Locally abundant round and to the west of Matlock in a circle about ten miles in diameter, and very generally associated with *Minuartia verna*.

The Derbyshire form has the fruits narrowly obovate, shallowly notched or even truncate, and with a long style usually much exceeding the notch. It has been named *T. virens* Jordan or *T. calaminare* Lejeune & Court. but is identical with neither of these Continental forms.

W. Bradford Dale 2063, 2164, Wensley 2660, 2561, Picory Corner 2365, Masson 2960, 2959 and 2858, Gratton Dale 1959, about the Black Rocks, Cromford 2955, Dean Hollow 2855, nr. Slaley 2757, Bonsall 2758, 2559, Heights of Abraham and High Tor 2958, Via Gellia, nr. the Pig of Lead 2857, Carsington Pastures 2553.

E.   Crich 3454.

Linton: Castleton 18.

TEESDALIA R.Br.

T. nudicaulis (L.) R.Br.   *Shepherd's Cress*
Native. Heaths, waysides, dry banks, etc., on sand and gravel. Very rare.
W. Dirtlow Rake 1481.
Linton: Middleton Dale 27.

CAPSELLA Medicus

C. bursa-pastoris (L.) Medicus   *Shepherd's Purse*
Native. Cultivated land, waysides and waste places. Abundant and widespread.
Recorded for all squares except **W.19**.
Linton: 'Common everywhere'.

HORNUNGIA Reichenb.

H. petraea (L.) Reichenb. (*Hutchinsia petraea* (L.) R.Br.)
                                                                              Rock Hutchinsia
Native. In shallow soil on limestone rocks and scree, commonly with a southerly aspect; usually associated with other annuals such as *Arabidopsis thaliana, Cardamine hirsuta* and *Saxifraga tridactylites* and often also with scattered plants of *Festuca ovina*. Locally frequent but restricted to the Carboniferous Limestone. See also maps (page 96).

W. Pindale 1582, Pictor Rocks, Ashwood Dale 0872, Lover's Leap, Buxton 0772, Topley Pike 0572, Cunning Dale 0773, Deep Dale 0971, Old Dale, Blackwell Mill 1372, Blackwell Dale railway bank 1072, Chee Dale 1273, Miller's Dale 1473, Monsal Dale 1871, Cressbrook Dale 1774, Ravensdale 1773, Tansley Dale 1775, Biggin Dale 1457, Wolfscote Dale 1357, Lindale 1551, in various spots throughout Dovedale 1453, Manystones Quarry, Brassington 2355, High Peak Railway nr. Longcliffe 2255, Wirksworth 2755.
Linton: 'Locally abundant'.

COCHLEARIA L.

C. officinalis group ('*C. alpina* (Bab.) H. C. Watson') *Scurvy-grass*

Native. On damp ledges of north-facing cliffs of Carboniferous Limestone and as a colonist of recently reworked spoil from old lead-mines on the limestone plateau. Locally frequent.

The Derbyshire plant has often been named *C. alpina* (Bab.) H. C. Watson but, like other inland populations in northern England, it does not agree exactly with descriptions either of the Scottish alpine (now believed to be indistinguishable from *C. pyrenaica* DC.) or of the maritime *C. officinalis* L. It is certainly closer to *C. pyrenaica* in the form of its fruits.

W. Winnats 1382, Cave Dale 1583, Bradwell Dale 1780, Pin Dale 1582, Dirtlow Rake 1481, Snitterton 2760, Deane Hollow 2856, Middleton-by-Wirksworth 2755, Brightgate 2659.

ALYSSUM L.

A. alyssoides (L.) L. (*A. calycinum* L.)

Introduced. A rare casual of grassy fields and arable land.

No recent records.

Linton: Yeldersley nr. Ashbourne 24.

BERTEROA DC.

B. incana (L.) DC. (*Alyssum incanum* L.)

Introduced. A casual in cultivated ground. Two plants seen in a field of clover and rye in late summer, 1946.

S. Ashbourne 1746.

Linton: Nr. Dale 43.

LUNARIA L.

L. annua L. *Honesty*

Introduced. Much cultivated in gardens and occasionally escaping.

S. Occasional garden escape, Bretby 2823.

Linton: Not recorded.

DRABA L.

D. incana L. *Hoary Whitlow Grass*

Native. Limestone rocks, cliff-ledges and spoil-heaps. Local and rare; confined to the Carboniferous Limestone.

W. Deep Dale 0971, Diamond Hill 0570, Blackwell Dale 1072, Peter Dale 1275, Miller's Dale 1672, Water-cum-Jolly Dale 1672, Monsal Dale 1771, Cressbrook Dale 1774.

Linton: Lover's Leap, Buxton 07, Monk's Dale 17, Great Rocks Dale 17.

D. muralis L. *Wall Whitlow Grass*

Native. Limestone rocks, rocky slopes and scree, walls, abandoned railway tracks. Locally frequent, but almost restricted to Carboniferous Limestone.

W. Winnats 1382, Cave Dale 1583, Harpur Hill 0571, Deep Dale 0971, Chee Dale 1172, Taddington Dale 1671, Ravensdale 1773/4, Red Rake Mine, Calver 2474, nr. Eyam 2076, Dowel Dale 0767, Hartington 1360, Monyash 1566, Ricklow Dale 1666, Lathkill Dale 1966, Alport 2264, Wolfscote Dale 1357, 1455/6, Dovedale 1453, Biggin Dale 1457, nr. Bradbourne 2051, Via Gellia 2757/8.

EROPHILA DC.

E. verna (L.) Chevall. (*Erophila vulgaris* DC.) *Whitlow Grass*

Native. Rocks, walls, dry places, especially on limestone. Locally common and widespread.

Most Derbyshire material comes under ssp. *verna*, with a rather narrow fruit often exceeding 5 mm. in length and with broadly lanceolate or elliptical leaves having many stellate and branched hairs. Ssp. *spathulata* (A. F. Lang) Walters, with broadly ovate to sub-orbicular fruit less than 5 mm. long and with obovate-spathulate leaves densely clothed with branched hairs, has been recorded from a few localities, as has its var. *inflata* (Bab.) O. E. Schultz, with linear-lanceolate leaves and an inflated fruit, circular in section, both from the Carboniferous Limestone. Ssp. *praecox* (Steven) Walters (E. *stenocarpa* Jordan), having an oblanceolate or elliptical fruit like that of ssp. *verna* but with leaves with many unbranched and few branched hairs, has been recorded only on one occasion.

W. Recorded for all squares except 09, 19, 28.

E. Ford, Eckington 4081, Cordwell Valley 3176, Markland Grips 5074, Ashover Butts 3562, Pleasley Vale 5265, Whatstandwell 3354.

S. Wyaston 1842, Ashbourne 1746, Snelston 1541, Windley 2945, Mackworth Church 3133, Ticknall 3624.

Linton: Pinxton 45, Trent Lock 43, Egginton Common 22.

Ssp. **praecox** (Steven) Walters.

S. Mapleton 1648.

Linton: Not recorded.

ARMORACIA Gilib.

A. rusticana Gaertn., Mey. & Scherb. (*Cochlearia armoracia* L.) *Horseradish*

Introduced. Formerly much cultivated and naturalized here and there on roadsides and in fields and waste places.

W. Hathersage 2381, Coombs Dale 2274, Alport 2164, Tissington 1752, Kniveton 2050.

E. Bradway 3380, Ford 4081, Brockwell 3472, Ashover 3463, Pleasley Vale 5165, Little Eaton 3641, Kimberley 4143.

S. Recorded for all squares except 43.

Linton: About Castleton 18, Via Gellia 25, Breadsall 33, nr. Repton 32.

CARDAMINE L.

C. pratensis L.   *Cuckoo Flower, Lady's Smock*

Native. Damp meadows, pastures and roadsides, by streams and in wet places and flushes of the gritstone moorlands up to 1,500 ft. or higher.

The Derbyshire plants vary considerably in size and habit but all have lilac flowers and all seem to fall within the range of *C. pratensis* L., *sensu stricto*.

Recorded for all squares.

Linton: 'generally distributed'.

C. amara L.   *Large Bitter-cress*

Native. Margins of streams, canals and ponds, flushes, fens, etc. Frequent and widespread.

Recorded for all squares except:

> **W.** 09, 19.
> **E.**   57, 35, 45.

S.   Bentley Brook 1748, Mugginton 2843, Shardlow 4530, Old Trent, Repton 3027, Weston-on-Trent 4027.

Linton: Stirrup Wood 99, Derwent Dale 19, Mackworth 33.

C. impatiens L.   *Narrow-leaved Bitter-cress*

Native. Shady woods, river-banks, moist limestone rocks and scree. Local, and chiefly on the Carboniferous Limestone.

W.   Ashwood Dale 0772, Cunning Dale 0872, Blackwell Dale 1072, Chee Dale 1373, Water-cum-Jolly Dale 1673, Monsal Dale 1771, Lathkill Dale 1865, Bradford Dale 2063, Wolfscote Dale 1456, Bonsall 2858, Via Gellia 2656.

E.   Duffield 3443.

S.   Dove Bank 1546, Eaton 1036, Darley Abbey 3538.

Linton: Taddington Dale 17, Langley Mill 44, Breadsall 33, Bretby 32.

C. flexuosa With.   *Wood Bitter-cress*

Native. Moist shady places, by streams and ditches, etc. Common and widespread.

Recorded for all squares except:

> **W.** 09.
> **E.**   44.
> **S.**   33, 43.

Linton: Stirrup Wood 99, Mackworth 33.

C. hirsuta L.   *Hairy Bitter-cress*

Native. Walls, bare ground, rocks, scree, cultivated land. Common and widespread.

Recorded for all squares except:

> **W.** 09, 08.
> **E.**   48, 47, 46, 56.

Linton: 'Generally distributed'.

C. bulbifera (L.) Crantz (*Dentaria bulbifera* L.)   *Coralwort*
Introduced.
E.   Graves Park, Sheffield 3582.
S.   Long Eaton 4832.
Linton: Between Strines and Marple 98.

BARBAREA R.Br.

B. vulgaris R.Br.   *Winter Cress, Yellow Rocket*
Native. Hedges, stream-banks, damp waysides, etc. Frequent.
Recorded for all squares except:
              W. 09, 18, 28.
              E.   47, 46, 35.
Linton: Bradwell 18, Ashover 36.

B. intermedia Boreau   *Intermediate Yellow Rocket*
Introduced. Arable and waste land. Rare.
W.   Chee Dale 1273, Middleton Dale 2275.
Linton: Yeldersley and Sturston 24.

B. verna (Mill.) Aschers.   *Early-flowering Yellow Rocket*
Introduced. Arable and waste land. Rare and only in the south of the county.
S.   A few plants in seed-hay, Osmaston 1944.
Linton: Shirley Village 24, Winshill 22, Calke 32.

ARABIS L.

A. caucasica Willd. (*A. albida* Stev.)   *Arabis*
Introduced. A garden escape extensively naturalized on limestone cliffs near
Matlock and elsewhere.
W.   Castleton 1483, Ashwood Dale 0772, Tideswell Dale 1573, Miller's Dale
1473, Blackwell Mill 1172, Stony Middleton 2375, Froggatt Bridge 2476,
Ashford By-pass 1969, Bradford Dale 2063, High Tor, Matlock 2958,
Cromford 2957.
E.   Cromford 3057, Crich 3553, Whatstandwell 35, Milford 3545.

A. hirsuta (L.) Scop.   *Hairy Rock-cress*
Native. Cliff-ledges, rocky slopes and dry grassland, especially on the lime-
stone, walls; not uncommon as a colonist of lead-mine spoil-heaps. Locally
common.
W.   Recorded for all squares except 09, 19.
E.   Markland Grips 5074, nr. Ashover 3563, Pleasley Vale 5165, recorded
for 35, Ilkeston 4540.
S.   Dale Abbey 4338, recorded for 32.
Linton: Creswell Crags 57, Breadsall 33.

A. glabra (L.) Bernh. (*A. perfoliata* Lam.) *Tower Mustard*

Native. Roadsides and dry hedgebanks. Rare and decreasing.

S. Roadside, Repton, towards Milton 3126.

Linton: Castleton 18, Dovedale 15, Matlock Bath 25, between Ashbourne and Okeover 14, Shirley 24, Drakelow and Bretby 22, roadsides and banks about Ticknall, Hartshorne and Milton 32.

NASTURTIUM R.Br.

N. officinale R.Br. *Water-cress*

Native. Streams, ditches and areas frequently flooded by moving water. Frequent. Cultivated as green or summer water-cress.

Earlier records may include some for *N. microphyllum* and *N. x sterilis*.

Recorded for all squares except:

W. 09, 19.

Linton: Tarden, nr. Mellor 98.

N. microphyllum (Boenn.) Reichenb. *One-rowed Water-cress*

Native. In similar habitats to *N. officinale* and distinguishable by the leaves turning purple-brown in autumn and the seeds in a single row in each cell of the fruit. Frequent.

W. Recorded for 27, Lathkill Dale 1765, Fenny Bentley 1850, Bonsall Moor 2559.

E. Totley Bents 3080, Scarcliffe Park Wood 5270.

S. Ashbourne Green 1948, Hanging Bridge 1546, Offcote 2048, Barton Blount 2234, Egginton 2628.

Linton: v. *microphyllum* (Reichb.) Shirley Wood 24.

N. officinale × N. microphyllum (*Rorippa × sterilis* Airy-Shaw)
Brown Water-cress

Native. In similar habitats to the parents. Frequent in the north-west of the county and should be looked for elsewhere. Cultivated as brown or winter water-cress.

Distinguishable from both parents by the dwarfed and misshapen fruits which set very few good seeds; resembles *N. microphyllum* in having stem and leaves turning brown in autumn.

W. Monsal Ponds and Taddington Dale 1770, Monk's Dale 1374, Lathkill Dale 1765, Chee Dale 1273, Hartington 1261, Bradbourne 2052.

E. Markland Grips 5074.

S. The By-flats, Hanging Bridge 1546, Hulland Moss 2546.

Linton: Not recorded.

RORIPPA Scop.

R. sylvestris (L.) Besser (*Nasturtium sylvestre* (L.) R. Br.) *Creeping Yellow-cress*
Native. Moist ground by streams and ponds, especially where water stands only in winter. Local.
W. Buxworth Canal 0282, Coombs Lake 0379, Water-cum-Jolly 1672, garden weed, Tissington Hall 1752.
E. Brookside 3470.
S. Garden weed, Ashbourne 1746, Cubley 1638, Trent banks nr. junction with River Erewash 5133, nr. Swarkestone Bridge 3728, nr. Twyford Ferry 3228, Kings Newton 3928.
Linton: Cromford 25, Beighton 48, Shipley 44, Derby and Breadsall 33.

R. islandica (Oeder) Borbás (*Nasturtium palustre* (L.) DC.)
Marsh Yellow-cress
Native. Moist places, especially where water stands only in winter. Local.
W. Recorded for 08, Mill Clough, Derwent 1888, stream nr. Buxton 0672, Combs Reservoir 0479, Miller's Dale 1573, Lathkill Dale 1765, 2066, Alport 2264, Via Gellia 2857.
E. Birley Hay 3980, Pebley Pond, Barlborough 4880, Press Reservoirs 3565, Stubbin 3667, Hardwick Ponds 4563, recorded for 35, Alfreton Brook 4056, Butterley Reservoir 4052, Stanley 4140, West Hallam 4341.
S. Recorded for all squares except 22, 21.
Linton: Tarden nr. Mellor 98, Birley Hay, Eckington Rd., Chesterfield 37, Burton 22.

R. amphibia (L.) Besser (*Nasturtium amphibium* (L.) R.Br.)
Great Yellow-cress
Native. Ponds, ditches, streams, canals. Occasional.
E. Canal nr. Ambergate 3452, Whatstandwell 3354, Alfreton Brook 4056, by railway, Duffield 3443, Kirk Hallam 4640.
S. Recorded for all squares except 14, 24.
Linton: 'Not uncommon'.

AUBRIETA Adanson

A. deltoidea (L.) DC.
Introduced. Much grown in gardens and occasionally escaping.
E. 'probably this sp.' Cromford 3057, on limestone cliff with *Arabis caucasica*.
Linton: Not recorded.

HESPERIS L.

H. matronalis L. *Dame's Violet*
Introduced. A garden escape occasionally naturalized in hedgerows, waysides, stream-banks and waste places.

W. Recorded for 08, Ashwood Dale 0872, Pictor Wood, Grin Wood, the Cavendish Golf Course all in 07, Hucklow 1878, Chee Dale, nr. Blackwell, Miller's Dale, High Rake all in 17, Longstone Edge 2073, Baslow and Calver Mill in 27, Alport 2264, Parwich 1953, Via Gellia 2857, Bonsall 25.

E. Grave's Park, Sheffield 3581, Barlow 3475, Brockwell Lane 37, Glapwell 4766, Doe Hole 3558, nr. Allestree 3540, Cumber Hills 34.

S. Banks of the Dove from Mappleton 1646, to Rocester 1139, Shirley Wood 2042, recorded for 32.

Linton: Between Castleton and Eyam 18, Milltown nr. Ashover 36.

ERYSIMUM L.

E. cheiranthoides L.   *Treacle-mustard*

Probably introduced. Arable land and waste places, stream-banks. Rare.

W. Grin Plantation 0572.

E. Brookside 3470, Stonegravels 3872, Matlock 3159, Breadsall 3740.

S. Clifton Tip 1644, Norbury Goods Yard 1242, Derby 3434, Spondon 3935, Dale Abbey 4338.

Linton: Pleasley Station 57, Derby and Mickleover 33, Draycott 43.

CHEIRANTHUS L.

C. cheiri L.   *Wallflower*

Introduced. Old walls, limestone cliffs and quarries. Rare and local.

W. Middleton Dale 2275.

E. Pleasley Vale 5165, old walls at Cromford 3057.

S. Mackworth Castle 3137.

Linton: Haddon Hall 26, Crich 35, Wingfield Manor 35, Codnor Castle 45, Derby 33, Repton 32.

ALLIARIA Scop.

A. petiolata (Bieb.) Cavara & Grande (*Sisymbrium alliaria* (L.) Scop.)
                                                          Hedge Garlic, Garlic Mustard

Native. Hedgerows, woods and wood-margins, wall-bases and waste ground, especially on nutrient-rich soils. Common and widespread.

Like *Anthriscus sylvestris* in benefiting from the increased availability of phosphate where farm animals pass to and fro.

Recorded for all squares except 09, 19.

Linton: 'Common throughout the county'.

SISYMBRIUM L.

S. officinale (L.) Scop.   *Common Hedge Mustard*

Native. Hedge-banks, roadsides, waste places. Common except in the extreme north of the county.

Recorded for all squares except:

W. 19, 18, 28.
E. 46, 45.

Linton: 'Common throughout the county except perhaps in the extreme north'.

S. irio L.   *London Rocket*

Doubtfully native. A rare casual; no recent records.

Linton: Wingfield Manor 35, between Burton and Ashby.

S. orientale L.   *Eastern Rocket*

Introduced. A casual of waste ground, much increased as a result of bombing during the war of 1939-45.

E.   Ashover 3463.

S.   Ashbourne 1746, Clifton 1644, Brailsford 2541, Derby 3536, Chaddesden 3637, Netherseal 2813.

Linton: Not recorded.

S. altissimum L.   *Tall Rocket*

Introduced. A casual of waste ground. Occasional, chiefly in the east and south of the county.

W.   Edale Church 1285, Diamond Hill 0570, Ashwood Dale, 0772, recorded for 17, Dove below Hartington 1259.

E.   North Wingfield 4064, Temple Normanton 4167, Stonegravels 3872, Crich 3554.

S.   Ashbourne 1746, Clifton 1644, Derby 3633, 3536, Baltimore Bridge 3630, Willington 2929, Swarkestone 3628, Netherseal 2813.

Linton: Not recorded

ARABIDOPSIS Heynh.

A. thaliana (L.) Heynh. (*Sisymbrium thalianum* J. Gay)   *Thale Cress*

Native. Walls, dry banks, hedgerows, sandy fields, waste places. Locally frequent, especially in shallow soil over limestone.

W.   Recorded for all squares except 09, 19.

E.   Recorded for 36, 35, Trent 3148, Sandiacre 3747.

S.   Ashbourne 1746, Brailsford 2541, Cubley 1638, recorded for 43, Tutbury Castle 2129, King's Newton 3826, walls, Smisby 3419.

Linton: Sutton-on-the-Hill 23, Mickleover and Derby 33.

CAMELINA Crantz

C. sativa (L.) Crantz   *Gold of Pleasure*

Introduced. A cornfield weed. Rare.

E.   Creswell 5374, Stanley 4140.

S.   Clifton 1545, Ashbourne 1746.

Linton: Heanor 43, Breadsall 33, Normanton by Derby 33.

DESCURAINIA Webb & Berth.

D. sophia (L.) Prantl (*Sisymbrium sophia* L.)   *Flix-weed*

Doubtfully native. A rare casual; no recent records.

Linton: Matlock 25, Markeaton Hill 33, Ticknall 32.

RESEDACEAE

RESEDA L.

R. luteola L.   *Dyer's Rocket, Weld*

Native. Banks, walls, arable land and disturbed ground, especially on calcareous substrata. Occasional.

W.   Recorded for 07, Topley Pike 1072, Middleton Dale 2175, Stoney Middleton, Calver New Bridge 27, nr. Alport 2264, Via Gellia 25.

E.   Brierley Wood, Sheepbridge 3775, Chesterfield 3871, Barlborough 4777, nr. Staveley 4174, Creswell 5273, Scarcliffe Park Wood 5270, Milltown Quarries, Ashover 3561, Pleasley 5164, Ambergate 3451, Little Eaton 3641, nr. Langley Mill 4446.

S.   Ashbourne Goods-yard 1746, Derby 3537, Trent Ballast Tips 4831, Burton 2323, Chellaston 3829, King's Newton 3826, Swadlincote 2919, Netherseal 2813.

Linton: Beighton 48, Pinxton 45, Codnor Park Station and railway, West Hallam 44.

R. lutea L.   *Wild Mignonette*

Native. Waste places, disturbed ground and arable fields, especially on calcareous soils. Infrequent.

W.   Ladmanlow Tip 0471, nr. Topley Pike and railway nr. Buxton 07, Friden 1660, nr. Parsley Hay 1463, Alsop Moor Brickworks 1656, Wirksworth 2854.

E.   Killamarsh 48, Holmesfield Church 3277, nr. Wingerworth 3667, Ambergate 3551, nr. Park Nook, Kedleston 3241, Stanley Brook 4140.

S.   Clifton goods-yard 1644, nr. Derby 3537, railway bank, Breaston 4433, Trent bank, Sawley, and ballast tips, Trent 43, Shobnall, Burton 2323, Netherseal Colliery 2715.

Linton: Nr. Bamford Station 28, between Rowsley and Bakewell 26.

VIOLACEAE

VIOLA L.

V. odorata L.   *Sweet Violet*

Native. Hedge-banks, plantations, etc., especially on calcareous soils; often near houses and probably a garden escape.

The white-flowered form is not uncommon.

Recorded for all squares except:

> **W.** 09, 19, 18, 28.
> **E.** 38, 48, 45.
> **S.** 22.

Linton: Hathersage 28, Burton 22.

forma *alba*   *White Violet*

W. Hassop 2272, Fenny Bentley 1750, Bonsall Moor 2559.

E. Windley 3045, Morley, Shottle 34, West Hallam 4341, Stanley 4140.

S. Ashbourne 1746, Hollington 2340, Breadsall 3739, Markeaton 3237, Dale 4338, Stanton-by-Dale 4638.

Linton: Alport, Lathkill Dale 26, Morton 38, Scarcliffe Park 57, Burton, Bretby 22, Repton 32.

V. hirta L.   *Hairy Violet*

Native. Grassy slopes, hedgerows, banks, chiefly on limestone. Locally frequent.

W. Recorded for 08, Conies Dale 1280, Diamond Hill 0570, Cunning Dale and Woo Dale 07, Chee Dale 1172, Miller's Dale and Monsal Dale 17, Sir William Hill 2177, recorded for 06, Lathkill Dale 1965/6, Alport 2264, Wolfscote Dale 1357, Brassington Rocks 2154, Via Gellia 2556.

E. Scarcliffe Park Wood 5170, Markland Grips 5174, Pleasley 5164.

S. Ticknall 3523, recorded for 43.

Linton: Ashover 36.

V. riviniana Reichenb.   *Common Dog Violet*

Native. Woods, hedgerows, rough pastures, moorland. Abundant and widespread.

Forms in more exposed habitats, with smaller leaves, flowers and capsules, than in woodland forms, may belong to ssp. *minor* (Gregory) Valentine.

Recorded for all squares except:

> **W.** 09.

Linton: 'Common. Generally distributed'.

Ssp. **riviniana**

W. Fox House 2680, Longshaw 2578, Baslow 2672, Wolfscote Dale 1455, Brassington 2154.

E. Beauchief Abbey 3381, Cordwell Valley 3176, Markland Grips 5074, Langwith Wood 5068, Ambergate 3451, Belper 3447, West Hallam 4341.

S. Ashbourne 1746, Ireton Wood 2748, Brailsford 2541, Hulland Moss 2446, Alkmonton 1838, Mackworth 3137, Dale 4338.

Ssp. **minor** (Gregory) Valentine

W. Upper Derwent by dam 19.

E. Recorded for 37.

S. Nr. Ashbourne 1647, Shirley Park 2042, Alkmonton 1838.

V. reichenbachiana Jord. ex Bor. (*V. silvestris* Reichenb.)   *Pale Wood Violet*
Native. Woods and hedgerows, chiefly on calcareous soils. Locally frequent.

W. Monk's Dale 1374, lower Peter Dale 1275, Longshaw 2578, Lathkill Dale 1966, Dovedale 1452.

E. Ford Valley 4080, Unthank Wood 3076, Pebley Pond 4978, Markland Grips 5074, Ashover 3562, Langwith Wood 5068.

S. Fenny Bentley 1849, Snelston Common and Darley Moor 14, Kniveton 2147, Yeaveley 2044, Alkmonton 1937.

Linton: Great Shacklow Wood 16, Wirksworth 25, Fritchley 35, Hilton Common 23.

The sterile hybrid of *V. riviniana* and *reichenbachiana* has been recorded for Hollington, nr. Longford (2239), growing with the parents.

V. canina L. (*V. ericetorum* Schrad.)   *Heath Violet*
Native. Dry grassland and heaths, probably always on non-calcareous soil even when over limestone. Rare.
The Derbyshire plants belong to ssp. *canina*.

W. Wolfscote Dale 1455, Middleton-by-Wirksworth 2754.

Linton: Shirley Park 24, Bentley Carr 13, Ockbrook 43, Seal Wood 21.

V. canina × **riviniana** (*V. ericetorum* × *riviniana*)

W. Wolfscote Dale 1455.

Linton: Shirley Park 24.

V. palustris L.   *Marsh Violet*
Native. Bogs, acid fens and wet places on acid soils, both in woods and in the open. Locally frequent and characteristic of flushes on the gritstone moorlands and of woods in gritstone cloughs.

The Derbyshire plants belong to ssp. *palustris*.

W.   Alport Dale 1390, recorded for 08, Kinder Scout 1187, Higger Tor Lane 2582, Burbage Edge 0372, Axe Edge 0370, Umberley Brook 2870, Froggatt 2476, Washgate Valley 0567.

E.   Dore 3081, Cordwell Valley 3176, Unthank 3075, Whitwell Wood 5278, Hate Wood 3463, Pleasley 5265, Shining Cliff Woods 3352.

S.   Bradley Brook 2244, Hulland Moss, Mugginton and Shirley Wood 24, Cubley 1539, Repton Rocks 3221.

Linton: Crowden Reservoir 09, Horsley Carr 34.

V. lutea Huds.   *Mountain Pansy*
Native. Characteristic of pastures on non-calcareous but only slightly acid soils, as on the brows of the limestone dales and locally on the shale-grits; often abundant on soils derived from igneous and volcanic substrata. Locally common, often with *Agrostis tenuis, Lathyrus montanus* and *Betonica officinalis*. Most of the Derbyshire populations are uniformly yellow-flowered, but some include plants with the flowers violet or violet and yellow.

W.   Recorded for 08, Pin Dale 1582, Bradwell Moor 1380, Woo Dale 0972, Ebbing and Flowing Well 0879, Brook Bottom Tideswell 1476, Chelmorton Low 1770, nr. Calton Nick Quarry 1171, Miller's Dale 1373, Priestcliffe Lees 1573, Middleton Dale 2175, Deep Rake 2273, Sir William Road 2177, Thirkelow Rocks 0469, High Low, Sheldon 1768, Ricklow Dale 1666, Lathkill Dale 1965/6, Bradford Dale 2063, Minninglow 2057, Harboro' Rocks 2354, Carsington Pasture 2453/4, Black Rocks 2955.

E.   Freebirch 3073.

Linton: Charlesworth Combes 98, Ashover 36.

V. tricolor L.   *Wild Pansy*
Native. Cultivated ground and waste places, chiefly on non-calcareous soils. Frequent.

W.   Woo Dale 0973, Black Edge 0576, Longstone Edge 1972.

E.   Ford Valley 4080, Holmesfield 3277, Kelstedge 3363, Pleasley Vale 5165, nr. Whatstandwell 3255, Morley 3940, West Hallam 4341.

S.   Breadsall 3739, Dale 4338.

Linton: Bradwell 18, Hathersage 28, Cromford and Matlock 25, Shirley and Yeldersley 24, Etwall 23, Repton and Calke 32.

V. arvensis Murray   *Field Pansy*
Native. Arable and waste places. Frequent and widespread.
Recorded for all squares except:

> **W.** 09, 19, 27, 26.
> **E.** 44.
> **S.** 23.

Linton: 'Common everywhere'.

V. arvensis × tricolor

S. Ednaston 2342, Hulland 2546, above Bradley Wood 2046.

Linton: Not recorded.

POLYGALACEAE
POLYGALA L.

P. vulgaris L.　　　　　　　　　　　　　　　　*Common Milkwort*

Native. Pastures, especially on limestone; banks, waysides, wood-margins. Locally frequent.

Polygala oxyptera Reichenb. grades into *P. vulgaris* and cannot be maintained as a distinct species.

W. Recorded for all squares except 09, 19, 18, 28.

E. Recorded for all squares except 45, 34, 44, 37, 38.

S. Ashbourne 1747, Osmaston 1944, throughout the Hulland-Bradley area 2045, Breadsall 3839, recorded for 22, Ticknall lime quarries 3523.

Linton: Allport Dale, Marebottom 19, Bradwell 18, Pinxton 45, Kedleston 34.

P. serpyllifolia Hose (*P. serpyllacea* Weihe)　　　*Thyme-leaved Milkwort*

Native. Pastures, heaths and moorlands on non-calcareous soils. Frequent. The records for the Carboniferous Limestone are for localities on acid plateau soil, where *P. serpyllifolia* is commonly associated with *Galium saxatile*.

W. Howden Reservoir 19, nr. Derwent 1788, above Hathersage 28, recorded for 07, 17, Longshaw 2578.

E. Ford 4081, Troway 3879, Whitwell Wood 5278, Kelstedge 3363, Chanderhill 3369, Pleasley Vale 5265, Alport Height 3051.

S. Hulland Moss 2546, between Shirley Wood & the Mill, and Mugginton 24, recorded for 22, Ticknall 3523, Overseal 2915.

Linton: Nr. Chapel 08, Via Gellia and Cromford Dale 25, Morley 34, Rodsley 14, Cubley Common 13, Mickleover 33, Long Eaton 43.

HYPERICACEAE
HYPERICUM L.

H. androsaemum L.　　　　　　　　　　　　　　　　*Tutsan*

?Formerly native. Woods and hedges. Very rare and perhaps now extinct. No recent records.

Linton: Matlock 35.

H. calycinum L.　　　　　　　　　　　　　　　*Rose of Sharon*

Introduced. Naturalized in plantations, shrubberies, etc. Rare.

W. Lovers' Walk, Matlock Bath 2958.

Linton: Knowle Hills 32.

H. perforatum L.   *Common St. John's Wort*

Native. Woods, hedgerows, grassland. Common, especially in disturbed soil over limestone.

Recorded for all squares except:

W. 09, 19.

H. maculatum Crantz (*H. dubium* Leers)   *Imperforate St. John's Wort*

Native. Damp hedge-banks and wood-margins. Rare, and chiefly on the Coal Measures.

W. Combs nr. Chapel 0378, Miller's Dale 1473, Monk's Dale 1374.

E. Whirlow 3182, Millhouses Sta. 3383, New Whittington 3975, Whittington Moor 3974,

Alfreton Brook 4056, Bargate nr. Belper 3645.

S. By the River Dove nr. Ashbourne 1646.

Linton: Buxton 07, Matlock 35.

H. × desetangsii Lamotte (*H. maculatum × H. perforatum*)

E. Brimington 37.

Linton: Not recorded.

H. tetrapterum Fries (*H. quadratum* Stokes)

Square-stemmed St. John's Wort

Native. Damp meadows and marshes, by ditches and streams. Frequent.

Recorded for all squares except:

W. 19.
E. 45.

Linton: Nr. the Snake Inn 19, between Tibshelf and Hucknall, 45.

H. humifusum L.   *Trailing St. John's Wort*

Native. Dry moorland, heath and open woods on non-calcareous soils. Local.

W. River Derwent 19, Hayfield 08, Yorkshire Bridge 1985, Derwent 1889, Jaggers Clough 1687, Ladybower 2086, Longshaw 2480, Hathersage 2280, 07, Miller's Dale 17, Bretton Clough 2077, Kirk Ireton 2650.

E. Freebirch 3173, Barlborough 4878, Wingerworth 36, between Breadsall and Morley Moor 3841.

S. Clifton 1744, Bradley Wood 2046, Marston Montgomery 1237, Repton Rocks 3221.

Linton: Buxton 07, Pinxton 45, Coxbench 34, Yeaveley 13, Mickleover 33, Egginton 22, Drakelow 21.

H. pulchrum L. *Slender St. John's Wort*

Native. Rough grassy places, heaths and hedge-banks, usually on non-calcareous soils; in the limestone areas on moist loamy soils, often with *Betonica officinalis*. Frequent and widespread.

Recorded for all squares except:

> **W.** 09, 19.
> **E.** 38, 36, 46, 45.
> **S.** 33, 22, 12.

Linton: Snake Inn 19, nr. Ripley, Pinxton, between Tibshelf and Hucknall 45, Mickleover 33, Stanton, Dale 43, Egginton Common, Burton 22.

H. hirsutum L. *Hairy St. John's Wort*

Native. Woods, hedges and grassland, chiefly on calcareous soils and commonly on stabilized scree. Locally common.

W. Recorded for all squares except 09, 19, 28.

E. Recorded for all squares except 38, 48, 37, 45, 34.

S. Limestone outlier, Norbury 1242, nr. Hollington 2240, recorded for 13, Longford 2137, Twyford 3228, Ticknall 3523.

Linton: Barlborough 47, Newton Wood 45, Kedleston Park 34, Mickleover, Breadsall and between Osmaston and Sinfin Moor 33, New Stanton 43, Burton, Bretby, Stapenhill and Egginton Common 22, Cauldwell 21.

H. montanum L. *Mountain St. John's Wort*

Native. Woods, cliffs, hedge-banks and south-facing grassy slopes on shallow loamy limestone soils. Local.

W. Duke's Drive, Buxton 0762, Cunning Dale 0972, Ashwood Dale on the north side 0772, Monk's Dale 1374, Deep Dale 1071, between Wormhill and Tideswell 1375, by the Wye nr. Blackwell Mill 1072, Topley Pike 1072, Chee Dale 1273, Cressbrook Dale 1772, Hartington 1360, Long Dale 1363, Biggin Dale 1457, Via Gellia 2757, Matlock Bath 2958.

E. Scarcliffe Park Wood 5270, Whitwell Wood 5277, Markland Grips 5074, also Creswell in 57, recorded for 35.

Linton: Stapenhill and Winshill 22, Ticknall Quarry 32.

H. elodes L. *Marsh St. John's Wort*

?Native. Bogs and wet places by ponds and streams on acid soils. Very rare and perhaps extinct.

E. Wessington 3757.

Linton: Tansley Moor 36.

CISTACEAE

HELIANTHEMUM Miller

H. nummularium (L.) Miller (*H. chamaecistus* Miller)   *Rockrose*

Native. Rock outcrops, stabilized scree and grassy slopes on limestone, on both black and loamy limestone soils; indifferent to aspect except in high and wet areas of the Carboniferous Limestone where it is restricted to south-facing habitats. Locally abundant.

W. Recorded for all squares except 09, 19, 28.

E. Creswell Crags 5374, Markland Grips 5074, Whitwell Wood 5279, Fallgate and Milltown 3561, Glapwell 4966, Pleasley Vale 5265, Crich Stand 3455.

Linton: Bolsover and Clowne 47

CARYOPHYLLACEAE

SILENE L.

S. dioica (L.) Clairv. (*Lychnis dioica* L., (in part))   *Red Campion*

Native. Woods, hedges, roadsides on a wide range of soils but avoiding podsolized or waterlogged soils; rock-ledges and boulder-scree in the limestone dales, especially where shaded; spoil-heaps of old lead workings. Common.

A dwarf form occurs on north-facing cliff-ledges of the Winnats.

Recorded for all squares except 19.

S. alba (Miller) E. H. L. Krause (*Lychnis alba* Miller)   *White Campion*

Native. Hedgerows, waste places and cultivated land. Frequent.

W. Duke's Drive, Buxton 0672, Ladmanlow Tip 0471, Tideswell 1575, Froggatt Edge 2476, Lathkill Dale 1966, Tissington 1752.

E. Brampton 3270, Pebley Pond 4878, Elmton 5073, Alton 3664, Astwith 4464, Pleasley Vale 5265, Alfreton Brook 4056.

S. Recorded for all squares except 13.

Linton: 'Generally distributed but scarcely plentiful anywhere'.

Hybrids between *S. dioica* and *S. alba*, with pink flowers and intermediate in other characters, are seen not infrequently in hedgerows.

S. noctiflora L.   *Night-flowering Campion*

Native. A weed of arable land. Rare.

W. Monsal Dale 17, Great Rocks Dale 1073.

E. Whirlow 3182, Allestree 33.

S. Hulland 2546.

Linton: Norton Lees 38, Church Broughton 13, Bretby Mill 32, White Lees, Ticknall 32, Foremark Hills 32, Repton Rocks 32, Gresley 21, Linton 21, Drakelow 21.

S. vulgaris (Moench) Garcke (*S. cucubalus* Wibel)   *Bladder Campion*
Native. Grassy slopes, roadsides, arable and disturbed land. Frequent. Varies
greatly in hairiness.
Recorded for all squares except:

> W. 09, 19, 15.
> S.   23.

Var. **puberula** Syme
W.   Cressbrook Dale 1773.
E.   Markland Grips 5074, Glapwell 4866, Press 3565, Wingerworth 3766.
Linton: Monk's Dale 17, Bakewell 26, Pleasley 56, Whatstandwell 35.

S. gallica L.   *Small-Flowered Catchfly*
S.   Sandybrook Hall 1748.
Linton: Not recorded.

S. nutans L.   *Nottingham Catchfly*
Native. Edges and ledges of limestone outcrops, usually with a southerly
aspect. Locally common but restricted to the Carboniferous Limestone.
See also maps (page 96).
There is a good deal of variation in leaf-width, etc., between one population
and another, but all Derbyshire forms come within ssp. *nutans*.
W.   Deep Dale 0971, Ashwood Dale 0772, Chee Dale 1273, Monk's Dale
1374/5, Miller's Dale 1473, 1573, Tansley Dale 1674, 1774, Middleton Dale
2275, Ashford 1869, Hartington 1260, Wolfscote Dale 1357, Dovedale 1551,
Biggin Dale 1457, recorded for 25.
Linton: Bakewell 26.

LYCHNIS L.

L. flos-cuculi L.   *Ragged Robin*
Native. Damp meadows, marshes, fens, wet woods. Common throughout
the county in suitable habitats.
Recorded for all squares except:

> W. 19, 09.

Linton: 'Common throughout the county'.

AGROSTEMMA L.

A. githago L. (*Lychnis githago* Scop.)   *Corn Cockle*
Introduced. Cornfields. Formerly not infrequent but now very rare.
W.   Buxton 0672, Blackwell Mill 1172, nr. Matlock 2858.
E.   Renishaw 4478, Whatstandwell 3354.
S.   Clifton 1545.
Linton: Brough 28, between Bakewell and Hathersage 27, Norton 38, Morley
34, Yeldersley 24, Mickleover and Breadsall 33, Ockbrook 43, nr. Burton 22,
Calke and nr. Repton Shrubs 32, Cauldwell 21.

CUCUBALUS L.

C. baccifer L.   *Berry Catchfly*
W. Recorded for 07.
Linton: Not recorded.

DIANTHUS L.

D. armeria L.   *Deptford Pink*
Doubtfully native. Dry pastures, waysides, disturbed ground. A very rare casual.
W. Ashbourne 1746.
Linton: Edensor 27, Renishaw 47.

D. caryophyllus L.   *Clove Pink, Carnation*
Introduced. A very rare casual. No recent records.
Linton: Nr. Edensor 27.

D. deltoides L.   *Maiden Pink*
Native. Grassy fields and banks on shallow loam over limestone, quarry spoil-heaps, old lead workings. Rare.
Very sensitive to rabbit-grazing.
W. Ladmanlow 0471, Peter Dale 1275, Hassop 2373, Glutton 0867, Lathkill Dale 1665, Alport 2264, Bradford Dale 2164, Brassington Rocks 2154.
Linton: Litton Dale 17, Bakewell-Chatsworth 26, nr. Winster 25, Codnor Castle 44.

SAPONARIA L.

S. officinalis L.   *Soapwort*
Doubtfully native. Hedgebanks and waysides, chiefly near villages and probably an escape from cultivation. Rare.
W. Whaley 0181, recorded for 07.
E. Pebley Pond 4878, Hasland 3969 Ambergate, 3552.
S. Ashbourne 1746, Chellaston 3830, Railway bank, Breaston Station 4433.
Linton: Elmton 57, Ednaston 24, Hulland Knob 24, Duffield 34.

CERASTIUM L.

C. arvense L.   *Field Mouse-ear Chickweed*
Native. Dry banks, waysides and grassland, especially on calcareous or slightly acid sands. Rare.
W. Matlock 2860, High Peak Quarry, Brassington 2355.
E. Creswell Crags 5374, Markland Grips 5074, Morley 4140.
S. Bradley Wood 2046, nr. Thrumpton 5031.
Linton: Breadsall Moor 34, Long Eaton 43, Repton 32.

C. tomentosum L.   *Dusty Miller*
Introduced. A garden escape, rapidly forming large patches on hedge-banks, waysides and waste places. Not infrequent.
W.   Litton 1675, Matlock 2959.
E.   Ashover 3463, Pleasley Vale 5165, Holloway 3356.
Linton: Not recorded.

C. fontanum Baumg. ssp. **triviale** (Link) Jalás (*C. triviale* Link)
                                              Common Mouse-ear Chickweed
Native. Meadows and pastures, waysides, walls, waste places. Common and widespread.
Recorded for all squares except:
                            E. 34.
Linton: 'Throughout the county'.

C. glomeratum Thuill.   *Sticky Mouse-ear Chickweed*
Native. Arable land, banks, walls and waste places. Common and widespread.
Recorded for all squares except:
                            W. 09, 19, 18.
                            E. 38, 48, 45, 44.
                            S. 22.
Linton: 'Generally distributed'.

C. semidecandrum L.   *Little Mouse-ear Chickweed*
Native. Dry open habitats chiefly on calcareous or sandy soils; rocks, walls. Locally frequent.
W.   Monsal Head 1871, Cressbrook Dale 1772, Sir William Hill 2177, Coombs Dale 2274, Chrome Hill 0767, Hartington 1360, Dovedale 1453, Brassington Rocks 2054.
E.   Markland Grips 5074, Creswell Crags 5374, Pleasley Vale 5265, Morley Moor 3841, Breadsall Moor 3742.
Linton: Axe Edge 07, Lathkill Dale 16,   Repton 32.

MYOSOTON Moench

M. aquaticum (L.) Moench (*Stellaria aquatica* (L.) Scop.)
                                                      Water Chickweed
Native. Marshes and fens, sides of streams and ditches, damp shady places. Local.
W.   Matlock 2960.
E.   Killamarsh 4581, Pebley Pond 4879, Alfreton Brook 4056, Butterley Reservoir 4052, Little Eaton 3641.
S.   Chaddesden 3736, Markeaton 3337, Darley Abbey 3537, River Dove‚ Scropton 1829, Marston-on-Dove 2329, Twyford 3228, Old Trent, Repton 3127, Swarkestone 3728.
Linton: Scarcliffe Park Wood 57.

STELLARIA L.

S. nemorum L.   *Wood Stitchwort*

Native. Damp woods and sides of streams. Local and rare; apparently decreasing.

The Derbyshire plant is ssp. *nemorum*.

W. Lathkill Dale 1966.

E. The Moss between Ford and Eckington 38, Allestree 3440, Darley Abbey 3538.

Linton: Charlesworth 98, Mellor 98, Strines 28, Eyam Dale 27, Stand Wood 27, Lindup Wood 26.

S. media (L.) Vill.   *Chickweed*

Native. A weed of cultivated ground and waste places.

Recorded for all squares.

Linton: 'Everywhere. Very common'.

S. pallida (Dumort.) Piré (*S. media* var. *boraeana* Jordan)   *Lesser Chickweed*

Native. A weed of cultivated ground and waste places, especially on light sandy soils. Rare.

W. The Nabbs, Dovedale 1453, Thorpe Pastures 1551.

E. Pleasley Vale 5265.

S. Lime Pits, Snelston 1541.

Linton: Miller's Dale 17, Newton Solney and Bretby 22.

S. neglecta Weihe (*S. umbrosa* Opiz & Rufr.; incl. *S. media* var. *major* Koch)
                                                                    Greater Chickweed

Native. Hedge-banks, wood-margins, streamsides, shady places. Locally frequent.

W. Thorpe 1551, Bradbourne 2052, Bonsall 2858.

E. The Moss, Eckington 4180, Troway 3879, Pebley Pond 4878, West Hallam 4341.

S. Recorded for all squares except 43, 32.

Linton: High Tor 25, Ambergate 35.

S. holostea L.   *Greater Stitchwort*

Native. Woods and hedge-banks on all but podsolized or waterlogged soils. Common.

Recorded for all squares except:

W.28.

Linton: 'Common'.

S. palustris Retz.   *Marsh Stitchwort*
Native. Marshes and fens. Rare, and only in the south of the county.
E.   River Trent at Sawley Junction 4832.
S.   Clifton 1544, Willington 2928.
Linton: Breadsall 33, Repton 32, Drakelow 21.

S. graminea L.   *Lesser Stitchwort*
Native. Pastures and waysides on non-calcareous but not strongly acid soils, commonly with bracken; often in moist places, including acid fens. Frequent and widespread.
Recorded for all squares except:
> **W.**   09.
> **E.**   34.

Linton: 'generally distributed and plentiful'.

S. alsine Grimm (*S. uliginosa* Murr.)   *Bog Stitchwort*
Native. Acid flushes and streamlets, on damp mud or in Sphagnum. Frequent.
Recorded for all squares except:
> **W.**   09.
> **E.**   57, 35, 34, 44, 56.
> **S.**   33.

Linton 'generally distributed'.

MOENCHIA Ehrh.

M. erecta (L.) P. Gaertner, B. Meyer and Scherb. (*Cerastium quaternellum* Fenzl)   *Upright Chickweed*
Native. Gravelly pastures. Very rare.
E.   Griff Wood, Ault Hucknall 4665.
Linton: Repton 32.

SAGINA L.

S. apetala Ard. (incl. *S. ciliata* Fries)   *Annual Pearlwort*
Native. Walls, rock-ledges, bare ground in dry places, etc.
There are two subspecies, formerly treated as distinct species:
Ssp. **erecta** (Hornem.) F. Hermann ('*S. apetala*'), with bluntish sepals wide-spreading in fruit. Locally frequent on walls, roadsides, waste places, etc., especially on sandy or gravelly ground.
W.   Recorded for 08, Stanage Edge 2384, recorded for 07, Calver New Bridge 2475, Darley Bridge 2762, Alsop Station 1554, Matlock 2858.
E.   Pleasley Vale 5265, Cromford 3056, Whatstandwell 3354.
S.   Ashbourne Churchyard 1746, Dale Abbey 4338.

Linton: Taddington Dale 17, Ashford 16, Haddon Hall 26, Millthorpe 37, Stubbing Court 37, Whitwell 57, Ashover 36, Marehay 34, Shirley 24, Breadsall 33, Sawley 43, Egginton 22, Ticknall 32, Drakelow 21.

Ssp. **apetala** (*S. ciliata* Fries), with more or less acute sepals appressed or only slightly spreading in fruit. Less frequent than ssp. *erecta* but in similar habitats except that ssp. *apetala* is more characteristic of rock-ledges on the Carboniferous Limestone.

W. Buxton 0573, Lathkill Dale 2063, Wolfscote Dale 1357, Dovedale 15, Brassington Quarries 2354, Carsington Pastures 2553.

E. Pleasley Vale 5265.

S. Clifton 1644, Weston-on-Trent church wall 3927.

Linton: Miller's Dale 17, Monsal Dale 17, Killamarsh 48, Barlow 37, Whatstandwell 35, Codnor Castle 45, Mugginton 24, Mansfield Road, Derby 33.

S. procumbens L.   *Procumbent Pearlwort*

Native. Paths, lawns, banks, waste ground, etc. Common and widespread. Recorded for all squares.

Linton: 'Common'.

S. subulata (Sw.) C. Presl   *Awl-leaved Pearlwort*

Probably introduced. Dry sandy, gravelly or rocky places. A single recent record, suggesting that this species is beginning to spread in Derbyshire as elsewhere.

E. Railway track, Pleasley 5265.

Linton: Not recorded.

S. nodosa (L.) Fenzl   *Knotted Pearlwort*

Native. Moist places on sand, peat or calcareous tufa. Local.

W. Recorded for 08, Macclesfield Old Road, Buxton 0372, Axe Edge 0270, Cat and Fiddle Road 0271, nr. Ladmanlow 0371, Congleton Road nr. Buxton 0171, Chee Dale 1273, Cressbrook Dale 1778, Monk's Dale 1374, Coombs Dale 2274, Middleton Dale 2175, recorded for 26, Lindale 1551, nr. Brassington 2355, Heights of Abraham 2958.

S. Ashbourne Green 1847.

Linton: Mellor 98, Castleton 18, Repton Rocks 32.

MINUARTIA L.

M. verna (L.) Hiern (*Arenaria verna* L.)   *Mountain Sandwort*

Native. Lead-mine spoil-heaps and limestone rock-ledges. Locally abundant on old lead-workings, rare elsewhere. See also map (page 95).

W. The Winnats 1382, Hartle Dale 1680, Hope Valley 1783, Deep Dale, Buxton 0971, Great Rocks Dale 1073, Chee Dale 1273, Miller's Dale 1573, Taddington Dale 1671, Cave Dale 1583, Tansley Dale 1774, Longstone Edge

2073, Middleton Dale 2175, Coombs Dale 2274, Long Dale 1860, Gratton Dale 2060, Lathkill Dale 2065, Black Rocks 2955, Via Gellia 2656, Bonsall 2758, Wirksworth 2854, Brassington Rocks 2154, Carsington Pastures 2553, Harborough Rocks 2455.

E. Milltown, Ashover 3561, Crich 3554.

M. hybrida (Vill.) Schischkin (*Arenaria tenuifolia* L.).
W. Lathkill Dale 1865.
Linton: Miller's Dale 17, Dovedale 15, Brassington 25.

MOEHRINGIA L.

M. trinervia (L.) Clairv. (*Arenaria trinervia* L.)   *Three-nerved Sandwort*
Native. Hedge-banks, wood-margins and shady places on loose dry soils; dry scree and rock-ledges on black limestone soil; often with *Glechoma hederacea* and *Sambucus nigra*. Locally frequent.
Recorded for all squares except:

W. 09, 19.

ARENARIA L.

A. serpyllifolia L.   *Thyme-leaved Sandwort*
Native. Walls, rock-ledges, arable land and bare ground; avoided by rabbits. Frequent.
There are two subspecies in Derbyshire:

Ssp. **serpyllifolia,** with sepals and ripe fruit more than 3 mm. long, the fruit swollen and rounded at the base. This is the more abundant and widespread form.
W. Recorded for all squares except 09, 19.
E. Recorded for all squares except 34, 45.
S. Recorded for all squares except 13, 22, 21.
Linton: South Normanton 45, Burton 22.

Ssp. **leptoclados** (Reichenb.) Guss., with sepals and ripe fruit less than 3 mm. long, the fruit straight-sided. In similar localities to ssp. *serpyllifolia* but more restricted to the Carboniferous Limestone. Frequent.
W. Roadside, Topley Pike 1072, Middleton by Wirksworth 2755.
E. Pleasley Vale 5165.
S. Ashbourne goods yard 1746, roadside grit-heap, Snelston 1543, corn-field nr. Anchor Church 3327.
Linton: About Matlock 35, South of Hulland 24, Weston Cliff 42.
Intermediates between the two subspecies are sometimes encountered.

SPERGULA L.

S. arvensis L.   *Corn Spurrey*
Native. An often troublesome weed of arable land on acid soils and an indicator of a need for liming. Frequent.

There are two varieties differing in characters of the seed:

Var. **arvensis,** with brownish-black seeds covered with pale club-shaped deciduous papillae and with the wing narrow or absent. This is the common form in the county.

W. Nr. Hayfield 08, Ringinglow 2983, Sladen 0772, nr. Totley 2979, Newhaven 1561, Wirksworth 2854, Carsington 2454.

E. Recorded for all squares except 57, 34.

S. Recorded for all squares except 43.

Var. **sativa** (Boenn.) Mert. & Koch, with black dull seeds, minutely tubercled and narrowly winged. Rare.

S. Hulland Gravel-pit 2645, Smisby 3419.

SPERGULARIA (Pers.) J. & C. Presl

S. rubra (L.) J. & C. Presl (*Buda rubra* (L.) Dumort.)   *Red Sand-spurrey*
Native. Dry open places on acid, sandy or gravelly soils, heaths, roadsides, etc. Local.

W. Grainfoot Clough 1888.

E. Beighton 4483, Top Lane nr. Holloway 3257, Gun Hill 3044, Breadsall 3742.

S. Hulland Gravel Pits 2645, gravel paths at Sudbury Hall 1532, Kirk Langley Hall 2839, Markeaton 3237, Dale Abbey 4338, sand quarry, Dale 4238.

Linton: Baslow 27, Troway 37, quarry by Matlock Church 35, Codnor Castle 45, Willington 22, Melbourne 32, Weston Cliff 42, Drakelow 21.

HERNIARIA L.

H. hirsuta L.   *Hairy Rupture-wort*
Introduced. There is an old record (1789) for this very rare casual, which has not been seen in the county since then.

Linton: Duffield 34.

SCLERANTHUS L.

S. annuus L.   *Knawel*
Native. Dry places and cultivated and waste ground on sandy or gravelly soil; avoiding calcareous soil. Local.

E. Barlborough Park 4778.

S. Hulland 2546, Ednaston 2442, Mugginton 2843, Locko 4138, Overseal 2815.

Linton: Ashford 16, Dovedale 15, Breadsall Moor 34, Ticknall 32, Calke Abbey 32, Repton 32.

PORTULACACEAE

MONTIA L.

M. fontana L.   *Blinks*

Native. Streamsides, springs, flushes, moist pastures and occasionally in gardens and arable fields; chiefly on non-calcareous substrata. Frequent in suitable habitats and reaching altitudes of 1,500 ft. or more on the gritstone moorlands.

A very variable species which has been divided into four more or less distinct subspecies. The two recorded for the county are:

Ssp. **amporitana** Sennen (ssp. *intermedia* (Beeby) Walters) with the ripe seed somewhat shining and having 3–4 rows of long acute tubercles on the keel, none elsewhere.

Ssp. **variabilis** Walters, also with the ripe seed somewhat shining but showing a variable development of broad low tubercles on the keel, none elsewhere.

More information is needed about the distribution of these two subspecies.

W.   Alport Dale 1390, Hayfield 08, Kinder Scout 0888, Derwent 1889, Hood Brook, Hathersage 2382, Foxhouse 2680, Cavendish Golf Course, Buxton 0473, Watford Wood 0474.

E.   Holmsfield 3277, Pratthall 3472, Unthank 3976, Donkey Racecourse, Chesterfield 3572, Stubbin Ponds 3667, Tansley Moor 3161, Holymoorside 3469, Alderwasley 3153.

S.   Sturston 1946, Lodge Farm, Ashbourne 1745, Mugginton 2843, Hulland Moss 2546, Bradley Brook 2245, nr. Dale 4338.

Linton: Froggatt 27, Umberley Brook 27, Sutton Scarsdale Ponds 46, Pinxton 45, Hilton Common 23, Repton 32, Cauldwell 21.

M. perfoliata (Willd.) Howell   (*Claytonia perfoliata* Donn ex Willd.)
Introduced. Cultivated, disturbed and waste ground, especially on sandy soil. Local.

No record before 1930.

W.   Darley 2961.

E.   Stone Edge Golf Course 3367, Tansley 3459, Babington Hospital, Belper 3446.

S.   Ashbourne 1746, Cubley 1638, Hoon Ridge, Hilton 2331.

Linton: Not recorded.

M. sibirica (L.) Howell (*Claytonia sibirica* L.)

Introduced. Damp woods and copses and shaded streamsides. Locally frequent and probably increasing.

Said to have been found first in the Edensor woods by Sir Joseph Paxton or one of his acquaintances. A specimen was sent to Baxter who published the record in his 'Genera of British Plants' (1837).

The colour of the petals varies from white to purple.

W.   Watford Bridge 0086, Hathersage 2580, Corbar Wood 0574, Stand Wood 2670, Tissington Hall 1752, Lindup Wood 2567, Calton Hill 2468, Beeley 2667.

E.   Beauchief 3482, Horsleygate Hall 3177, Eckington Woods 47, Quarndon 3341, West Hallam churchyard 4341.

S.   Edlaston Rectory 1842, Bradley 2246, Yeldersley 2144, Alder Moor, Sudbury 1835, Overseal 2915.

Linton: Hayfield 09, Mickleover 33.

AMARANTHACEAE

AMARANTHUS L.

A. retroflexus L.

Introduced. A rare casual of arable land and waste places.

No recent records.

Linton: 'Casual at New Brampton, near Ashgate Plantation', 37.

CHENOPODIACEAE

CHENOPODIUM L.

C. bonus-henricus L.   *Good King Henry, All-good, Wild Spinach*

Probably introduced. Roadsides, farm-yards, waste places; usually near buildings. Frequent.

W.   Recorded for all squares except 09, 19, 28.

E.   Beauchief Abbey 3381, Whittington 3974, Eckington 47, nr. Scarcliffe Park Wood 5270, Holymoor 3468, Pleasley Vale 5265, Duffield 3542, Stanley 4141.

S.   Bradley Wood 1946, Brailsford 2541, Etwall 2631, Chaddesden 3736, Church Gresley 2719.

Linton: New Mills 09, Killamarsh 48, Norbury Quarry 14, Burnaston 23, Sawley 43, Chellaston 32, Ticknall 32.

C. polyspermum L.   *All-seed*

Native. Waste places, roadsides, field borders. Rare, and chiefly in the south and east of the county.

W.   Ladmanlow refuse tip 0471.

E.   Nr. Whitwell 57, Little Eaton 3640.

S.   Derby 3535, Littleover 3233, Markeaton 3238.

Linton: Heanor 44, Borrowash 43, Calke 32.

C. vulvaria L.   *Stinking Orache, Stinking Goosefoot*
Possibly native before 1930. Now only casual in waste places. Rare.
S.   Derby canal bank 3636.
Linton: Dethick 35.

C. album L.   *Fat Hen, White Goosefoot*
Native. Arable land, waysides and waste places, especially on richly manured soil. Common.
Recorded for all squares except:
W. 09, 19.
Linton: 'General throughout the county'.

C. suecicum J. Murr ('*C. album* var. *viride* (L.)')
Introduced. Waste ground and rubbish-tips. Very rare.
No recent records.
Linton: Beighton 48.

C. murale L.   *Nettle-leaved Goosefoot*
Native. Waste places, chiefly on light soils. Very rare.
S.   Sturston tip 1946. 'Once, about 1935'.
Linton: Not recorded.

C. urbicum L.   *Upright Goosefoot*
Possibly native before 1930. Now casual and rare.
S.   Trent 4831, Coton in the Elms 2514.
Linton: Not recorded.

C. rubrum L.   *Red Goosefoot*
Native. Farmyards, manure-heaps, waste ground. Rare, and chiefly in the south and east of the county.
W. Beeley 2268.
E.   Pebley Pond 4878, Scarcliffe Park Wood 5270, Pleasley Vale 5265.
S.   Ashbourne 1746, Baltimore Bridge 3630, Trent 4931, Cauldwell 2517.
Linton: Longford 23, Church Broughton 23, Sandiacre 43, Swarkestone 32, Repton 32.

C. glaucum L.   *Glaucous Goosefoot*
Doubtfully native. Rich waste ground. Very rare.
S.   Hulland 2545.
'Two plants on a manure heap in a field in 1949. In 1950 there was a large patch; none seen in 1953'.
Linton: Not recorded.

ATRIPLEX L.

A. patula L.   *Spear Orache, Common Orache*
Native. Roadsides, arable fields, waste places. Common.

W. Moorfield nr. Glossop 0492, Watford Bridge 0086, Bamford 2083,
High Rake 1677, Eyam 2276, Lathkill Dale 1966, Youlgreave 2165, Thorpe
1550, Kniveton 2050.

E. Recorded for all squares except 45.

S. Recorded for all squares.

Linton: Alport Dale 19, Mellor 98, Edale 18.

A. hastata L.   *Hastate Orache, Fat Hen*
Native. Roadsides, arable fields, waste places. Local, and chiefly in the south
and east of the county.

W. Hartington 1260, recorded for 26, Bradbourne 2052.

E. Whittington 3974, Pebley Pond 4878, Pleasley Vale 5165, Bramley Vale
4665, Crich 3454, Pinxton 4454, Little Eaton 3641, nr. Morley 4040.

S. Ashbourne 1746, Alkmonton 1838, Derby 3537, nr Scropton 1829,
Swarkestone 3628, Cauldwell 2517.

Linton: Burbage 07, Norton 38, Beighton 48, Long Eaton, Sawley, Great
Wilne and Trent Lock 43, Thrumpton 53, Burton and Drakelow 22.

TILIACEAE

TILIA L.

T. platyphyllos Scop.   *Large-leaved Lime*
Probably native. In old woods on limestone screes and cliffs, often with
T. cordata and intermediates; but undoubtedly planted in some localities.
Occasional.

W. Tissington 1751, Matlock Bath 2958, Via Gellia 2757, Wirksworth 2854.

E. Markland Grips 5074, Scarcliffe Park Wood 5270, Pleasley Vale 5265,
Whatstandwell 3354, Duffield 3344.

S. Copley nr. Ashbourne 1947, Lodge Farm, towards Tinkers' Inn 1844,
Marston Montgomery 1238, Kirk Langley 2939.

Linton: Mickleover 33, Dale 43, Burton 22, Calke 32.

T. cordata Miller   *Small-leaved Lime*
Native. In old woods on calcareous soil and on cliffs and boulder-screes both
of the Magnesian and Carboniferous Limestones; less frequently on base-rich
but non-calcareous soils, as on the Magnesian Limestone plateau, and
occasionally in cloughs of the gritstone or shale-grits; on limestone often with
T. platyphyllos and intermediates; probably planted in some localities.
Occasional, but locally frequent on the Carboniferous Limestone.

Seedlings are very rarely seen in Derbyshire.

W. Shatton Village 1982, Taddington Dale 1770, Coombs Dale 2274, Tissington 1751, Bonsall Moor 2559.

E. Nr. Overton 3472, Pleasley Vale 5265.

S. Copley, Ashbourne 1947, Yeldersley 2144, Boylestone Churchyard 1835, Meynell Langley 2938.

Linton: King Sterndale 07, Fritchley 35, Mickleover and Mackworth 33, Calke 32.

T. × **vulgaris** Hayne (*T. platyphyllos* × *T. cordata*)   *Common Lime*
Probably native in a few localities, but much planted elsewhere. In some of the old woods on limestone cliffs and screes, where there is good reason to suppose limes to be native, a range of intermediates between *T. cordata* and *T. platyphyllos*, the parents of the Common Lime, can be found with one or both parents. These include what appear to be first-crosses and also back-crosses with both parents. Occasional.

W. Ashopton 2086, recorded for 07, 16, Tissington 1751, Wirksworth 2854.

E. Povey 3881, Ault Hucknall 4665, Pleasley Vale 5261, Morley 3940, Stanley 4140, West Hallam 4341.

S. Ashbourne Green 1947, Bradley 2245, Doveridge 1133, Longford 2138, nr. Derby 3437, Chaddesden 3837, recorded for 43, Swarkestone 3728, Walton 2318.

Linton: Chelmorton 17/16, Bretby, Burton and Brizlincote 22.

MALVACEAE

MALVA L.

M. moschata L.   *Musk Mallow*
Native. Grassy pastures, hedgebanks, waysides, etc., on all but the most acid soils; also on sunny limestone outcrops. Local and not uncommon.

W. Hathersage 2380, Monsal Dale, Taddington Dale, Monk's Dale in 17, Coombs Dale 2274, Chatsworth 2670, Hollinsclough 0867, Lathkill Dale 1865, Robin Hood's Stride 2262, Calton Lees 2568, Dovedale 1453.

E. Povey Farm 3880, Ford Valley 4180, Moorhall 3175, Linacre 3573, Fallgate 3562, The Butts, Ashover 3463.

S. Recorded for all squares except 23.

Linton: Renishaw 47, Sutton Scarsdale 46, Cromford Canal 35, Morley 34.

M. sylvestris L.   *Common Mallow*
Native. Roadsides, waste places; often in rank vegetation at the foot of limestone cliffs in the dales. Frequent.

W. Hathersage 2381, Monsal Dale 1771, Ballcross 2470, Chatsworth 2670, Lathkill Dale 1966, 2165, Thorpe Cloud 1550, Cromford 2957.

E.   Eckington 4480, Dronfield 3678, Brimmington 4073, Scarcliffe Park Wood 5170, Milltown 3561, Pleasley Vale 5265, nr. Langley Mill 4546.

S.   Recorded for all squares except 14.

Linton: Bradwell 18.

M. neglecta Wallr. ('*M. rotundifolia* L.')   *Dwarf Mallow*

Native. Roadsides, waste places. Frequent, but chiefly in the south of the county.

W.   New Bridge, Calver 2475, entrance to Via Gellia 2957.

E.   Nr. Barlborough 4777, Creswell Crags 5374, Scarcliffe Park Wood 5270, Glapwell 4866, Upper Langwith 5169, nr. Whatstandwell 3354, West Hallam 4341.

S.   Wyaston Grove 1842, Mugginton 2943, Sutton-on-the-Hill 2333, Littleover 3334, Dale Abbey 4338, Spread Eagle 2929, Repton 3026, Overseal 2914.

Linton: Miller's Dale 17, Matlock Bath 25, Beighton 48, Markland Grips 57, New Brampton 37, Clay Cross Station 36.

M. parviflora L.

Introduced. A rare casual, not seen in recent years. No recent records.

Linton: Pinxton 45.

LINACEAE

LINUM L.

L. bienne Miller (*L. angustifolium* Hudson)   *Pale Flax*

Introduced. A rare casual.

S.   Nr. Sudbury 1632, Hilton 2430, Etwall 2532.

Linton: Renishaw 47, between Willington and Etwall 23.

L. usitatissimum L.   *Common Flax*

Introduced. Cultivated and waste ground. An occasional escape.

E.   Cutthorpe 3573, Pebley Pond 4878.

S.   Mappleton, casual 1647, Derby 3537, Dale Abbey 4338, Swarkestone Bridge 3628.

Linton: Wormhill and Great Rocks Dale 17, between Baslow and Hassop 27, Rowsley and Haddon 26, Matlock 25, Beauchief and Millhouses 38, Spital 37, Pinxton 45.

L. perenne L.   *Perennial Flax*

Almost certainly introduced. The reference in the Buxton Guide (1854) was probably to a casual occurrence.

No recent records.

Linton: 'Casual? Buxton Guide 1854'.

L. catharticum L.   *Purging Flax*

Native. Pastures, heaths, moors and rock-ledges, to 1,500 ft. Common, especially in the limestone areas.

Recorded for all squares except:

> **W.** 09, 19.
> **E.** 46, 44, 48.
> **S.** 22.

Linton: West Hallam and Loscoe 44, nr. Burton and Bretby 22.

GERANIACEAE

GERANIUM L.

G. pratense L.   *Meadow Cranesbill*

Native. Moist meadows and roadsides on fertile soils, especially over Carboniferous Limestone and along the sides of roads metalled with limestone; often on stream-banks and near gateways. Local, but often very abundant and then a beautiful sight in full flower.

Recorded for all squares except:

> **W.** 09, 19.
> **E.** 38, 57.
> **S.** 22.

Linton: Drakelow 21.

G. sylvaticum L.   *Wood Cranesbill*

Probably introduced. 'Woods in mountain districts. Very rare'.

S. Breadsall (probably an escape) 33.

Linton: Matlock 25, Ashwood Dale 07, Chatsworth 27.

G. endressii Gay

Introduced. A garden escape. Very rare.

W. Chinley (probably garden escape) 0482.

Linton: Not recorded.

G. nodosum L.

Introduced. An escape from gardens which may occasionally become more or less naturalized. Rare.

S. Darley Abbey Park 3538.

Linton: Not recorded.

G. phaeum L.   *Dusky Cranesbill*

Introduced. A garden escape, occasionally naturalized in hedgebanks and bushy places.

W.   Stubbins Lane, Chinley 0382, recorded for 07, Calver 2373.

E.   Wingerworth 3867, Duffield 3344, Morley 3941.

S.   Callow, Ashbourne 1747, Ashley, Ashbourne 1746.

Linton: Matlock Bath 25, Baslow 27, Calke 32.

G. sanguineum L. *Bloody Cranesbill*

Native. Cliff-ledges and stabilized scree-slopes of the Carboniferous Lime-stone, in the open or in light shade; often with *Melica nutans, Rubus saxatilis* and *Convallaria majalis* on the site of ancient woodland. Local.

W.   Nr. Chinley 0482, Ashwood Dale 0772, Deep Dale, Buxton 0971, Great Rocks Dale 1073, Chee Dale 1273, Churn Hole 1072, Miller's Dale 1373, Monk's Dale 1374, Lathkill Dale 1865, Haddon Hall 2366, Beresford Dale 1258.

E.   Hallam (an escape) 4346.

Linton: Hartington 16, Matlock 25.

G. pyrenaicum Burm. fil. *Mountain Cranesbill*

Doubtfully native. Hedgebanks, waysides, waste places. Local and uncommon.

W.   Cowdale 0872, Ashford 1970, nr. Hassop 2272, nr. Baslow 2572, Parwich Hall 1854.

E.   Pleasley 5264.

S.   Ashbourne Hall 1846, Snelston 1442, recorded for 43.

Linton: Litton and Monsal Dales 17, Osmaston by Ashbourne 24, Markeaton Road 33, Stapenhill 22.

G. columbinum L. *Long-stalked Cranesbill*

Native. In shallow soil on south-facing rocky limestone slopes and in old limestone quarries. Local, and restricted to the Carboniferous Limestone.

W.   Deep Dale 0971, Blackwell Mill 1272, Monk's Dale 1374, Tideswell Dale 1574, 1375, Taddington Dale 1770, Coombs Dale 2374, Wolfscote Dale 1455, Dovedale 1451, Via Gellia 2457.

Linton: Ashover 36, Matlock Bath 25.

G. dissectum L. *Cut-leaved Cranesbill*

Native. Cultivated and waste ground, waysides, rocky limestone slopes. Common.

W.   Hathersage 2380, Ladmanlow 0471, Deep Dale 0970, Cowdale 0872, Chee Dale 1272, Miller's Dale 1373, 1573, Taddington Dale 1670, Coombs Dale 2274, Alport 2164, Beresford Dale 1259, Alsop-en-le-Dale 1655, Bonsall Moor 2559.

E.   Norton 3682, Ford 4081, Cutthorpe 3372, Barlow Grange 3173, recorded for 47, Markland Grips 5075, Creswell 5374, Ashover, the Butts 3463, Astwith 4464, Pleasley Vale 5265, Ilkeston 4642.

S. Atlow 2348, Scropton 1930, Barton Blount 2134, Mickleover 3134, Trent Lock 4831, Thrumpton Ferry 5031, Egginton 2729, Willington 3028, Walton 2318, Overseal 2712, Measham 31.

Linton: Chapel-en-le-Frith 08, Shirland 35, Pinxton 45.

G. rotundifolium L. *Round-leaved Cranesbill*

Probably introduced. A casual of hedgebanks and cultivated land. Very rare.

No recent records.

Linton: Ashwood Dale 07, Ashbourne 14, Derby 33, Repton 32.

G. molle L. *Dove's-foot Cranesbill*

Native. In short open turf of dry grassland and waysides, in shallow soil on south-facing limestone outcrops, in cultivated land and in waste places. Common.

Recorded for all squares except:
W. 09, 19, 28.

Linton: 'Frequent'.

G. pusillum L. *Small-flowered Cranesbill*

Native. Cultivated and waste ground, field borders and open turf in dry grassland; especially in loamy soil over limestone. Rather local.

W. Cave Dale 1483, Monsal Dale 1772, Coombs Dale, Calver 2274, Stoney Middleton 2275, Dovedale 1453, Hall Dale 1353.

E. Markland Grips 5074, High Peak Railway, nr. Black Rocks 3056, Cocky Dumbles, Nutbrook 4442.

S. Snelston Limepits 1541, Canal, Trent Lock 4831, Thrumpton 5031.

Linton: Norton Lees 38, Ashwood Dale 07, Creswell 57, Yeldersley 24, Egginton 22, Findern 33, Repton 32, Milton 32, Calke 32. Rosliston 21.

G. lucidum L. *Shining Cranesbill*

Native. Rocks, walls and banks, especially on limestone; abundant on shaded limestone scree. Locally abundant.

The white-flowered form is occasionally seen.

W. Recorded for all squares except 09, 19.

E. Fallgate 3562, Cromford 3057, Windley 3045.

S. Cubley Common 1639, recorded for 43.

Linton: Norton 38, Pleasley Park 56, Anchor Church 32, Repton 32.

G. robertianum L. *Herb Robert*

Native. On moist rich soils in woods, hedgebanks, etc., over a wide range of substrata and often abundant on boulder scree and rocks in limestone woods; a pioneer colonist of limestone scree, shaded or unshaded. Common.

The Derbyshire form is the common inland ssp. *robertianum.*

Recorded for all squares.

Linton: 'Common'.

ERODIUM L'Hér.

E. cicutarium (L.) L'Hér.   *Common Storksbill*
Native. Arable fields, waysides and waste places on sand and on loamy
downwash over limestone, often with *Aira praecox* and *A. caryophyllea*.
Rather rare.
The Derbyshire form is ssp. *cicutarium*.
W.   Middleton Dale 2275, Dovedale 1451, Lin Dale 1551, Via Gellia 2757.
E.   Nr. Baslow 3072, Creswell Crags 5374, Stanley 4140, West Hallam 4340.
S.   Sudbury 13, Old Sawley 4731, Ticknall 3523, Weston-on-Trent 4027.
Linton: Charlesworth 98, Ashwood Dale 07, Baslow 27, Renishaw 47, Bread-
sall Moor 34, Shirley 24, Stanton-by-Bridge 32, Repton Rocks 32.

OXALIDACEAE

OXALIS L.

O. acetosella L.   *Wood-sorrel*
Native. Moist, shady places on soils covering a wide range in acidity but
infrequent on strongly podsolized soils. Common and widespread.
Recorded for all squares except:
>   **W.** 09.
>   **E.** 45.
>   **S.** 22.

Linton: 'Generally distributed'.

O. europaea Jord. ('*O. stricta* L.')   *Upright Yellow Sorrel*
Introduced. Waste places and gardens. A rare casual, not seen in recent years.
No recent records.
Linton: Cultivated ground, Etwall 22.

BALSAMINACEAE

IMPATIENS L.

I. noli-tangere L.   *Touch-me-not*
Introduced. A rare casual.
W.   Roadside, Matlock Bath 2958.
Linton: Osmaston Park 14 (once only).

I. capensis Meerb.   *Orange Balsam*
Introduced. Naturalized on the banks of rivers and canals. Local and in-
frequent but probably spreading.
S.   Canal bank nr. Clay Mills 2526, Findern, canal bank 3129, Canal bank,
Swarkestone 3629, Canal east of Shardlow 4530.
Linton: Not recorded.

I. parviflora DC.   *Small Balsam*
Introduced. Woods and shady waste places. Local but increasing.
W. Hassop 2272, Matlock 2960, Via Gellia 2656, 2857, Cromford 2957, Matlock 2959.
E. Barlow 3475, Shining Cliff Woods 3354, Ambergate 3352, 3551, Belper 3447, Duffield 3543, Stanley 4041.
S. Etwall 2632, Derby 3536, Darley Abbey 3538, Wilne Toll Bridge 4431.

I. glandulifera Royle   *Policeman's Helmet*
Introduced. Well established on stream-banks and moist or shaded roadsides and in waste places. Locally abundant.
First recorded about 1930.
W. Tideswell Dale 1573, Cressbrook Dale 1772, Bubnell 2471, Darley Dale 2762, Matlock 2959.
E. Ecclesall Woods 3282, Beighton 4483, Unstone 3777, Renishaw Park 4478, Cromford 3057, Hazelwood 3344.
S. Spinnyford Brook 2441/3, Darley Abbey 3538, Swarkestone 3728, Weston-on-Trent 4127, River Trent, Walton 2218.
Linton: Not recorded.

ACERACEAE

ACER L.

A. pseudoplatanus L.   *Sycamore*
Introduced. Well established in woods, hedges and plantations on all but very poor soils, but thriving best on deep, moist, well-drained rich soils and therefore most characteristic of the lower parts of valley-sides; very tolerant of wind and often planted for shelter in exposed places; reaching c. 1,400 ft. Common and increasing.
Recorded for all squares.
Linton: 'Frequent, especially abundant in the north'.

A. platanoides L.   *Norway Maple*
Introduced. Planted as an ornamental tree and occasionally establishing itself from seed. Rare.
E. Whirlow 3182, nr. Creswell Crags 5274.
S. Shirley Park 2143.
Linton: Not recorded.

A. campestre L.   *Field Maple*
Native. A characteristic component of old woods on calcareous or other nutrient-rich soils but infrequent in recent secondary ashwoods on limestone; abundant as a hedgerow plant on the Coal Measures. Locally frequent.
Recorded for all squares except:
                    W. 09, 19, 08.
Linton: Chapel 08.

HIPPOCASTANACEAE
AESCULUS L.

A. hippocastanum L.   *Horse Chestnut*

Introduced. Generally distributed as a planted tree and occasionally reproducing from seed.

Recorded for all squares except:

> **W.** 09, 19.
> **E.**   57, 34, 56, 45.

Linton: Not recorded.

AQUIFOLIACEAE
ILEX L.

I. aquifolium L.   *Holly*

Native. In woods over a wide range of soil types including the most acid, in limestone scree-woods and on limestone and gritstone cliffs; often surviving from old woodland; ascending to 1,400 ft. or higher. Common.

Recorded for all squares except 09.

CELASTRACEAE
EUONYMUS L.

E. europaeus L.   *Spindle-tree*

Native. In woods, hedges and scrub, chiefly on limestone. Local.

W. Monk's Dale 1275, 1374/5, Demons Dale 1670, Lathkill Dale 1865, Dovedale 1451, Via Gellia 2556.

E. Pebley Pond 4880, Scarcliffe Park Wood 5270, Whitwell Wood 5279, Pleasley Vale 5265.

S. Nr. Kirk Langley 3037.

Linton: Castleton 18, Great Shacklow Wood, Ashford 26, Duckmanton 46, Wood nr. Lea 35, Clifton 14, Shirley 24, Winshill 22, Repton Rocks 32, Drakelow 21.

RHAMNACEAE
RHAMNUS L.

R. catharticus L.   *Buckthorn*

Native. In older woods, hedges and scrub on limestone, but infrequent in recent secondary ashwoods. Rather local.

W. Cunning Dale 0872, rocks above railway, Ashwood Dale 0772, Chee Tor 1273, also Monsal Dale, Cressbrook Dale, Peter Dale and Priestcliffe Lees in 17, Coombs Dale 2274, Lathkill Dale 1966, Conksbury Bridge 2165, Dovedale 1451, Via Gellia 2656, also Griffe Grange Valley and Brassington in 25.

E. Pebley Pond 4880, nr. Beighton 4883, nr. Spinkhill 4578, Scarcliffe Park Wood 5170, also Whitwell Wood and Markland Grips in 57, Pleasley Vale 5265, between Stanley and Morley 4040.

S. Nr. Kniveton 1949, Bradley 2146, Alkmonton 1937, nr. Littleover 3334.

Linton: Hartle Dale and Dirtlow Rake 18, Barton Fields 23, between Stanton and Dale Abbey 43, Drakelow and Walton 21.

FRANGULA Miller

F. alnus Miller (*Rhamnus frangula* L.) *Alder Buckthorn*
Native. In woods, hedges and scrub, usually on damp soils which are some-what acid and peaty. Local:

W. Lathkill Dale 16.

E. Ecclesall Wood 3282, Scarcliffe Park Wood 5170, Markland Grips 5074, Whitwell Wood 57, Wingerworth Great Pond 3667, Pleasley Vale 5265.

S. Recorded for 43, Repton Rocks 3221.

Linton: Froggatt 27, Mossborough Moss 48, nr. Renishaw 47, Glapwell 46, Horsley Carr 34, Breadsall Moor 34, Drakelow 21.

LEGUMINOSAE
LUPINUS L.

L. nootkatensis Sims *Lupin*
Introduced. A rare casual.

S. Mansfield Road, Breadsall 3739.

Linton: Not recorded.

GENISTA L.

G. anglica L. *Needle Furze, Petty Whin*
Native. Heaths and unburnt moorland on the less poor types of moorland soil: Rare.

W. Old railway track, nr. Yorkshire Bridge 1886, between Grindleford and Froggatt 2471/6, between Dale Head Farm and Brand Side 0469.

E. Tansley 3259, Dethick Common 3359.

Linton: Nr. Wythen Lache 07, Fernilee-Coombs 07, Ladybower 18, Baslow 27, Beeley Moor 27, Cromford Moor 25, Pinxton 45, Bradley Brook 24, Breadsall Moor 34, Egginton 22, Willington 32.

G. tinctoria L. *Dyer's Greenweed*
Native. Rough pastures, hedgebanks and field borders, especially on basalt, shale or gritstone, but also on non-calcareous plateau-loam over the Carboni-ferous Limestone. Locally frequent.

W. Elton 2161, Fenny Bentley 1750, Froggatt 2457, Hognaston 2350, Grange Mill 2557.

E. Troway 3880, Ford 4081, Linacre 3372, Eckington 4479, Ashover 3462, Whatstandwell 3254, Windley 3045, Denby Common 4046.

S.   Cubley Common 1640, Bradley Brook 2244, Radbourne 2836, Dale Abbey 4338.

Linton: Bradwell 18, Hathersage 28, Marston Montgomery 13.

ULEX L.

U. europaeus L.   *Gorse, Furze*

Native. Rough pastures, heaths and waysides especially on the lighter non-calcareous soils and avoiding shallow limestone or very poor and acid soils; in the limestone dales restricted to non-calcareous plateau-loam. Common and widespread.

Recorded for all squares except:

E. 46.

Linton: 'Frequent and general'.

U. gallii Planchon   *Western Dwarf Gorse*

Native. Heaths, moorlands and waysides on non-calcareous soils, especially after disturbance; in the limestone dales only on non-calcareous plateau-loam. Rather local but very conspicuous when in flower.

Derbyshire is close to the eastern limit of this species.

W.   Moorfield 0492, recorded for 08, Yorkshire Bridge 1985, Hathersage 2283, Corbar Hill 0574, Burbage Edge 0373, Froggatt 2476, Brand Side 0468, Beeley Moor 2967, Parwich Moor 1757, between Matlock and Bonsall 2858, Longcliffe 2356, Grange Mill 2457.

E.   Beauchief 3381, Holmesfield 3277, Pebley Pond 4978, Tansley Moor 3261, Loads, Chesterfield 3169, Bilberry Knoll, Cromford 3057, Breadsall Moor 3742.

S.   Darley Moor 1641, Spinnyford Brook 2445.

Linton: Alport Dale 19, Cresswell 57, Boylestone 13, Spread Eagle 22, Calke 32, Gresley 21.

CYTISUS L.

C. scoparius (L.) Link   *Broom*

Native. Rough pastures, field borders, hedgebanks, banks by roads and railways, etc., on non-calcareous soils. Locally frequent. 'Wholly absent from the limestone' (Purchas).

W.   Nr. Glossop 09, Alport Castles 1490, Chapel 0580, Yorkshire Bridge 1985, Bamford 2083, Buxton 0573, Tideswell 17, Calver New Bridge 2475.

E.   Twentywell 3280, Troway 3880, Killamarsh 4680, Brierley Wood 3775, Railway Bank, Creswell 5274, Ashover 3463, Whatstandwell 3254, Stanley 4141.

S.   Recorded for all squares.

Linton: Birchover 26, nr. Renishaw 47, Pinxton 45.

ONONIS L.

O. repens L.   *Restharrow*

Native. Dry pastures, roadsides and disturbed ground on sandy and limestone soils. Local.

W. Hathersage 2283, Ashwood Dale 0772, Miller's Dale 1473, 1574, Lathkill Dale 1966, Bradford Dale 2264, railway nr. Alsop 1653, Kniveton Quarry 2150, Brassington Rocks 2154.

E. Povey Farm, Troway 3880, Beighton 4482, Barlow Brook 3475, Whitwell 5278/9, Walton 3568, Stony Houghton 4966, Pleasley Vale 5265.

S. Scropton 1930, nr. Dale Abbey 4239, Ticknall 3523, Bretby Castle Field 2923.

Linton: Baslow 27, Renishaw 47, Markland Grips 57, Pleasley 56, Clifton 14, Osmaston 24, Shirley 24, Repton 32.

O. spinosa L.   *Restharrow*

Native. Pastures and roadsides, especially on heavy soils. Local.

W. Monk's Dale 1373/4, Taddington 1571, above Grange Mill 2457.

E. Chaddesden 4041.

S. Mackworth 3137, roadside, Dale 4338, nr. Trent Station 4931, Thrumpton Ferry 5031, Swarkestone 3828, Overseal 2916.

Linton: Wormhill 17, Lathkill Dale 16, Pleasley 56, Ockbrook 43, Ticknall 32, Willington 32.

O. alopecuroides L.

Introduced. A casual of waste and cultivated ground, perhaps from bird-seed. Rare.

E. Crosshills nr. Bolsover, on roadside tip 5169.

Linton: Not recorded.

MEDICAGO L.

M. falcata L.   *Sickle Medick*

Introduced. A very rare casual.

S. Darley Abbey 3538.

Linton: Not recorded.

M. sativa L.   *Lucerne, Alfalfa*

Introduced. In waste places and on roadsides as an escape from cultivation. Occasional.

W. Brough 1882, Taddington by-pass 1471, Calver Sough 2374, Alport 2065.

E. Old Brampton 3372, Barlow 3475, Creswell 5274, Pleasley Vale 5165, nr. Ambergate 3550, Ilkeston 4642, Stanley Pit 4240.

S. Ashbourne 1846, Osmaston 1943, Bradley 2246, Chaddesden 3836, Railway banks, Darley 3635, Netherseal 2718, Swadlincote 2919.

Linton: Renishaw 47, Railway Works, Derby 33.

M. lupulina L.   *Black Medick*
Native. Pastures, reseeded meadows, roadsides and crevices of limestone outcrops; very tolerant of drought. Common and widespread.
Recorded for all squares except:

W. 09, 19, 08.

Linton: 'Generally distributed and plentiful everywhere'.

M. polymorpha L. (incl. *M. denticulata* Willd.)   *Hairy Medick*
Introduced. A rare casual.

S. Sturston 1946, Snelston 1543, Canal, Nottingham Rd. Station, Derby 3536, Thrumpton 5131.

Linton: Mickleover 33.

M. arabica (L.) Hudson   *Spotted Medick*
Probably introduced. A rare casual of cultivated ground.

E. Dore 3081.

S. Stores Road, Derby 3537.

Linton: The Hague, Renishaw 47, Shirley 24, Hollington 24.

MELILOTUS Miller

M. altissima Thuill. ('*M. officinalis* Lam.')   *Tall Melilot*
Probably introduced. Waysides, grassy fields, railway banks, waste places. Occasional.

W. Great Rocks Dale 1073, Tideswell Dale 1573, roadside nr. Owler Bar 2977, Hindlow 0867.

E. Abbeydale Hall 3180, nr. Norwood 4682, Bramley Vale 4566, nr. Ilkeston 4642.

S. Recorded for all squares except 23.

Linton: Renishaw 47, Codnor Park 45.

M. officinalis (L.) Desr. (*M. arvensis* Wallr.)   *Common Melilot*
Introduced. Cultivated and waste ground. Rare.

W. Between Edale and Hope, casual on roadside 1285, 1783.

S. Recorded for 43.

Linton: Shirley Common and Ednaston 24, railway, Derby-Borrowash 33/43, Repton 32.

M. alba Desr.   *White Melilot*
Introduced. A casual of cultivated and waste ground. Rare.
E.   Abbey Lane 3382, Brockwell 3772, Birdholme 3668, Rowthorn 4764,
Little Eaton 3640, Stanley 4142.
S.   Mapleton 1648, Ashbourne 1745, between Bradley and Brailsford 2245,
Shirley 2041, Stores Road, Derby 3636, nr. Trent Lock 4932, Swarkestone
3628, Netherseal Colliery 2715.
Linton: Spital, Chesterfield 37.

M. indica (L.) All.   *Small-flowered Melilot*
Introduced. A rare casual of waste places.
E.   Birdholme 3868, Rowthorn 4764.
S.   Clifton goods yard 1644, Egginton railway bank 2629, Burton 2523.
Linton: Mickleover 33.

TRIGONELLA L.

T. foenum-graecum L.   *Fenugreek*
Introduced. A rare casual.
S.   Waste ground between Derby and Little Eaton 3639 (last seen 1939).
Linton: Not recorded.

TRIFOLIUM L.

T. ornithopodioides L. (*Trigonella purpurascens* Lam.)   *Birdsfoot Fenugreek*
Probably introduced. A rare casual, not seen in recent years.
No recent records.
Linton: Breadsall 33.

T. micranthum Viv. ('*T. filiforme* L.')   *Slender Trefoil*
Native. In short open turf of pastures, commons, waysides and lawns,
especially on sandy or gravelly soils. Occasional.
W.   Recorded for 07, Chee Tor 1273, nr. Blackwell Mill 1373, nr. Dovedale
1651.
S.   Sandybrook Hall nr. Ashbourne 1748, Offcote Hurst, in the lawn 1948,
The Laurels, Ashbourne, 1746, nr. Sawley Church 4731, Thrumpton 5031.
Linton: Middleton Dale 27, Sapperton and Boylestone 13, Hilton 23, Burton
22, Calke and Milton 32.

T. dubium Sibth.   *Lesser Yellow Trefoil*
Native. In short turf of pastures, commons, waysides and lawns. Common
and widespread.
Recorded for all squares except:
                              W. 09, 19.
Linton: 'Common. Generally distributed'.

T. campestre Schreber ('*T. procumbens* L.')   *Hop Trefoil*
Native. Pastures, waysides and arable land. Frequent.

W. Recorded for 07, Peter Dale 1275, Middleton Dale 2175, Lathkill Dale 1966, Dovedale 1451, Wolfscote Dale 1455.

E. Eckington 4480, recorded for 47, Creswell 5374, Markland Grips 5074, Wingerworth 3667, Pleasley Vale 5261, Upper Langwith 5169, recorded for 35.

S. Snelston, on the limestone outlier 1541, Alfreton Road, Derby 3537, Thrumpton 5031, recorded for 32, Swadlincote 2919.

Linton: Between Whaley Bridge and Chapel 08, Peveril Castle and Edale 18, Tibshelf and Pilsley 46, Codnor and Loscoe 44, Turnditch, Shirley and nr. Osmaston 24, Marston Montgomery and Doveridge 13, Dale and Stanton 43, Burton 22.

T. aureum Pollich ('*T. agrarium* L.')   *Hop Trefoil*
Introduced. A rare casual of cultivated fields.

S. One plant, roadside, Ednaston 2342.

Linton: Bradley Park Farm 24.

T. hybridum L.   *Alsike Clover*
Introduced. An escape from cultivation and frequently establishing itself on roadsides, field borders and waste ground. Frequent.

W. Recorded for 08, Burbage 0473, Chelmorton 1170, recorded for 26, Alsop 1554, Kniveton 2150, Wirksworth 2854.

E. Whirlow 3182, Cutthorpe 3473, Markland Grips 5074, Glapwell 4866, Astwith 4464, Cromford 3057, Nutbrook, Ilkeston 4541.

S. Recorded for all squares.

Linton: Edale 18, Bamford and Hathersage 28, Baslow 27, Ashover, Press and Wingerworth 36.

T. repens L.   *White Clover, Dutch Clover*
Native. Pastures, meadows and waysides on a wide range of soils but not on the poorer acid soils; a rapidly spreading colonist of bare waste ground. Common and widespread.

Recorded for all squares.

Linton: 'Common. Generally distributed'.

T. fragiferum L.   *Strawberry Clover*
Native. Damp meadows and waysides. Local and rare.

S. Ashbourne Green 1847.

Linton: Between Fenny Bentley and Tissington 15, Duffield 34, nr. Shirley, by Yeldersley Lane 24, nr. Stanton-by-Dale 43.

T. medium L. *Zigzag Clover*

Native. Pastures, banks and roadsides on neutral and weakly acid soils, especially on Coal Measures and non-calcareous plateau-loam over limestone and on soil derived from igneous rocks. Locally frequent.

Recorded for all squares except:

> **W.** o6, 26.
> **S.** 33.

Linton: Findern and Breadsall 33.

T. arvense L. *Haresfoot Trefoil*

Native. Open habitats on light sandy non-calcareous soils; dry fields, banks, roadsides. Local.

W. Monk's Dale 1374, Peter Dale 1275.

E. Ford 4081, Wingerworth Great Pond 3667, Upper Langwith 5068, West Hallam Station 4240.

S. Recorded for 23 and 33, Marston-on-Dove 2329.

Linton: Breadsall Moor 34, nr. Borrowash 43, Burton 22, Repton and nr. Foremark 32.

T. scabrum L. *Rough Trefoil*

Introduced. A very rare casual; no recent records.

Linton: Bakewell 26, Burton 22.

T. striatum L. *Soft Trefoil*

Native. Open habitats on well-drained soils, calcareous or non-calcareous; dry pastures, banks, south-facing limestone outcrops. Very local but often abundant.

W. Peter Dale 1275, Hassop Mines 2373, Lathkill Dale 1966, Thorpe Cloud and Thorpe Pasture 1550/1, Parwich Hill 1954.

S. Nr. Trent Station 4931, nr. Sawley Church 4731, by River Trent nr. Thrumpton 5031.

Linton: Drakelow 22, Repton 32.

T. incarnatum L. ssp. **incarnatum** *Crimson Clover*

Introduced. An occasional escape from cultivation.

W. Calver Sough 2374.

E. Renishaw 4478, Clowne 4975.

S. Mackworth 3137, Derby 33.

Linton: Not recorded.

T. pratense L. *Red Clover*

Native. Grassy places on all except waterlogged or very acid soils. Common and widespread.

M

The wild plant is var. *pratense*, a long-lived perennial with toothed leaflets and usually solid stems. The cultivated form, var. *sativum* Schreber, larger but shorter-lived, with usually entire leaflets and a hollow stem, frequently escapes from cultivation and is found on roadsides, field borders, etc.

Recorded for all squares.

Linton: 'Common. General throughout the county'.

Var. **sativum** Schreber

E.   Ambergate 3452, Whatstandwell 3354, Morley 3940, Little Eaton 3741.

S.   Hulland 2546, also Mugginton and Brailsford in 24, about Derby 33, Locko 4138, Swarkestone 3728.

Linton: Miller's Dale 17, Burton 22.

T. subterraneum L.   *Subterranean Clover*

Doubtfully native. Dry pastures on sandy or gravelly soils. Very rare.

S.   Nr. Sawley Church 4731, by River Trent nr. Thrumpton Ferry 5031.

Linton: Nr. Sandiacre Church 43, Repton 32.

The following species of *Trifolium* have been recorded as rare casuals, from waste ground between Derby and Little Eaton 3639/3640 and seen last in 1939: *T. uniflorum* L. and *T. monanthum* A. Gray.

ANTHYLLIS L.

A. vulneraria L.   *Kidney Vetch, Ladies' Fingers*

Native. In open turf on dry banks and slopes, especially over limestone or by roads metalled with limestone. Local.

W.   The Winnats 1382, Sir William Road, Grindleford 2680, nr. Topley Pike 0971, also Cunning Dale 0872, Cressbrook Dale 1773, also Monk's Dale, Peter Dale, Miller's Dale, Chee Dale, and Monsal Dale in 17, Longstone Edge 2072, also Middleton Dale and Sir William Hill in 27, Ashford 1868, Milldale 1455, Via Gellia 2556, also nr. Grange Mill 2457, Cromford 2956, Brassington 2254, Harborough Rocks 2455.

E.   Quarry nr. Barlborough 4777, Markland Grips 5074, Pleasley Vale 5165, recorded for 35, West Hallam 4246.

Linton: Killamarsh 48, South Normanton 45, Yeldersley 24, Dale 43, Mickleover 33, Ticknall 32, between Repton and Willington 22, Linton 21.

LOTUS L.

L. corniculatus L.   *Birdsfoot-trefoil*

Native. Pastures, waysides and waste places on all but the poorest soils. Common and widespread.

Recorded for all squares.

Linton: 'Common, general and everywhere abundant'.

A depauperate form is recorded for Longstone Edge 2073. Linton also remarks
on this from the same locality 'a lane-side between Curbar and Far End,
Baslow' 27.

L. tenuis Waldst. & Kit.   *Slender Birdsfoot-trefoil*
Native. Fields and grassy waysides on a wide variety of soil types but
especially where there has been some disturbance. Rare.
W. Longstone Edge 1973, Coombs Dale 2274.
Linton: Dovedale 15, Nether Langwith 57, Osmaston-by-Ashbourne 24/14,
Caldwell 21.

L. uliginosus Schkuhr   *Large Birdsfoot-trefoil*
Native. Damp meadows, marshes, stream-banks and hedgerows. Frequent.
W. Derwent Reservoir 1691, recorded for 08, Derwent 1889, Bamford
2083, Hathersage 2381, Cavendish Golf Course, Buxton 0473, Taddington
1671, also High Dale and Chee Dale 1273, Calver New Bridge 2475, also
White Edge, Froggatt Edge and Leash Fen in 27, nr. Ashford 1769, Thorpe
1650, Brassington 2355.
E. Beauchief 3381, Whirlow 3182, Linacre 3272, also Chesterfield, Brampton
and Cordwell in 37, Whitwell Woods 5277, Wingerworth Great Pond 3667,
Kelstedge 3363, Ault Hucknall 4665, Hardwick 4563, Pleasley Vale 5265,
nr. Crich 3654, West Hallam 4140.
S. Nr. Ashbourne 1745, Hulland Moss 2546, Marston Common 1439,
Recorded for 43, Spread Eagle, Findern 2929, Grange Wood, Lullington
2714.
Linton: Renishaw 47, Matlock and Cromford Canal 35, Newton Wood 45,
Church Broughton and Barton Fields 23, Breadsall 33, Ockbrook and New
Stanton 43, Repton and Calke 32.

GALEGA L.

G. officinalis L.   *Goat's Rue, French Lilac*
Introduced. Naturalized in waste places. Occasional.
S. Old railway track, Ashbourne 1746.
Linton: Not recorded.

ROBINIA L.

R. pseudoacacia L.   *Locust, False Acacia*
Introduced. Much planted and occasionally establishing itself from seed.
S. Odd specimens planted at Ashbourne 1846, and in Longford Park 2138,
Burton, in garden 2623.
Linton: Not recorded.

ASTRAGALUS L.

A. glycyphyllos L.   *Milk-vetch*

Native. Hedgebanks, wood-margins and quarry spoil-heaps on the Magnesian
Limestone. Rare and local.

E.   Roadside nr. Whitwell Wood 5279, Whitwell Wood 5278, Steetley
Quarry 5479.

Linton: Not recorded.

ORNITHOPUS L.

O. perpusillus L.   *Birdsfoot*

Native. A colonist of bare non-calcareous sand or gravel on waysides, field
borders and disturbed ground. Local.

E.   Breadsall Moor 3742, Locko Park 4041, Morley 34.

S.   Bull Hill 2046.

Linton: Edensor 27, Dethick 35, Edlaston Coppy 14, Egginton 22, Findern 32,
Melbourne 32, Willington 32, Repton 32.

CORONILLA L.

C. varia L.   *Crown Vetch*

Introduced. Naturalized in a few places. Rare.

W. Hathersage, River Derwent 2281.

Linton: Not recorded.

HIPPOCREPIS L.

H. comosa L.   *Horse-shoe Vetch*

Native. Dry sunny limestone slopes. Rare.

W. Deep Dale 0971, Deep Dale, Topley Pike, Blackwell Mill, Wye Dale all
in 1072, Ravensdale 1573, Chee Dale 1273, The Nabbs, Dovedale 1453.

Linton: Monsal Dale, Topley Pike, and the wooded slopes of Chee Tor 17,
Dovedale 15.

ONOBRYCHIS Scop.

O. viciifolia Scop.   *Sainfoin*

Introduced. A casual escape from cultivation. Rare.

No recent records.

Linton: Matlock.

VICIA L.

V. hirsuta (L.) S. F. Gray   *Hairy Tare*

Native. Arable fields, roadsides, waste places. Locally frequent.

W. Buxton 0673, Miller's Dale 1473, Lathkill Dale 1966, Minninglow 2057.

E.   Beauchief Hall 3381, Beighton 4482, Linacre 3372, Breadsall Moor 3742, Stanley 4142.

S.   Recorded for all squares.

Linton: Baslow 27, Pinxton 45.

V. tetrasperma (L.) Schreber (*V. gemella* Crantz)   *Smooth Tare*
Native. Arable land, hedges, waysides. Infrequent.

W.   Miller's Dale 1473.

E.   Nr. Beauchief Hall 3381, High Lane, Holloway 3256.

S.   Sturston Tip 1946, Radbourne Lane 2836, by canal, Burton 2323, Melbourne 3825.

Linton: Clowne 47, behind 'Phoebe's Cottage', Matlock 25, Shirley 24, Mackworth 33, Walton 21.

V. cracca L.   *Tufted Vetch*
Native. Hedges, wood-margins, damp meadows, etc. Common and widespread.

Recorded for all squares except:

W.   19, 09.

Linton: 'Common'.

V. sylvatica L.   *Wood Vetch*
Native. Rocky woods and scrub, especially on limestone. Local and rather rare.

W.   Cressbrook Dale 1773, roadside between Bradbourne and Fenny Bentley 1950, by the colour works, Via Gellia 2857, Heights of Abraham 2858.

S.   Breadsall Village 3739.

Linton: King Sterndale 07, Lathkill Dale 16/26, Duffield 34, Coldwall Bridge 14, Repton Rocks 32.

V. sepium L.   *Bush Vetch*
Native. Hedges, thickets, wood-margins and grassy places on a wide range of soils but avoiding strongly acid soils; a frequent colonist of limestone scree. Common and widespread.

Recorded for all squares.

Linton: 'Common'.

V. sativa L. (incl. *V. angustifolia* (L.) Reichard)   *Common Vetch*
Native and introduced.

There are two subspecies in the county:

Ssp. **sativa,** with oblong to obovate leaflets and flowers at least 15 mm. long, is the cultivated form which is found as a casual of arable land, railway banks and waste places. Occasional.

W. Hayfield 0386, High Rake 1677, Middleton Dale 2275, Lathkill Dale 1966, nr. Oaker 2860, Bonsall Moor 2559.

E. Recorded for all squares except 45, 34.

S. Recorded for all squares except 22, 43.

Linton: Egginton Common, Burton and Willington 22.

Ssp. **angustifolia** (L.) Gaudin, with linear to oblong leaflets and flowers usually less than 15 mm. long, is the wild plant of grassy places and wood-margins. Frequent.

W. Derwent 1789, High Dale 1571.

E. Linacre 3572, Whitwell Wood 5278, Pleasley Vale 5265, Little Eaton 3640, Chaddesden cross-roads 4040.

S. Railway bank, Ashbourne 1747, Cross o' th' Hands 2846, Ednaston 2442, Etwall 2632, Littleover 3334, Egginton 2628, Swarkestone 3728, Stanton-by-Bridge 3727.

Linton: Edale 18, Miller's Dale 17, Baslow 27, Morley 34, Draycott 43.

V. lathyroides L.   *Spring Vetch*

Doubtfully native. There is a single old record.

No recent records.

Linton: Repton 32.

V. pseudo-cracca Bertol.

Introduced. A rare casual.

E. Little Eaton 3640.

Linton: Not recorded.

LATHYRUS L.

L. aphaca L.   *Yellow Vetchling*

Introduced. A very rare casual.

W. Garden weed, Belvedere Terrace, Buxton 0572.

S. From bird seed, Clifton 1545.

Linton: Pinxton 45, Burton 22.

L. nissolia L.   *Grass Vetchling*

Native. Disturbed ground of field borders, banks and waysides. Rare and decreasing.

No recent records.

Linton: Nr. Alfreton 45, Nutwood, Darley by Derby 33, between Breaston, Risley and Draycott 43.

L. hirsutus L.   *Hairy Vetchling*
Introduced. Casual. Fields and waste places. Very rare.
S.   Little Eaton Road, Darley by Derby 3638.
Linton: South Normanton 45, Burton 22.

L. pratensis L.   *Meadow Vetchling*
Native. Meadows, hedges, grassy banks and waysides. Common and
widespread.
Recorded for all squares.
Linton: 'Common and abundant'.

L. tuberosus L.   *Earth-nut Pea*
Introduced. A rare casual.
E.   Egginton Sewage Farm 2631, railway bank between Etwall and Mickle-
over 2632.
S.   Catton Park 2015 (in pheasant food: persisted three years).
Linton: Not recorded.

L. sylvestris L.   *Narrow-leaved Everlasting Pea*
Native. Woods, thickets, bushy banks and hedgerows. Local and rare.
S.   By canal tow-path, Weston-on-Trent and Weston Cliff, abundant
3927, Weston-on-Trent railway bank 4128.
Linton: Weston-on-Trent and casual below Chellaston 32.

L. palustris L.   *Marsh Pea*
Perhaps formerly native, but this seems doubtful.
No recent records.
Linton: Pinxton 45.

L. montanus Bernh.   *Bitter Vetch*
Native. Pastures, heaths, waysides, banks, wood-margins, etc., on neutral
or slightly acid soils but neither on calcareous nor strongly acid soils; very
characteristic of the non-calcareous plateau-loam on the brows of the limestone
dales, where it is commonly associated with *Agrostis tenuis, Viola lutea* and
Betonica officinalis. Locally frequent.
Recorded for all squares except:
S.   13, 43.
Linton: New Stanton and Ockbrook 43.

L. annuus L.
Introduced. A rare casual.

E.   One mile south of Little Eaton 3640.

S.   Stores Road, Derby 3537.

Linton: Not recorded.

ROSACEAE
SPIRAEA L.

S. salicifolia L.   *Willow Spiraea*

Introduced. Frequently planted in hedges and occasionally naturalized. Rare.

W. Buxton 0573, Miller's Dale 17, nr. Rowsley 26.

Linton: Not recorded.

PHYSOCARPUS (Cambess.) Maxim.

P. opulifolius (L.) Maxim.

Introduced. Planted in gardens and elsewhere and occasionally becoming established.

W. Lathkill Dale (planted) 1966.

Linton: Not recorded.

FILIPENDULA Miller

F. ulmaria (L.) Maxim. (*Spiraea ulmaria* L.)   *Meadow-sweet*

Native. Swamps, marshes, wet woods and meadows, streamsides and wet rock-ledges. Common and widespread.

Recorded for all squares except:

W. 09, 19.

Linton: Nr. Snake Inn 19.

F. vulgaris Moench (*Spiraea filipendula* L.)   *Dropwort*

Native. Dry grassland, especially on south-facing limestone slopes with loamy soils. Local.

W. Recorded for 08, Back Dale 0870, Cressbrook Dale 1773, Miller's Dale 1373, Litton Dale 1574, Tideswell Dale 1573, Hand Dale 1460, Lathkill Dale 1765, Gratton Dale 2060, nr. Aldwark 2358, nr. Grange Mill 2357.

S Knowle Hills 3525, Swarkestone 3628, between Chellaston and Weston 32, Drakelow 2318.

Linton: Youlgreave 26.

RUBUS L.

R. chamaemorus L.   *Cloudberry, Knoutberry*

Native. On high-level blanket bogs at altitudes upwards of about 1,500 ft. and forming extensive patches on peat bared by erosion. Locally frequent on the northern moorlands. See also map (page 91).

Male and female plants are distinguishable by their leaves, those of the male plant being more deeply lobed. Ripe fruit is very rarely seen owing to grazing by sheep.

W. Ashop Head 0690, Bleaklow 1196, Derwent Edge 1991, Kinder Scout 0988, nr. Edale Cross 0786, Colbourne Moors 0983, Jaggers Clough 1487, Chapel-en-le-Frith 07, Axe Edge 0071.

Linton: 'Plentiful locally'; but no records outside the area covered by those listed above.

R. saxatilis L. *Stone Bramble*

Native. Rocky slopes and stabilized scree on the Carboniferous Limestone, both in the open and in light shade; often associated with *Convallaria majalis*, *Melica nutans* and *Geranium sanguineum* and then probably the site of ancient woodland. Local and rather uncommon; very rarely elsewhere than on the Carboniferous Limestone.

W. Recorded for 07, Monsal Dale 1871, Cressbrook Dale 1773, Monk's Dale 1374, Miller's Dale 1372, Chee Dale 1172, Deep Dale and Topley Pike 1072, above Chee Tor 1273, Fin Cop 1770, Coombs Dale 2274, Longstone Edge 2073, Lathkill Dale 16, Hurt's Wood, Dovedale (Staffs. side) 1353, Via Gellia 2757.

E. Markland Grips 5074.

Linton: Nr. the Snake Inn 19.

R. idaeus L. *Raspberry*

Native. Woods, hedgebanks, waysides, bushy slopes. Common and widespread.

Recorded for all squares except:

> **W.** 09.
> **E.** 45, 56, 48.
> **S.** 33, 22.

White-fruited forms occur.

Linton: Mackworth 33, Drakelow 22.

R. caesius L. *Dewberry*

Native. Moist woods, thickets, thickets, hedges and rough pasture, especially on limestone or base-rich clay. Locally frequent.

W. Recorded for 08, 18, Monk's Dale 1374, also Monsal Dale and Miller's Dale in 17, recorded for 06, Ashford 1769, Alport 2264, Dovedale 1453, Via Gellia 2656, also Wirksworth and Ible Wood in 25.

E. Brierley Wood 3775, Scarcliffe Park Wood 5170, also Creswell Crags in 57, Langwith Wood 4967, East of Astwith 4564, Pleasley Vale 5265, Crich 3455, Morley Moor 3941.

S. Ashbourne 1746, Sturston 2146, Shirley 2141, Cubley 1638, Etwall 2732, Trent 4931, Marston-on-Dove 2329, recorded for 32.

Linton: Between Hassop and Baslow 27, Killamarsh 48, Barlborough and Bolsover 47, above Allestree 34, Newton Wood 44, Mickleover 33, Thrumpton Ferry 53, Calke and Repton 32.

R. fruticosus L., *sensu lato*   *Blackberry, Bramble*
Native. Woods, hedges, thickets, plantations, heaths, etc. Common and
widespread.
The forms of this aggregate, commonly treated as separate species, are very
numerous and difficult to determine. Linton was himself an expert on the
group, and we are greatly indebted to Dr. E. S. Edees for his help in bringing
Linton's treatment up-to-date. Almost all the recent records are his and we
have quoted freely from his paper on Derbyshire brambles (Proc. B.S.B.I.,
5, 13-19) and from letters.

R. scissus W. C. R. Watson ('*R. fissus* Lindl.')
Native. 'Heathy places in woods and by roadsides, especially in hilly districts'.
Frequent and widespread.
W. Nether Booth 1486, Dore 2980, Smeekley Wood 2976, Eyam 2277.
E. Brierley Wood 3676, Holymoorside 3368, Wirksworth 3053, Mapperley
4343, West Hallam Station 4240.
S. Bradley Wood 1946, Hulland Carr 2645, Dale Abbey 4338.
Linton: Derwent Dale 19, Edale and by River Noe, Hope 18, Hathersage 28,
Eyam, Ramsley Moor nr. Baslow, Grindleford Bridge, Froggatt and Smeekley
Wood 27, Holmesfield 37, Harthill Moor and nr. Stanton 26, Moor Lawn
Coppice, Holymoorside 36, Wirksworth and Kirk Ireton, Scow Brook 25,
Alderwasley 35, Shirley Wood, The Holt, Ednaston, Drabble Carr, Bradley
Brook, Cross o' the Hands and Breward's Carr, Mugginton 24.

(**R. plicatus** Weihe & Nees)
Linton probably took a wider view of this species than is taken today and it
is uncertain to what his records refer.

R. fissus Lindl. ('*R. rogersii* Linton')
Native. Rare.
S. Shirley 2141.

R. eboracensis W. C. R. Watson
Native. This species has been overlooked and will probably prove to be
quite common.
W. Cromford Moor 2955.

R. sublustris Lees (*R. corylifolius* var. *sublustris* (Lees) Rogers)
Native. 'Widely distributed in hedgerows, especially on the limestone'.
W. Hope 1684, Hathersage 2281, Miller's Dale 1473, Tideswell Dale 1574,
Froggatt 2476.
S. Thorpe Cloud Station 1650, Tissington, by the church 1752, between
Tissington and Bradbourne 1952, Hopton Wood 2656, Mapleton 1748,
Yeldersley Hollies 2243, roadside north of Marston Montgomery 1439.
Linton: Ednaston 24, Pleasley 56, etc.

(R. balfourianus Bloxam ex Bab.)
The presence of this species in Derbyshire, though reported by Linton, is not yet confirmed.

R. warrenii Sudre (*R. dumetorum* var. *concinnus* Baker ex Warren)
Native. Hedges, road-margins, etc. Common and widespread.
W. Eyam 2276, Owler Bar 2977.
E.   Unthank Lane 3075, Hardwick Wood 3666, nr. Belper 3246.
Linton: Charlesworth 09, Mills nr. Mellor 98, Chapel-en-le-Frith and Whaley Bridge 08, Ashopton 18, Bamford and Hathersage 28, Ashwood Dale 07, Miller's Dale 17, Eyam Moor, Shatton and Stoke Moor Wood, Grindleford 27, Bar Moor Clough 37/6, Bolsover and Duckmanton 47, Kirk Dale, Ashford 16, Matlock and Callow nr. Wirksworth 25, Shirland 35, Alfreton 45, Bradley and Shirley 24, Allestree Park, Belper, Denby and Duffield 34, Marston Montgomery 13, nr. Derby 33/43, Dale and Risley Park 43, Willington 22.

R. tuberculatus Bab. (*R. dumetorum* var. *tuberculatus* (Bab.) Rogers)
Native.
S.   Roadside between Thurvaston and Culland Hall 2438.

R. rubriflorus Purchas (*R. dumetorum* var. *rubriflorus* (Purchas) Rogers)
Native. Hedges. Locally frequent.
E.   Hazelwood nr. Belper 3246.
S.   Nr. Osmaston 1943, road to Ashbourne north of Osmaston 1944, Yeldersley Hollies 2243, roadside south of Culland Hall 2438.
Linton: Killamarsh and Spink Hill, Eckington 48, Bagthorpe Farm, Old Brampton 37, Bolsover 47, Creswell 57, Langwith 56/7, Walton 36, Newton Wood 45/6, Wirksworth 25, Fritchley 35, Yeaveley 14, Longford and Shirley 24, Boylestone, Cubley and Somersal Herbert 13, Barton Blount, Church Broughton and Etwall 23, Markeaton 33, Egginton Common 22.

(R. scabrosus P. J. Muell.)
'There is an abundant bramble in Staffordshire and Cheshire and no doubt in Derbyshire too, which does not tally with any of the published descriptions. It is nearest *R. scabrosus* P. J. Muell. but it has white flowers' (E. S. Edees, *in litt.*).

R. myriacanthus Focke (*R. dumetorum* var. *diversifolius* Rogers, in part).
Native. 'Often found on low ground near water'.
S.   Edlaston 1743, Repton and Repton Shrubs 3024.

R. gratus Focke
Native. Heaths, woods, copses. Rare.
Linton: By the brook in Shirley Wood 24.

R. calvatus Lees ex Bloxam (*R. villicaulis* var. *calvatus* Lees ex Bloxam)
Native. 'Frequent, especially in the south of the county'.
S.   Edlaston 1743, Brailsford 2441, Dawson's Rocks 3322.
Linton: Low Leighton, New Mills 08, Hathersage Booths 28, The Holt, Edlaston 14, Ednaston Hall, Shirley and Snelston 24, Allestree Park 34, Findern 23, Sudbury, Repton Rocks, Dimminsdale, Calke, and lane to Milton just above Repton Park Road 32.

R. carpinifolius Weihe & Nees
Native. Hedges and bushy places, open woods. Local.
W.   Glossop 0291, Dore 2980.
S.   Lane to Spinnyford Brook 2445, Dawson's Rocks 3322.
Linton: Derwent Dale 19, New Mills and Hague Bar 98, Whaley Bridge 08, Ashopton and nr. Hope 18, Upper Hurst Brook, Bamford, nr. Hathersage Booths and Otterdale below Shatton Moor 28, Beauchief Abbey and Dore 38, Cressbrook School Church 17, Bar Brook below Baslow, Calver, Froggatt and Abney Clough 27, Youlgreave and Robin Hood's Stride 26, Wingerworth 36, wood south of Bradley, Shirley Wood and Brook, and copse north of Ednaston Lodge 24, Allestree Park and Belper 34, Sudbury 13, Markeaton 33, Repton Rocks 32.

R. nemoralis P. J. Muell. (*R. villicaulis* var. *selmeri* (Lindeb.) Rogers)
Native. Hedges, carrs, thickets. Frequent.
W.   Glossop 0291, Hayfield 0388, New Mills 0185, Barber Booth nr. Edale 1184.
S.   Dawson's Rocks 3322.
Linton: By the Etherow nr. Dinting 99, Charlesworth and Mills nr. Mellor 98, Chinley and between New Mills and Bugsworth 08, Edale and Ashopton 18, Hathersage Booths 28, Stony Middleton, Froggatt, Curbar and Hay Wood, Grindleford 27, Kelstedge 36, The Holt, Edlaston, Blackwall, nr. Nether Biggin and Breward's Car, Mugginton 24, Burnaston and east of Church Broughton 23, Egginton Common 22, Repton Rocks, Repton Road, Ticknall, and White Lees, Ticknall 32.

R. laciniatus Willd.
Introduced. Hedges and roadsides nr. gardens; an escape from cultivation. Rare.
E.   Holmesfield 3273.

R. durescens W. R. Linton
Native. Hedges, heaths, bushy banks, woods. Local.
S.   Bradley Wood 1946, nr. Cross o' the Hands ('an excellent bush where the roads and grid lines cross') 2846.

Linton: Between Ambergate and Wirksworth 25/34, Hillcliff Lane, Shottle, Hulland Moss, Breward's Car, Mugginton, Wood Lane, Brailsford, between Shirley and Ednaston, Bradley Wood 24, Belper, Duffield, Holbrook, Daypark 34, between Church Broughton and Sutton 23.

R. lindleianus Lees

Native. Hedges, woods, heaths. 'Very common in south Derbyshire. Perhaps less common in the north, but generally distributed and probably to be found in every part of the county'.

S.   Brailsford Green, by the side of the path to the church and all round the field 2441.

Linton: By the Etherow nr. Dinting 99/09, Charlesworth 09/98, Low Leighton nr. New Mills 08, Woodseats, Castleton and Shatton east of Bradwell 18, Bamford, Hathersage and Grindleford Bridge 28, Beighton and Coal Aston 48, Miller's Dale 17, Creswell Crags 57, Matlock Bath, Cromford and Wirksworth 25, Whatstandwell and Cromford Canal 35, Bradley Wood, Shirley 24, Boylestone 13, Church Broughton 23, Allestree Park and Denby 34, Mapperley and Loscoe 44, Dale, New Stanton, Sandiacre and Long Eaton 43, Sudbury, Calke and Bretby 32.

R. muenteri Marss.

Native. Probably not uncommon, but confused by Linton with *R. polyanthemus* Lindeb.

W. New Mills 0185.

S   Lane to Spinnyford Brook 2445.

Linton: Nr. Shirley 24.

R. amplificatus Lees (*R. macrophyllus* var. *schlechtendalii* (Weihe) Rogers)

Native. Woods, hedges, etc. Occasional.

W.   By Bar Brook nr. Baslow 2673.

E.   Between the ponds west of Wingerworth 3567.

Linton: Spinnyford Brook nr. Hulland 24.

R. danicus (Focke) Focke

Native.

S.   Ednaston 2342.

(R. rhodanthus W. C. R. Watson ('R. rhombifolius Weihe'))

It is not clear what Linton meant by *R. rhombifolius* Weihe, but it may be *R. rhodanthus* W. C. R. Watson, recorded for the county by Watson.

(R. incurvatus Bab.)

Possibly native, but the true plant has not so far been confirmed for Derbyshire.

R. polyanthemus Lindeb. (*R. pulcherrimus* Neum.)

Native. Hedges, commons, heaths. Frequent.

W. Ladybower 1986, Dore 2980, Holymoorside 2469.

E. Between the two ponds west of Wingerworth 36.

S. Nr. Bradley Wood 14, Cross o' the Hands, Ednaston, lane to Spinnyford Brook and Yeldersley Old Hall 24, roadside south of Culland Hall 23, Dawson's Rocks 32.

Linton: The numerous records doubtless include some for *R. muenteri*, but the species is certainly frequent in the county.

R. cardiophyllus Muell. & Lefèv. ('*R. rhamnifolius* Weihe & Nees')

Native. Hedges, heaths. 'Apparently rather rare'.

S. Cross o' the Hands 2846, between Thurvaston and Culland Hall 2438.

Linton: Monsal Dale 17, Breward's Carr, Mugginton, and Long Lane towards Brailsford 24, Belper 34, Thurvaston Stoop and Osliston 23, Calke 32.

R. dumnoniensis Bab. (*R. rotundatus* P. J. Muell. ex Genev. *sec* W. C. R. Watson)

Native. Hedges, thickets.   Rare.

S. Dawson's Rocks 3322.

Linton: Nr. Repton towards Willington and between Repton Park and the Hayes at the corner of the roads to Burton and Hartshorn 32.

R. lindebergii P. J. Muell.

Native. Heaths, bushy banks, fields, 'A local species of dry sunny hillsides which may be commoner in the limestone dales than our records show'.

W. Glossop 0291, roadside descending Rushup Edge 1184.

E. Unthank Lane 3074.

Linton: Briergrove, Rowarth 99, Wash and Malcoff nr. Chapel and Chinley 08, Edale and Hartle Dale, Bradwell 18, Bamford 28, Taddington Dale 17, Froggatt, Abney Clough, Shatton and Otterdale 27, Fox Lane nr. Unthank 37, Edensor 26, Wadshelf 36, Masson Mill, Matlock Bath, and Wirksworth 25, Hulland Moss, The Knowl, Bradley, Breward's Car, Mugginton, Mercaston and Yeldersley Lane 24, Stenson and Findern 33, Repton Rocks 32.

R. ulmifolius Schott (*R. rusticanus* Merc.)

Native. 'This species has a peculiar distribution in Derbyshire . . . It likes clay and calcareous soils, but shuns sandstone, so that it is often abundant in districts where other species are rare . . . .'

W. Alsop-en-le-Dale 1554, Fenny Bentley 1650, Thorpe 1550, Tissington 1852.

E. Notts-Derbyshire border nr. Trent 43.

S. Ashbourne Green 1948, nr. Shirley 2141, nr. Marston Montgomery 1439, nr. Radbourne 2735, Sutton-on-the-Hill 2333, nr. Hollington 2338, Swarkestone 32.

Linton: Spink Hill, Eckington, by pool east of Killamarsh and Beighton 48, Monsal Dale 17. Calver 27, Millthorpe nr. Holmesfield and Crowhole, Barlow 37, Barlborough 47, Calton Lees, Chatsworth 26, Shottle Gate and Hazelwood 34, Kirk Hallam 44, Sutton 23, New Stanton and Dale 43.

R. winteri P. J. Muell. ex Focke ('*R. robustus* P. J. Muell.')
Native. Known only from Linton's specimen collected at Stydd in 1896.
Linton: Yeaveley, Stydd and between Snelston and Darley Moor 14.

R. falcatus Kalt. ('*R. thyrsoideus* Wimm.')
Native. Hedges. Locally plentiful.
S. Swarkestone 3728. 'There are fine examples in the lane to the church and younger bushes in the churchyard'.
Linton: Etwall 23, Findern, Elvaston, Chellaston Station and Staker Flats, Mickleover 33, nr. Trent Station, nr. Sawley, Thrumpton Ferry and Trent Lock 43, Egginton Common and between Stapenhill and Cauldwell 22, nr. Swarkestone Church, Repton, and nr. Stanton-by-Bridge 32.

R. sprengelii Weihe
Native. 'A beautiful woodland species, widely distributed throughout the county, but perhaps particularly common in the north'.
W. Glossop 0291, Hayfield 0388, New Mills 0185, Dore 2980, Owler Bar 2977, Sherriff Wood 2378, Copy Wood 2665.
E. Brierley Wood 3676, Fox Lane 3076, Wood nr. Holymoorside 3469, Upper Loads 3169, Walton Wood 3668, Wood nr. Ambergate 3252.
S. Edlaston 1743, Bradley Wood 2046, Bradley 2245, Cross o' the Hands 2846, lane to Spinnyford Brook 2445, Ednaston 2342, Dawson's Rocks 3322.
Linton: Crowden Reservoir 09/00, Glossop 09, Chinley 08, Edale and Ashopton 18, Upper Hurst Brook, Bamford, Hood Brook, Hathersage, moors nr. Ringinglow and Dore and Hathersage Booths 28, Coal Aston and Beauchief Abbey 38, Park Hall Woods nr. Barlborough 48, Stony Middleton, Abney and Highlow Hall 27, Linacre Wood 37, Hackenthorpe and Moorhole 47, Stanton Moor and Birchover 26, between Whatstandwell and Wirksworth 25, and Blackwall, Shirley, Yeldersley, Bradley and Mercaston 24, Belper and brookside above Denby 34, Boylestone 13, Dale 43, east of Egginton Common 22, Repton Rocks, between Calke and Melbourne and near Southwood Farm 32.

R. vestitus Weihe & Nees ('*R. leucostachys* Schleich.')
Native. Hedges, banks, woods. Frequent.
E. Smalley 4243, Mapperley 4343, West Hallam 4342.
Linton: Etherow and Mills nr. Mellor 98, Chinley 08, Ashopton 18/28, Bolsover 48, Monsal Dale and Cressbrook Mill 17, nr. Eyam, Grindleford and Abney

Clough 27, Duckmanton 47, between Greenhill and Bradway 38, Elmton, Markland Grips, Langwith, Creswell and Pleasley 57, Great Shacklow Wood 16, Birchover 26, Dovedale 15, Masson Wood, Matlock, and Pig of Lead, Cromford 25, Cromford Canal 35, Shirley and Brailsford 24, Belper, Duffield, Shottle Gate, Denby and Allestree Park 34, Sudbury 13, Hatton 23, West Hallam Station 43, between Calke and Pistern Hill 32.

R. criniger (E. F. Linton) Rogers

Native. Lane-sides, woods, banks, 'Plentiful where brambles grow in south Derbyshire, especially about Ashbourne and Shirley'.

S.   Bradley Wood 1946, Cross o' the Hands 2846, Ednaston 2342, lane to Spinnyford Brook 2445, Yeldersley Old Hall 2144, roadside north of Marston Montgomery 13, Bretby 2923, Dawson's Rocks 3322.

Linton: Bradley, Yeldersley Lane, Shirley, Snelston and Hulland Ward 24, Hatton and towards Heathtop, Little Derby, Etwall, and nr. Findern 23, Milton and Repton Rocks 32.

(R. mucronulatus Bor. (R. *mucronatus* Blox.))

Probably native, but there has been no recent confirmation.

R. taeniarum Lindeb. ('R. *infestus* Weihe')

Native. Open woodland, hedges. Rather rare.

W.   Froggatt 2476.

Linton: Low Leighton nr. New Mills 08, Hathersage Booths and Nether Hurst, Bamford 28, Froggatt and Baslow 27, Kelstedge, Brierley Wood nr. Sheepbridge (not typical) and Lawn Moor Copse, Holymoorside 37, Ambergate 35, between Rodsley and Yeaveley 14/24, nr. Duffield 34.

R. radula Weihe ex Boenn.

Native. Hedges, woods, commons. Rather rare. A specimen collected by Purchas in Dovedale in 1883 is correctly named.

Linton: Millhouses and Coal Aston 38, pool east of Killamarsh 48, High Dale, Taddington 17, Walton Wood 37, Dovedale 15, Wirksworth 25, Denby 34, Loscoe and Kirk Hallam 44.

R. echinatus Lindl.

Native. Hedges, woods, Not infrequent.

S.   Roadside north of Thurvaston 2438, hillside between Repton & Repton Shrubs 3024.

Linton: Creswell and Creswell Crags 57, Ashover Hay 36, Edlaston 14, Shirley, Brailsford 24, Kedleston 34, Boylestone, Cubley and Yeaveley 13, Longford, Hatton and Church Broughton to Sutton 23, Risley Park 43, Calke, Bretby and by the canal, Swarkeston Junction 32, Linton 21.

R. echinatoides (Rogers) Sudre (*R. radula* var. *echinatoides* Rogers)

Native. Hedges, woods, etc. 'A widely distributed and evidently common species in Derbyshire'.

W. Glossop 0291, Ladybower 1986, Dore 2980, Smeekley Wood 2976, Via Gellia 2757.

E. Fox Lane 3076, Unthank Lane 3075, Upper Loads 3169, roadside nr. Wingerworth 3666, Belper Lane End 3349.

S. Edlaston 1743, nr. Shirley 2141, Ednaston 2342, Ireton Wood 2847, Yeldersley Old Hall 2144, Thurvaston 2438.

Linton: Ashopton 18/28, by River Derwent below Mytham Bridge 28, Hackenthorpe 48, Taddington Wood and Miller's Dale 17, Calver, between Abney and Highlow Hill, Shatton and Manners Wood, Chatsworth 27, Unstone 37, Bolsover 47, Stanton Moor 26, Masson Woods 25, Cromford Canal 35, Snelston 14, Longford, Hillcliff Lane, Blackwall, Shirley, Yeldersley 24, Allestree Park 34, between Mickleover and Etwall 23, hedgebank of the Ashby road on Pistern Hill 32.

(R. rudis Weihe & Nees)

Possibly native, but there has been no recent confirmation for Derbyshire.

R. granulatus Muell. & Lefèv. ('*R. oigoclados* var. *bloxamianus* Coleman')

Native. Local.

S. Between Repton and Repton Shrubs 3124, on a hillside above a house nr. Repton 3024.

Linton: Nr. Burnaston House, nr. Spread Eagle Inn and on roadside between Little Derby and Egginton Station 23, Willington and between Willington and Egginton Common 22, Milton, Bretby, Repton and between Ticknall and Stanton-by-Bridge 32, nr. Rosliston and nr. Cauldwell 21.

R. foliosus Weihe & Nees

Native. Bushy places, hedges. Rare.

Linton: Bryan's Copse, Smisby 13, Knowle Hills, Repton Shrubs and Repton Rocks 32.

(R. rubristylus W. R. C. Watson ('*R. oigoclados* var. *newbouldii* Bab.'))

Possibly native, but there is no recent confirmation of Derbyshire records.

R. pallidus Weihe & Nees

Native. Copses and busy places, especially nr. streams. Rare.

W. By the stream in Smeekley Wood 2976.

Linton: Owler Car nr. Coal Aston and wooded ravine between Owler Lees and Totley 38.

N

R. newbouldii Bab. ('*R. podophyllus* P. J. Muell.')
Native. Bushy places, ravines, hedges. Local.
W. Roadside between Chisworth and Charlesworth 99.
E.   Cordwell 3176.
Linton: Charlesworth and Stirrup Benches 99/09, New Mills and Tarden nr.
Mellor 98, Hackenthorpe and Moorhole 48, Owler Bar 27, Unstone and
Dronfield 37, Willington 22, Milton, Repton and Repton Shrubs 32.

(R. scaber Weihe & Nees)
Possibly native, but all specimens so labelled seem to be of an undescribed
species.

R. rufescens Muell. & Lefèv. (*R. rosaceus* var. *infecundus* Rogers)
Native. Hedges, woods. Locally common.
S.   Between Rodsley and Yeaveley 1940, nr. Shirley 2142, Bretby 2923,
Dawson's Rocks 3322, Repton Shrubs 3123.
Linton: Brierley Wood, Unstone 37, Salter Wood nr. Ripley 45, Shirley,
Yeldersley and Cross o' the Hands, Hulland Ward 24, Cubley 13, Calke,
between Ticknall and Stanton-by-Bridge, Repton Shrubs and Repton Rocks
32.

(R. purchasianus (Rogers) Rogers (*R. rosaceus* var. *purchasianus* Rogers))
Possibly native, but the old records need confirmation.

(R. apiculatus Weihe (*R. anglosaxonicus* Gelert))
Possibly native, but the old records need confirmation.

R. retrodentatus Muell. & Lefèv. ('*R. borreri* Bell Salt.')
Native. Open woodland, bushy places. Rare.
S.   Dawson's Rocks 3322.
Linton: Woodland about Unstone and Sheepbridge 37.

R. griffithianus Rogers
Native. Known only from a single locality in the county.
S.   Dawson's Rocks 3322.

R. leightonii Lees ex Leighton (*R. radula* var. *anglicanus* Rogers)
Probably native.
Linton: Between Willington and Findern 23, Bretby 32.

(R. furvicolor Focke ('*R. melanoxylon* Muell. & Wirtg.'))
Possibly native, but the old records need confirmation.

R. hylocharis W. C. R. Watson ('*R. rosaceus* var. *hystrix* Weihe & Nees')
Native. 'Widely distributed in woods and hedges'.
W. Dore 2980, Sherriff Wood 2378, Smeekley Wood 2976.
E. Fox Lane 3076, Unthank Lane 3075, Holymoorside 3469, Upper Loads 3169.
S. Bradley Wood 1945, Edlaston 1743, Cross o' the Hands 2746, lane to Spinnyford Brook 2445, Dawson's Rocks 3322.
Linton: Mills nr. Mellor and New Mills 98, Whaley Bridge and Chinley Station 08, nr. Dore, Totley and Upper Hurst Brook, Bamford 28, Froggatt 27, between Unstone and Dronfield, between Dronfield and Dore Station, Bole Hill, Barlow, north-east of Coal Aston and Brierley Wood, Unstone 37, Birchover 26, Walton Wood 36, Wirksworth 25, Edlaston and Snelston 14, Cross o' the Hands, Nether Biggin, between Bradley and Hulland, Shirley 24, Belper and Shottle Gate 34.
'Most of Linton's records for *R. rosaceus* var. *hystrix* probably belong here'.

R. dasyphyllus (Rogers) Rogers ('*R. koehleri* ssp. *dasyphyllus* Rogers')
Native. Woods, hedges, commons. 'This is certainly one of the commonest species in Derbyshire . . . there is no doubt that it can be found wherever brambles grow. It is particularly common in hilly districts and grows equally well on limestone, sandstone and clay.'
Linton: 'Everywhere abundant'.

(R. marshallii Focke & Rogers)
'There is only one record based on a large bush between the two ponds west of Wingerworth. It is not typical *R. marshallii* and may well be of hybrid origin.'

(R. fuscoater Weihe & Nees)
The plant to which this name was given by Linton is a distinct and beautiful species which is fairly widespread in the north of the county, but it is probably an undescribed species. The records below are of this plant.
W. Dore, plentiful about the ganister mine 2980, hillside at Eyam Firs 2276, Froggatt 2476.
Linton: Ambergate 34, Cross o' the hands, Hulland Ward 24.

POTENTILLA L.

P. palustris (L.) Scop.   *Marsh Cinquefoil*
Native. Fens, bogs, wet heaths and moors, usually on moderately mineral-rich but non-calcareous peat in places where water stands at least in winter. Locally abundant, often with *Menyanthes*.
W. North of Nuns Brook, Fairfield 0674, Leash Fen 2972, Moss Carr 0665, Carsington Pastures 2454.

S. Quilow, Edlaston 1843, Darley Moor 1642, nr. Cubley Wood Farm 1640, Yeaveley 1839.

Linton: Tideswell 17, Tansley Moor 36, Yeldersley Lane and Mugginton 24, Repton Rocks 32, Gresley 21.

P. sterilis (L.) Garcke (*P. fragariastrum* Pers.)   *Barren Strawberry*

Native. Open woods, wood-margins, scrub and in open turf on moist grassy slopes, usually on calcareous or slightly acid soils. Common and widespread. Recorded for all squares except:

W. 09, 19.
E. 45.

Linton: 'Common'.

P. anserina L.   *Silverweed*

Native. Roadsides, waste ground, damp pastures and commons. Common and widespread.
Recorded for all squares except:

W. 09, 19.

Linton: 'Common'.

var. **viridis** Koch.

S. Spondon 3934.
Linton: Not recorded.

P. argentea L.   *Hoary Cinquefoil*

Native. Dry sandy grassland on banks and waysides and in pastures. Rare.
W. A small colony on limestone outcrop, Parwich Hill 1954.
Linton: Markeaton Road, nr. Derby Lodge 33, nr. Sandiacre Church 43.

P. recta L.

Introduced. A garden escape sometimes becoming established. Rare.
E. Lees Common 3476.
S. Garden escape, Ashbourne 1746; v. *sulphurea* occurred on ex-aerodrome land above Bradley Wood 2046, not persisting in either locality.
Linton: Not recorded.

P. norvegica L.

Introduced. Naturalized on waste ground in a few places. Rare.
S. Derby, in a timber-yard 3136, canal side, Trent 4931.
Linton: Not recorded.

P. tabernaemontani Ascherson (*P. verna* L., in part)   *Spring Cinquefoil*
Native. Ledges and top edges of limestone outcrops on south-facing slopes
of the dales; occasionally spreading into grassland at the foot of an outcrop.
Local, and almost totally confined to the Carboniferous Limestone.

W. Conies Dale 1280, Cunning Dale 0872, Back Dale 0870, Deep Dale and
Topley Pike 1072, Miller's Dale 1473, also Monk's Dale, Chee Dale, Cressbrook
Dale, Tansley Dale, Great Rocks Dale, and nr. Monsal Dale Station in 17,
Middleton Dale 2175, also Stoney Middleton, Longstone Edge and Coombs
Dale in 27, Hartington 1360, Lathkill Dale 1865, Bradford Dale 2063, Dovedale
1451.

E. Markland Grips 5074.

Linton: 'Local'.

P. erecta (L.) Raüschel (*P. sylvestris* Neck.)   *Common Tormentil*
Native. Pastures, heaths, moors, roadsides, bog-margins, open woods, etc.;
chiefly on well-drained acid soils and avoiding strongly calcareous or very
heavy soils. Common and widespread.

Recorded for all squares except:

                              E.   47.

Linton: 'Common'.

P. anglica Laicharding (*P. procumbens* Sibth.)   *Trailing Tormentil*
Native. Wood-margins, field borders, hedge-banks, roadsides; usually on
well-drained acid soils but avoiding strongly podsolized soils. Locally frequent
except on limestone.

Hybrids between *P. sterilis* and *P. reptans*, or between *P. anglica* and either
of these, may be difficult to distinguish from *P. anglica* but are usually highly
sterile.

W. Recorded for all squares except 19, 15.

E. Cinderhill Lane, Norton 3681, nr. Povey Farm, Troway 3880, The Moss
4280, nr. Peakley Hill 3376, nr. Eckington 4379, Whitwell Wood 5278, Press
Reservoirs 3565, Nethermoor 46.

S. Shirley Wood 2042/3, Brailsford 2541, recorded for 43 and 32.

Linton: Codnor 44, Burton 22.

P. × suberecta Zimmet. (*P. anglica* × *P. erecta*)
S. Sandpit by Slack Lane, Brailsford 2442.

Linton: Nr. Woodhays, Chisworth 09, Sheffield Road, Ashopton 18.

P. × mixta Nolte ex Reichenb. (*P. anglica* × *P. reptans*)
W. Beeley Moor nr. Wragg's Quarry 2866.

Linton: About Buxton 07, Dovedale 15, north side Kedleston Park 34.

P. reptans L.   *Creeping Cinquefoil*

Native. Hedge-banks, waysides and waste places on all but very poor soils; a rapidly spreading colonist of bare ground. Common and widespread.

Recorded for all squares except:

> **W.** 09, 19,

Linton: 'Common. General'.

FRAGARIA L.

F. vesca L.   *Wild Strawberry*

Native. Woods, scrub and rough grassland on calcareous or slightly acid soil, often becoming very abundant in wood-clearings. Common and widespread.

Recorded for all squares except:

> **W.** 09, 19.
> **E.** 46.
> **S.** 33, 21.

Linton: 'Generally distributed'.

F. moschata Duchesne (*F. elatior* Ehrh.)   *Hautbois Strawberry*

Introduced. An escape from cultivation, formerly established in a few localities but no recent records.

W. Grin Plantation, Buxton 0472, Ashwood Dale 0672, below Pig of Lead, Bonsall 2857.

E. West Hallam Station, by allotments 4242. [*F. ananassa*]

Linton: Nr. Haddon 26, Edgehill Wood, Duffield 34, Green Lane, Ockbrook 43, Egginton Station 22.

GEUM L.

G. urbanum L.   *Wood Avens, Herb Bennet*

Native. Woods, thickets, hedgebanks and shady places on calcareous or good non-calcareous soils but not on strongly acid or water-logged soils. Common and widespread.

Recorded for all squares except 19.

Linton: 'Common'.

G. rivale L.   *Water Avens*

Native. Streamsides, marshes, moist valley-bottom pastures and damp woods on base-rich soils. Common and widespread except on the gritstone moors.

W. Recorded for all squares except 09, 19, 28.

E. Freebirch 3072, Scarcliffe Park Wood 5270, Whitwell Woods 5278, Ault Hucknall 4665. Langwith Wood 5068.

S. Ashbourne 1746, Bradley 2245, recorded for 43.

Linton: Killamarsh 48, Breadsall 33, Calke 32, Repton 32.

G. x intermedium Ehrh. (*G. urbanum* x *G. rivale*)

Native. Hybrid swarms of intermediates between G. *urbanum* and G. *rivale* usually with both parent species, are not uncommon, especially in woods in the limestone dales.

W. Grin Plantation 0472, Miller's Dale 1373, Monk's Dale 1375, Hassop 2170, Lathkill Dale 1865, above Conksbury 2066, Parwich 1754.

E. Ault Hucknall 4665.

S. Spend Lane nr. Ashbourne 1748, Atlow 2249.

Linton: Nr. Calver 27, nr. Carsington 25, Osmaston Park 24.

DRYAS L.

D. octopetala L.   *Mountain Avens*

Doubtfully native. early last century, but no record since then.

No recent records.

Linton: Ashwood Dale 07.

AGRIMONIA L.

A. eupatoria L.   *Common Agrimony*

Native. Wood-margins, hedgebanks, waysides, field borders and banks on a wide range of calcareous and non-calcareous soils, but avoiding strongly acid soils. Frequent.

Recorded for all squares except:

W. 09, 19, 18, 28.
E. 45, 38.

Linton: 'Frequent'.

A. procera Wallr. ('*A. odorata* Miller')   *Fragrant Agrimony*

Native. Wood-margins, roadsides, banks, etc.; usually on light soils which are neither calcareous nor strongly acid. Local and rather scarce.

W. High Dale 1571, Monsal Dale 1771, Coombs Dale 2274, Froggatt Edge 2476, nr. Middleton 1762, Alport 2264, also Over Haddon 2066, nr. Rowsley 2566, Beeley 2667.

E. Totley Bents 3080, Ashgate nr. Chesterfield 3571.

S. Breadsall 3937, nr. Egginton 2727, Castle Donington 4025.

Linton: Mellor 99, Duckmanton 47, Ladyhole Farm 24.

ALCHEMILLA L.

A. vulgaris agg. ('*A. vulgaris* L.')   *Lady's Mantle, Bearsfoot*

Native. Pastures, meadows and waysides. Generally distributed.

Recorded for all squares except: **E.** 48.

Linton: 'Common'.

Three species of the aggregate occur in Derbyshire.

A. filicaulis Buser, *sensu lato* ('*A. vulgaris* var. *filicaulis* (Buser)'), is represented in the county by its two subspecies:

Ssp. **vestita** (Buser) M. E. Bradshaw, with stems, leaf-stalks and flower-stalks densely covered with spreading hairs, is the commonest of the Derbyshire Lady's Mantles found in pastures and short grassland throughout the county.

W. Recorded for all squares except 09. 19, 08, 28.

E. Troway 3880, Linacre 3472, Brockwell Tip 3771, Whitwell Wood 5278, Pleasley Vale 5165, Alfreton Brook 4056.

S. Ashbourne 1747, Sturston 2047, Cubley Covert 1539, Kirk Langley 2939, Dale 4438, Cauldwell 2517.

Ssp. **filicaulis**, like ssp. *vestita* but with the upper half of the stem and the flower-stalks glabrous, has been recorded only from the limestone dales.

W. Taddington Dale 1769, Lathkill Dale 1865, Monsal Dale 17.

The two subspecies are not always readily distinguishable and more careful search is needed.

Linton: Did not distinguish the two subspecies.

A. xanthochlora Rothm. ('*A. vulgaris* var. *pratensis* (Schmidt)')

Roadsides, meadows, banks. Frequent, especially in the north and west of the county.

W. Recorded for all squares except 19.

E. Beauchief 3381, Ford nr. Eckington 4081, Press Reservoirs 3565, Alfreton Brook 4056.

S. Ashbourne 1747, Sturston 2146.

Linton: Hood Brook, Hathersage 28, Great Shacklow Wood 26.

A. glabra Neygenfind ('*A. vulgaris* var. *alpestris* (Schmidt)')

Roadsides and rough grassland, chiefly in the north-west of the county and less frequent than *A. xanthochlora.* See also map (page 90).

W. Recorded for all squares except 09, 19.

E. Linacre 3372, Whitwell Wood 5278.

S. Offcote 1948, nr. Bradley Dam 2245.

Linton: Repton 32.

APHANES L.

A. arvensis L. (*Alchemilla arvensis* (L.) Scop.)   *Parsley Piert*

Native. Bare places on light dry soils, calcareous or non-calcareous; locally abundant about rock outcrops in the limestone dales; elsewhere a weed of arable land and sandy places. Common.

W. Recorded for all squares except 09, 19, 08, 28.

E. Creswell Crags 5274, Whitwell Woods 5278, Stonedge 3467, Astwith 4464, Pleasley Vale 5165, nr. Little Eaton 3642, Stanley 4142.

S. Edlaston 1842, Shirley Wood 2042, Mickleover 3034, recorded for 43, Repton Rocks 3222, recorded for 21.

Linton: 'Common. Generally distributed'.

A. microcarpa (Boiss. & Reuter) Rothm.   *Parsley Piert*

Native. In similar situations to *A. arvensis* but apparently restricted to acid soils. Rare, but probably under-recorded.

W. Ladmanlow Tip 0471.

S. Grange Wood, Lullington 2714.

Linton: Not recorded.

SANGUISORBA L.

S. officinalis L. (*Poterium officinale* (L.) A. Gray)   *Great Burnet*

Native. Damp meadows and moist grassy slopes. Frequent.

Recorded for all squares except:

> **W.** 09, 19, 26.
> **E.** 36, 35.
> **S.** 43.

Linton: Bakewell 26, Whatstandwell and Cromford Canal 35, Dale and Sandiacre 43.

POTERIUM L.

P. sanguisorba L.   *Salad Burnet*

Native. Grassland on calcareous or slightly acid soils. Common on the limestone, rare elsewhere.

Observations on plateau-loams of the Carboniferous Limestone suggest that successful establishment from seed is unusual on non-calcareous soils, though established plants may persist after the surface-soil has become leached.

W. Recorded for all squares except 09, 19.

E. Killamarsh 4581, Markland Grips 5074, Fallgate Quarries 3562, Pleasley Vale 5265, Crich Stand 3455, West Hallam 4142.

S. On the limestone outlier, Birchwood Quarry 1541, Trent 4932, Sawley 4731, Bretby Castle Field 2923, Swarkestone 3628, recorded for 21, Weston-on Trent 4128.

Linton: Bolsover 47, Palterton 46, Breadsall 33, nr. Thrumpton Ferry 53.

P. polygamum Waldst. & Kit.

Introduced. Formerly grown for fodder and occasionally escaping as a casual. Very rare.

E. Unthank 3075.

Linton: Mickleover 33.

ROSA L.

R. arvensis Hudson   *Field Rose*

Native. Wood clearings and wood margins, hedges, wayside scrub, etc.
Common and widespread.

Recorded for all squares except:

> **W.** 09, 19.
> **E.** 45.
> **S.** 32.

Linton: Mellor 99, nr. The Snake Inn 19.

R. pimpinellifolia L.   *Burnet Rose*

Native. A colonist of limestone scree, whether shaded or in the open; heaths
and hedges. Locally frequent on the limestone, rare elsewhere.

W. Horseshoe Dale 0970, Cressbrook Dale 1773, Monk's Dale 1374, Monsal
Dale 1771, Middleton Dale 2175, nr. Bonsall 2758.

Linton: Chapel 08, Rough Heanor 33, Dale 43.

R. canina group   *Dog Rose*

The Canina roses are very difficult taxonomically, and there has unfortunately
been no recent study of the Derbyshire forms. Most of our records are of
R. canina L. and *R. dumalis* Bechst. (incl. *R. glauca* Fries), but Linton lists also
some for *R. obtusifolia* Desv.

R. canina L.

Native. Woods, hedges, scrub, waysides, etc. Common and widespread.
Recorded for all squares.

This includes the common forms with leaflets glabrous on both sides or
pubescent beneath and eglandular or nearly so, with sepals reflexed after
flowering and falling before the fruit ripens and with a wide disc, so that the
fruit has a narrow aperture.

R. dumalis Bechst. (incl. *R. glauca* Fries)

Native. Woods, hedges, scrub. Occasional.

W. Kniveton 2050.

S. Thorpe 1549, Snipesmoor Lane, Ashbourne (now destroyed) 1945,
Rough Lane, Yeldersley 2243, also Bradley and Hulland in 24.

This includes forms differing from *R. canina* L. in having usually smaller
prickles, sepals erect after flowering and persistent at least until the fruit
reddens, and a narrow disc completely hidden by the flat villous head of
stigmas; the aperture of the fruit is wider than in *R. canina*, usually more than
1 mm. in diameter.

Linton lists localities for both *R. canina* var. *dumalis* and *R. glauca* Vill. as
well as for many other varieties and forms of *R. canina* and *R. glauca*.

R. obtusifolia Desv.
Native. Hedges. Not uncommon.
This includes Canina roses with strongly hooked spines, leaflets usually pubescent and glandular, sepals reflexed after flowering and soon falling, and fruit with a wide disc and narrow aperture.
No recent records, but should be looked for.
Linton lists several records for two varieties of *R. obtusifolia*.

R. villosa group   *Downy Rose*
Downy roses are not uncommon in Derbyshire, but as with the dog roses there is need for further study. The three British members of the group have all been recorded for the county.

R. tomentosa Sm.
Native. Woods, hedges, scrub. Occasional, but probably under-recorded.
W. Nr. Hathersage 2381, Middleton Dale 2175, Longstone Edge 2173.
Linton: 'Frequent'. Several localities are also given for var. *scabriuscula* Sm.

R. sherardii Davies
Native. Woods, hedges, scrub. Rare, but perhaps under-recorded.
W. Elton 2261.
Linton: Not recorded.

R. villosa L. (*R. mollis* Sm.)
Native. Woods, hedges, scrub. Not uncommon.
W. Recorded for 08, Cunning Dale 0872, Miller's Dale 1573, Cressbrook Dale and Monk's Dale 17, Bretton Clough 2078, recorded for 16, Lathkill Dale 2066, Dovedale 1453, Brassington 2154, Via Gellia 2556.
E. Linacre 3372, Cartledge 3277, Railway Mapperley 4342.
S. Thorpe Rough 1549, nr. Osmaston Fields Farm 1844, Kingsgrove, Mercaston 2542.
Linton: Bradwell, Ashopton, and nr. Hope 18, Mickleover 33, Calke 32

R. rubiginosa group   *Sweet-briar*
Sweet-briars are not common in Derbyshire, but there are records for both of the widespread British species.

R. rubiginosa L.
Native. Scrub, chiefly on limestone. Rare.
W. Recorded for 17, 27.
Linton: Snelston 14, nr. Mickleover 33, Spondon 43, Breadsall 34, Repton 32.

R. micrantha Borrer ex Sm.

Native. Hedges, heaths.

No recent records.

Linton: 'Not uncommon'.

PRUNUS L.

P. spinosa L.   *Blackthorn, Sloe*

Native. Woods, scrub, hedges, wayside thickets, etc., on a wide variety of soils. Common and widespread, locally suckering freely to form extensive thickets.

Linton states that blackthorn fruits well only in exceptional seasons.

Recorded for all squares except:

W. 09. 19, 08.

Linton: 'Common. Generally distributed'.

P. domestica L.   *Plum, Bullace, Greengage*

The numerous forms may be grouped into two subspecies:

Ssp. **domestica**   *Plum*

The true plum is a small tree without thorns and with the first-year shoots sparsely hairy or glabrous. The fruit is 4-7.5 cm. long with a flattened and sharply keeled stone.

Introduced. Naturalized in hedges nr. houses. Rare.

W. Roadside hedge, Carsington 2452.

S. Hedges and scrubby banks about Hulland 2446, etc., Repton, possibly an escape 3026.

Linton: Staveley 47, Heanor 44, Humbleton Farm, Derby 33, Ockbrook 43.

Ssp. **insititia** (L.) C. K. Schneider   *Bullace and Greengage*

The bullace is usually a somewhat thorny shrub with densely pubescent first-year shoots. The fruit is small, commonly 2–3 cm. long, purple or yellow and with the stone only slightly flattened and hardly keeled. The greengage is a small non-thorny tree with green fruit 3–5 cm. long. Its first-year shoots are finely hairy.

Probably introduced but more thoroughly established than ssp. *domestica*. Hedges and copses. Frequent.

W. Hathersage 2282, Washgate Valley 0567, Darley Bridge 2762, Parwich Hill 1854.

E. Cordwell Valley 3176, Dronfield Woodhouse 3477, old orchard, Creswell Crags 57, nr. Stretton 3760, Ambergate 3452, Windley 3045, Morley Lime 3940, nr. Denby Common 4046, between Stanley and Morley 4040.

S.   Atlow 2348, also Brailsford, Osmaston and Shirley in 24, roadside, Yeaveley 1839, in quantity in a wood at Kirk Langley 2939, Littleover 3333, also Mackworth and Breadsall in 33, Swarkestone 3728.

Linton: Edale, Pindale and Shatton 18, Via Gellia 25, Renishaw and Calow 47, Palterton 46, Pinxton 45, Drakelow 22.

P. cerasifera Ehrh.   *Cherry Plum*

Introduced. Often planted in hedges and sometimes found a long way from houses.

W.   Hathersage 2282, Great Longstone 2071, Bradway 3280, Dronfield 3377.

S.   Hedges north-east of Ashbourne, notably on Boothby Farm 1847, nr. gardens and orchards at Norbury 1242, etc., and also at Osmaston 2044.

Linton: Not recorded.

P. avium (L.) L.   *Gean, Wild Cherry*

Native. Woods and hedges on fertile loamy soils. Frequent, but rarely on limestone.

W.   Recorded for 08, Hood Brook, Hathersage 2382, Chee Dale 1273, Chatsworth 2670, Middleton Dale 2275, Bakewell 2169, Fenny Bentley 1750, Longcliffe 2154.

E.   Delves Wood, Povey 3880, Ford Valley 4080, Chanderhill 3370, above Creswell Crags 57, Wingerworth Great Pond 3667, Pleasley Vale 5265, Breadsall Moor 3742, nr. Duffield Church 3442, Stanley 4240, Mapperley 4343.

S.   Ashbourne Green 1947, Bradley 2246, Cubley 1638, Longford 2137, Locko 4038, Bretby 2922, Repton Rocks 3221, Overseal 2915, Grange Wood 2714, recorded for 31.

Linton: Mellor 98, Pinxton 45, Derby. 33.

P. cerasus L.   *Dwarf Cherry, Sour Cherry*

Probably introduced. Hedges. Local and rather rare.

Usually a shrub rather than a small tree, and probably always descended from cultivated Sour and Morello Cherries.

W.   Hathersage 2282.

S.   By main road, Osmaston 1944, Ladyhole Lane 2044, about Mercaston and Brailsford 2541/2, Hulland Ward 2645, Ingleby 3426.

Linton: Etherow district 09, between Hassop and Baslow 27, Clowne 47, Glapwell 46, Mackworth 33.

P. padus L.   *Bird Cherry*

Native. Woods and scrub, chiefly on Carboniferous Limestone and Edale Shales and commonly in valley-bottoms or on slopes facing north to west; occasionally on limestone cliffs. Local. See also map (page 90).

W.   Recorded for all squares except 09, 19.

E.   Cordwell Valley 3176, Totley Bank Wood 38, Upper Langwith 5169.

S.   Bradley 2245, Atlow Top 2348, about Mercaston 2543, etc., Cubley 1737, Peathayes 2139.

Linton: Stirrup 99, Alport Dale 19, Whatstandwell 35, Wyaston Brook 14, between Repton and Burton 22.

COTONEASTER Medicus

C. microphyllus Wall. ex Lindl.

Introduced. Commonly grown in gardens and occasionally naturalized, especially on limestone. Rare.

W.   Railway cutting north of Hartington 26, rocks nr. Longcliffe 2356.

Linton: Not recorded.

CRATAEGUS L.

C. laevigata (Poiret) DC. (*C. oxyacanthoides* Thuill.)   *Hawthorn*

Probably native in the extreme south of the county, introduced elsewhere. Woods, occasionally in hedges and then likely to have been planted. Local and rare.

W.   Hathersage 2381, Stand Wood 2670.

E.   Nr. Troway 3979, Quarndon 3241.

S.   Ashbourne Green 1847, Snelston 1644.   Intermediate plants are fairly frequent in this district.

Linton: Chapel 08, Buxworth 08, Cressbrook Dale 17, Wardlow Hay Cop 17, Miller's Dale 17, Renishaw 47, Blackwall 24, Yeldersley Lane 24, Doveridge 13.

C. monogyna Jacq. (*C. oxyacantha* L., in part)   *Hawthorn*

Native. Woods, scrub, hedges, abandoned fields, cliff-ledges, etc. Common and widespread.

Recorded for all squares.

Linton: 'Common'.

SORBUS L.

S. aucuparia L. (*Pyrus aucuparia* (L.) Ehrh.)   *Rowan, Mountain Ash*

Native. Woods, hedges, scrub, streamsides, mountain rocks and cliff-ledges, on a very wide range of soil types and rocks, from calcareous to strongly acid and heavily podsolized soils, and ascending to over 1,500 ft. Frequent.

Recorded for all squares except:

S.   33, 43.

Linton: 'Frequent'.

S. domestica L. (*Pyrus domestica* (L.) Ehrh.)   *Service Tree*
Introduced. Woods and plantations. Rare.
No recent records.
Linton: Crich Woods 35.

S. aria (L.) Crantz (*Pyrus aria* (L.) Ehrh.)   *Whitebeam*
Probably native in the limestone dales, introduced elsewhere. Rare.
W. Napkin Piece 9987, Snake Inn 1190, recorded for 08, Buxton 07, Chee
Dale 1273, King Sterndale 0972, Big Moor 2876, nr. Longnor 0964, Flagg
1368, Lathkill Dale 1966, Dovedale 15.
E. Barlborough 4777, Duffield 3443.
S. Wyaston 1941, Bradley Dam 2245, Bretby 2922.
Linton: Matlock, Osmaston Park 24.

S. rupicola (Syme) Hedlund (*Pyrus aria* var. *rupicola* Syme)
Native. Cliff-tops and cliff-ledges of the Carboniferous Limestone, rarely
elsewhere. Local.
W. Cave Dale 1482, Ashwood Dale 0772, Cressbrook Dale 1773, Miller's
Dale 1773, Monsal Dale 1772, Chee Dale 1273, Litton Mill 1772, Dovedale
1453, Matlock High Tor 2958.
Linton: Drakelow 22, Foremark 32.

S. torminalis (L.) Crantz (*Pyrus torminalis* (L.) Ehrh.)   *Wild Service Tree*
Very doubtfully native. Woods and plantations. Rare, and never beyond
suspicion of having been planted.
No recent records.
Linton: Repton 32.

PYRUS L.

P. pyraster Burgsd. ('*P. communis* L.')   *Wild Pear*
Almost certainly introduced. Very rare.
Cultivated pears belong to this species.
E. Nr. Ambergate 3551.
S. Two old trees, probably relic of orchard, Snelston 1642.
Linton: Matlock 35, Chesterfield 37, Morley Lime 34, Anchor Church 32.

MALUS Miller

M. sylvestris Miller (*Pyrus malus* L.)   *Crab Apple*
Native or sometimes introduced. Woods, hedges and scrub.
Recorded for all squares except:
W. 09, 19, 08, 16, 26.
Linton: 'Common'.

There are two subspecies in Derbyshire:

Ssp. **sylvestris** (*M. acerba* Mérat), usually thorny, with leaves soon becoming glabrous, flower-stalks and calyx glabrous or nearly so and fruit small and sour. Common everywhere.

Ssp. **mitis** (Wallr.) Mansfeld (*M. domestica* Borkh.), usually thornless, with leaves persistently pubescent beneath, flower-stalks and calyx tomentose and fruit large and often sweet. Descendants of cultivated apples. Occasional.

S.   Occurs on railway banks nr. Ashbourne 1746/7.

CRASSULACEAE
SEDUM L.

S. telephium L.   *Orpine, Live-long*
Native. Woods and hedgebanks, chiefly on Carboniferous Limestone.
W. Cave Dale 1482, Cow Dale 0872, Fairfield 0875, Tideswell Dale 1573, High Dale, Brushfield 1571, Ravensdale 1773, Coombs Dale 2274, Glutton Dale 0867, Lathkill Dale 1765, 2066, Buxton-Ashbourne disused railway track 16/15, Beresford Dale 1456, Dovedale 1455, Brassington Rocks 2154.
E.   Ashover 3463, Milltown 3561.
S.   Cubley Common 1639.
Linton: Chapel-en-le-Frith 08, Dethick 35.

S. dasyphyllum L.   *Thick-leaved Stonecrop*
?Introduced. On old walls. Rare.
No recent records.
Linton: Dethick 35, Pinxton 45.

S. anglicum Huds.   *English Stonecrop*
Native. Rocks and dry grassy banks. Rare.
W. Middleton Dale 2275, rocks nr. Calver 2374, Via Gellia 2556.
Linton: Nr. Chatsworth 27.

S. album L.   *White Stonecrop*
Introduced and naturalized. Walls, rocks, roofs, quarries. Occasional.
W. Recorded for all squares except 09, 19, 28.
E.   Wadshelf 3170, Kelstedge 3363, recorded for 35.
S.   Kirk Langley 2838.
Linton: Milton 32.

S. lydium Boiss.
Introduced.
W. Alport 2164.
Linton: Not recorded.

S. acre L.   *Wall-pepper, Biting Stonecrop*

Native. Dry grassy slopes, scree, cliff-ledges, rocks, walls, roofs, especially on limestone. Locally abundant.

W. Recorded for all squares except 09, 19.

E. Wadshelf 3170, Milltown 3561, Kelstedge 3463, Hazelwood 3244, West Hallam 4142.

S. Brailsford 2541, Kirk Langley 2838, Allestree 3439, Ticknall 3524.

Linton: Creswell Crags 57, South Normanton 45, Ashbourne 14, Makeney 34, Sandiacre 43.

S. forsteranum Sm. (*S. rupestre* L., in part)   *Rock Stonecrop*

Introduced. Formerly naturalized in a few places, but not seen for many years.

No recent records.

Linton: Castleton 18, Bradley Wood 24, Linton 21.

S. reflexum L.

Introduced and naturalized. Old walls and roofs. Rarely seen in recent years.

S. Repton 3027.

Linton: Buxton 07, Froggatt 27, Matlock 35, Edlaston 14, Horsley Castle 34, Dale Abbey 43.

SEMPERVIVUM L.

S. tectorum L.   *House Leek*

Introduced but hardly naturalized. Walls and cottage-roofs. Rare.

W. New Smithy 0582, Miller's Dale 1773, Over Haddon 2066.

E. Morley Almshouses 3941, Stanley 4140.

S. Ashbourne 1846, Yeaveley 1840, Breadsall 3639.

Linton: Rodsley 24, Mickleover 33, Repton and Bretby 32.

UMBILICUS DC.

U. rupestris (Salisb.) Dandy (*Cotyledon umbilicus* L.)   *Navelwort, Pennywort*

Native. Crevices of cliffs and walls; ruins. Rare.

W. Grindleford 2479, Froggatt Edge 2476.

S. Anchor Church 3327.

Linton: Bradwell 18, Chatsworth 27, Dovedale 15, Barlow 37, Anchor Church 32.

o

SAXIFRAGACEAE

SAXIFRAGA L.

S. umbrosa L.

Introduced. Limestone rocks. Rare.

W. Leadmill Bridge 2380, Sherriff Wood 2378.

S. Lodge Farm 1745, Yeldersley Brook 2144.

Linton: Ashwood Dale 07, Poole's Hole 07, Lathkill Dale 16.

S. spathularis x umbrosa   *London Pride*

Introduced. Commonly grown in gardens and sometimes escaping and becoming naturalized. Occasional.

W. Pictor Wood 0872, Ashwood Dale 0772, Sherriff Wood 2778.

Linton: Not recorded.

S. tridactylites L.   *Rue-leaved Saxifrage*

Native. Walls, rocks, roofs, dry places; common on limestone outcrops and cliff-ledges in the dales, where it is associated with several other annuals. Locally frequent.

W. Recorded for all squares except 09, 19, 08, 28.

E. Markland Grips 5074, Ashover 3463, Pleasley Vale 5265, Morley 3940.

S. Birchwood Quarry 1541, Bradley Churchyard 2245, Breadsall 4037, Ticknall 3524.

Linton: Chapel-en-le-Frith 08.

S. granulata L.   *Meadow Saxifrage*

Native. Pastures on calcareous or slightly acid soils, especially on moist slopes and banks; very abundant locally on loamy soils in the limestone dales.

W. Recorded for all squares except 09, 19.

E. Baslow 3572, Ashover 3462, nr. Duffield 3244, Stanley 4141.

S. Ashbourne 1646, Sawley 4832, Swarkestone 3628, Newton Solney 2725, Netherseal Churchyard 2812.

Linton: Hayfield 09, Charlesworth 98, Shirley 24, Doveridge 13, Mickleover 33.

S. hypnoides L.   *Mossy Saxifrage, Dovedale Moss*

Native. Limestone rocks and walls, chiefly in shade or on north-facing slopes. Locally frequent, but restricted to the limestone dales.

W. Pin Dale 1582, Cowdale 0872, Deep Dale 0972, Monk's Dale 1374, Cressbrook Dale 1773, Miller's Dale 1773, Middleton Dale 2175, Coombs Dale 2274, Earl Sterndale 0966, Lathkill Dale 1966, Bradford Dale 2063, Lode Mill 1454, Thorpe Cloud 1451, Hipley Hill 2054.

S. aizoides L.   *Yellow Mountain Saxifrage*
Possibly native in the past, but the evidence for this is rather slight.
No recent records.
Linton: 'Native, but probably extinct'. One record, Howard in Bot. Guide.

CHRYSOSPLENIUM L.

C. oppositifolium L.   *Golden Saxifrage*
Native. Stream-sides, springs, flushes, wet rocks and wet ground in woods;
commonly in the shade and tolerant of both calcareous and acid substrata.
Frequent.
Recorded for all squares except 45.

C. alternifolium L.   *Golden Saxifrage*
Native. In similar places to *C. oppositifolium* but more characteristic of moist
rocky woods on the limestone. Local but not infrequent.
W. Conies Dale 1280, Coombs 0278, Cowdale 0872, Ashwood Dale 0772,
Miller's Dale 1473, Chee Dale and Flag Dale 1273, Ashford 1869, Rowsley
2565, Woodeaves 1850, Bradbourne 2052, Griffe Grange 2556.
E. Kelstedge 3363.
S. Clifton 1544, Ashbourne 1946, Bradley Brook 2244, Atlow Rough
2247, Marston Montgomery 1237, Kirk Langley 2939.
Linton: Baslow 27, Dethick 35, Coxbench 34, Wyaston 14, Darley Abbey 33,
Knowle Hills and Repton 32.

PARNASSIACEAE
PARNASSIA L.
P. palustris L.   *Grass of Parnassus*
Native. Marshes, fens and moist grassy slopes; frequent on north-facing sides
of the limestone dales. Locally frequent.
W. Cave Dale 1482, Conies Dale 1280, Grin Plantation 0472, Harpur Hill,
Cunning Dale, Ashwood Dale in 07, Miller's Dale 1473, also Cressbrook,
Ravensdale, Tansley Dale, Monk's Dale, Chee Dale, Burfoot, Deep Dale,
Blackwell Mill and Wheston in 17, Calver New Bridge 2475, Brierlow 0969,
nr. Arborlow 1663, Dovedale 1454, Biggin Dale 1458, Grange Mill 2457,
nr. Slaley 2657, also Griffe Grange, Brassington Rocks and nr. Ballidon in 25.
E. Whitwell Wood 5278.
Linton: Hayfield 08, Pinxton 45, Copse Hill 14, Mugginton and Bradley 24,
Mackworth 33.

GROSSULARIACEAE
RIBES L.
R. rubrum L.   *Red Currant*
Native. Woods, hedges, river-banks. Not infrequent, but doubtless originat-
ing from gardens in some of its localities.

W. Old railway, Bamford 2083, on rocks, Lovers' Leap, Buxton 0772, between Baslow and Hassop 2372, by river, Alport 2264, nr. Lode Mill 1455, Tissington 1751, Brassington Rocks 2154.

E. Morley 3040, Allestree 3640, West Hallam 4340, Stanley 4240.

S. Riverside, Mapleton 1648, Fenny Bentley Brook 1646, Dale Moor 4538.

Sometimes white-fruited.

Linton: Mellor 98, Chee Tor to Wormhill 17, Pinxton 45, Radbourne 23, Mickleover, Breadsall and between Mackworth and Markeaton 33, Foremark Bottoms and Repton 32.

R. nigrum L.   *Black Currant*

Native. Damp woods and copses and shady stream-banks. Not uncommon.

W. Recorded for 07, Chee Dale 1172, also Blackwell Dale, Wormhill, Miller's Dale and Monsal Dale in 17, roadside, Newhaven 1660, Haddon 2266, Dovedale 15.

E. By spring, Gulley's Wood, Beauchief 3381, Lea Mills 3156, West Hallam 4340, Stanley 4240.

S. Fenny Bentley Brook 1646, also Henmore Brook and pondside, Tinkers' Inn, in 14, ditch nr. Locko Park 4238.

Linton: Stirrup Wood 99, Hood Brook 28, Baslow 27, Breadsall 33, Repton 32.

R. alpinum L.   *Mountain Currant*

Native. Limestone ravines and rocky woods; often hanging over vertical rocks like a curtain. Very local.

W. Cowdale 0872, King Sterndale 0971, Wormhill 1272, Flag Dale 1273, Miller's Dale 1573, Monk's Dale 1374, Ravensdale 1773, Eyam Dale 2176, Washgate Valley 0567, Hartington 1360, Lathkill Dale 1865, Bradford Dale 2063, Darley Bridge 2661, Dovedale and Milldale 1454, plantations at Newhaven 1659, and Tissington 1752, hedge nr. Kniveton Wood 2050.

S. Riverside, Thorpe 1549, streamside nr. Oxclose, Offcote 2047, roadside hedge, Somersal Herbert 1335.

This species is occasionally seen as a hedge-plant enclosing old cottage gardens, e.g. at Carsington, Kniveton and Hulland.

Linton: Derwent Chapel 18?, Creswell Crags 57.

R. sanguineum Pursh   *Flowering Currant*

Introduced. Much grown in gardens and occasionally escaping and establishing itself. Rare.

W. Old railway, Bamford 2083.

Linton: Not recorded.

R. uva-crispa L. (*R. grossularia* L.)   *Gooseberry*
Native, but often an escape from cultivation. Woods and hedges and shady
stream-banks. Not uncommon.
Recorded for all squares except:

> **W.** 09, 19, 08.
> **E.** 47, 46, 48, 38.
> **S.** 33, 22, 32, 23.

Linton: Burton 22, Calke, Foremark and Repton 32.

DROSERACEAE
DROSERA L.

D. rotundifolia L.   *Round-leaved Sundew*
Native. Bogs and wet peaty places on heaths and moors, commonly with
Sphagnum. Rather rare.

W. Upper Derwent 1696, Kinder Scout 08, Grindsbrook 1187, Shatton
Moor 1982, Jaggers Clough 1487/1587, Longshaw 2480, Stanage Edge 2284,
Hathersage 2581, Wild Moor 0274, Abney Moor 1979, Longshaw 2579,
Umberley Brook 2970. Raven Tor 2867, Relly Woods, Wirksworth 2853.

E. Hipper Sick 3068.

S. Hulland Moss 2546. (Not since 1951).

Linton: Moors nr. Dore 28, Tansley Common 36, Quarndon Common 33,
Foremark Park and Repton Rocks 32.

D. anglica Hudson ('*D. longifolia* L.')   *Great Sundew*
Native. Usually amongst *Sphagnum* in the wetter parts of bogs. Very rare
and perhaps now extinct.
No recent records.
Linton: Abney Moor 18, Moors nr. Buxton 07, East Moor 27.

LYTHRACEAE
LYTHRUM L.

L. salicaria L.   *Purple Loosestrife*
Native. Margins of ponds and slow streams and in marshes and fens. Frequent
at low altitudes in the east and south of the county.

W. Marsh at Woodeaves 1850, marsh below Thorpe Station 1650.

E. Recorded for 48, Whittington 3974, Linacre Top Reservoir 3272,
Renishaw Park 4478, Little Eaton canal, 3641, nr. Duffield Church 3442,
Belper 3447, Mapperley 4343, West Hallam 4341.

S. Recorded for all squares except 14.

Linton: Bakewell 26, Thrumpton Ferry 53.

PEPLIS L.

P. portula L.   *Water Purslane*

Native. Muddy margins of ponds and pools and in bare wet places; avoiding
calcareous substrata. Local.

W.   South side of Combs Lake 0379.

E.   Wingerworth 3867.

S.   Hulland Moss 2546, Willington 2928.

Linton: East of Killamarsh 48, Sutton Scarsdale 46, Morley Common 34,
between Rodsley and Yeaveley, and Tinkers' Inn 14, above Eaton Woods 13,
Repton 32.

THYMELAEACEAE

DAPHNE L.

D. mezereum L.   *Mezereon*

Native. Woods on calcareous soil and limestone scree. Very rare.

W.   Cressbrook Dale 1772, Lathkill Dale 1966, Griffe Grange Wood 2556,
Via Gellia 2857, Willersley Park 2957.

Linton: Chee Tor 17, thickets in Lathkill Dale 26, copse nr. Bakewell 26,
Matlock 25, Dove Valley 15.

D. laureola L.   *Spurge Laurel*

Native. Woods, thickets and hedges, chiefly on calcareous soils. Local and
rather rare.

W.   Cressbrook Dale 1773, Willersley 2957, Via Gellia below Bonsall 2857,
High Tor, Matlock Bath 2958.

E.   Markland Grips 5074, Whitwell Wood 5278, Pleasley 5265, 5164.

S.   Radbourne Park 2835.

Linton: Hardwick 46, Morley 34, Spondon Fields 43, Ockbrook 43, Foremark
32.

ONAGRACEAE

EPILOBIUM L.

E. hirsutum L.   *Great Hairy Willow-herb, Codlins & Cream*

Native. Streamsides, ditches, marshes, fens. Common and widespread.

Recorded for all squares except 19.

Linton: 'Generally distributed'.

E. hirsutum × montanum

W.   Chee Dale 1273.

Linton: Mytham Bridge, Bamford 18, Via Gellia Colour Works 25.

E. hirsutum × roseum
E.   Hasland 36 (Drabble, J. Bot. 1911).
Linton: Not recorded.

E. hirsutum × obscurum
E.   Tapton 37 (Drabble, J. Bot. 1911).
Linton: Not recorded.

E. parviflorum Schreber   *Lesser Hairy Willow-herb*
Native. Streamsides, ditches, marshes and fens. Common and widepread.
Recorded for all squares except:
>   **W.**   09, 19, 08, 07.
>   **E.**   34.

Linton: Buxworth 08, Ecclesbourne nr. Duffield 34.

E. montanum L.   *Broad-leaved Willow-herb*
Native. Woods on the more base-rich soils, hedgebanks, walls and rocks
and as a weed in gardens. Common and widespread.
Recorded for all squares.
Linton: 'Common'.

E. montanum × parviflorum
W.   Via Gellia 2757.
Linton: Chee Dale 17, nr. Hassop Station 27, Shirley 24.

E. montanum × obscurum
E.   Brockwell 37 (Drabble, J. Bot. 1911).
Linton: Nr. Hassop Station 27, boggy place below Bradley Wood and Edlaston
Coppy 14, Brailsford 24.

E. roseum Schreber   *Small-flowered Willow-herb*
Native. Damp roadsides, banks of streams and canals, ditches, shaded damp
places. Occasional.
W.   Cressbrook Dale 1774, Alport 2164, Thorpe 1651, Via Gellia 2857.
E.   Renishaw 4478, Pleasley Vale 5265, Moorwood Farm, Crich 3556,
West Hallam 4142.
S.   Fenny Bentley 1646, Clifton Station 1644, Ireton Wood 2847, Derby
3537, nr. Scropton 1829, Marston-on-Dove 2329.
Linton: Stoney Middleton 27, Froggatt 27, Oaker Hill 26, Barlow Brook 37,
Kelstedge 36, Little Eaton 34, Heanor 44, Cubley 13, Borrowash 43, Repton
32, Calke 32.

E. adenocaulon Hausskn.   *American Willow–herb*

Introduced. Streambanks, gardens, waste places, etc. Rare, but increasing. The first record for Great Britain was in 1891.

W.   Cressbrook Dale 1774.

E.   Ambergate 3551.

S.   Eaton 0936, Clifton, cinder tips 1644, Osmaston 2042, Etwall 2631.

Linton: Not recorded.

E. tetragonum L. (*E. adnatum* Griseb.)   *Square-stemmed Willow–herb*

Native. Sides of streams and ditches, damp hedgebanks, wet places. Rare.

E.   Matlock Moor 3062, Hardwick Ponds 4563, Coxbench 3743.

Linton: Mellor 99, Holymoorside 36, Breadsall 33.

E. obscurum Schreber   *Dull-leaved Willow–herb*

Native. Sides of streams and ditches, marshes, moist gardens and waste places. Common and widespread.

Recorded for all squares except:

W.   17, 15.
E.   34.
S.   33.

Linton: Miller's Dale 17, Fenny Bentley 15, nr. Duffield 34, Breadsall 33.

E. obscurum x **parviflorum**

No recent records.

Linton: Edale 18, Shatton 18, Brailsford, Bradley and Peat Hays, Shirley 24, Burnaston 23.

E. obscurum x **palustre**

No recent records.

Linton: Malcoff 08, east of Shirley Wood 24.

E. palustre L.   *Marsh Willow–herb*

Native. Marshes, acid fens, moorland flushes and boggy places by moorland streamlets; confined to acid substrata. Frequent.

Recorded for all squares except:

E.   45, 34.
S.   33.

Linton: Breadsall 33.

E. palustre x **parviflorum**

E.   Recorded for 57.

Linton: Nr. Robin Hood's Inn 27, nr. Bradley Brook 24.

E. nerterioides A. Cunningham

Introduced. Moist stony ground, gardens and waste places. Rare, but increasing.

W. Ladmanlow tip 0371, Thornbridge Hall 1970.

E. Garden weed in Ashover 36.

W. Garden weed in Ashbourne 1746.

Linton: Not recorded.

CHAMAENERION Seguier

C. angustifolium (L.) Scop. (*Epilobium angustifolium* L.)

Rose-bay Willow-herb

Native. Wood-margins and woodland clearings (especially after fire), disturbed ground and waste places, gardens; also cliff-ledges and scree slopes, especially on north-facing slopes. Locally very abundant and common everywhere except on very acid moorland soils.

Recorded for all squares.

Linton: 'Hedges, cars, woods, locally abundant'.

OENOTHERA L.

O. biennis L.
Evening Primrose

Introduced. Roadsides, railway banks, waste places. Locally frequent.

W. Via Gellia 2857, Coombs Dale 2374.

E. Donkey Race Course, Chesterfield 3672, Crich 3454, Allestree Ford 3540, Breadsall Moor 3741, railway banks, Mapperley 4342.

S. Clifton 1644, Derby 3436, ballast tip, Trent Lock 4932, Ticknall 3624, Netherseal Colliery 2715.

Linton: Breadsall Moor 34.

O. erythrosepala Borbás
Lamarck's Evening Primrose

Introduced. Waysides, banks and waste places. Rare.

E. Canal, Ambergate 3551.

Linton: Not recorded.

CIRCAEA L.

C. lutetiana L.
Enchanter's Nightshade

Native. Woods and shady places on moist base-rich soils. Locally frequent.

Recorded for all squares except 19, 08.

Linton: Frequent.

C. × intermedia Ehrh. (*C. lutetiana* × *C. alpina*)
Intermediate Enchanter's Nightshade

Native. Shady rocky places and damp woods. Rather rare. Formerly confused with *C. alpina*, which is not found in Derbyshire. A hybrid of *C. lutetiana* and *C. alpina* and rarely setting seed.

W. Chew Wood 9992.

E. Ecclesbourne Brook 3344.

S. Ashbourne 1746, Shirley Wood 2041, Boylestone 1835, Peathayes 2139, Derby 33.

Linton: Records under *C. alpina* probably all refer to this hybrid: Marple 98, Matlock Bath 25, Matlock 35.

HALORAGACEAE

MYRIOPHYLLUM L.

M. verticillatum L.   *Whorled Water-milfoil*

Native. Canals, ponds and slow streams. Rare.

S. Swarkestone 3628.

Linton: Canal nr. Killamarsh 48, backwater west of Sawley 43, Repton 32, Swarkestone Bridge 32.

M. spicatum L.   *Spiked Water-milfoil*

Native. Ponds, reservoirs, canals, streams and ditches, especially in calcareous water. Frequent.

W. Chatsworth 2669.

E. Canal, Killamarsh 4782, Pebley Pond 4879, Scarcliffe Park 5170, Pond, Hardwick Hall 4563, canal, Whatstandwell 3354, Ambergate 3551, canal, Ilkeston 4442.

S. Recorded for all squares.

Linton: Chapel Reservoir 08, Lathkill 26, reservoir at Cromford 25, Nether Langwith 57, Scarcliffe Park Wood 57, Pebley Pond 56, Codnor Castle 45, Mapperley 44.

M. alterniflorum DC.   *Alternate-flowered Water-milfoil*

Native. Pools, canals, streams and ditches, especially in acid or peaty water. Rare.

No recent records.

Linton: Pool at Kelstedge 36, canal between Renishaw and Staveley 47, Wingerworth 36, Sutton Scarsdale 45.

HIPPURIDACEAE

HIPPURIS L.

H. vulgaris L.   *Mare's-tail*

Native. Ponds, canals and slow streams, especially in base-rich water. Rather rare.

W. Lathkill Dale 1966, Tissington Dam 1751, Via Gellia 2957.

E. Canal, Norwood, Killamarsh 4782, The Walls, Whitwell 5078, Creswell Crags 5374, Markland Grips 5274, Scarcliffe Park Wood 5270, Hardwick Ponds 4563/4.

Linton: About Chatsworth 26/27, Melbourne, White Hollows, nr. Ticknall, Bretby Park, Swarkestone Bridge 32.

CALLITRICHACEAE

CALLITRICHE L.

C. stagnalis Scop.   *Common Star-wort*

Native. Shallow ponds, ditches, streams, etc., and on wet mud. Common and widespread.

Recorded for all squares except:

> **W.** 09.
> **E.** 38, 45, 34.

C. platycarpa Kütz. ('*C. verna* L.')

Native. Ponds, ditches, streams, etc., in lowland areas. Frequent; perhaps under-recorded owing to past confusion.

W. Lathkill Dale 2066.

E. Pebley Pond 4878.

S. Brailsford 2441, Mercaston 2643, Darley Abbey 3538, Marston-on-Dove 2329, Ticknall 3523.

Linton: Chapel-en-le-Frith 08, Burton 22.

C. obtusangula Le Gall

Native. Ponds, streams and ditches. Rare.

E. Hardwick Great Pond 4563.

Linton: River Derwent above Allestree 34, canal, Sandiacre 43.

C. hamulata Kütz.

Native. Ponds, ditches and slow streams. Infrequent.

W. Combs Reservoir 0379, Yorkshire Bridge 2085.

E. Pleasley Vale 5265.

Linton: Rushup Edge 18, Wild Moor 07, Middleton Dale 27, Chatsworth Park 26/27, Beauchief Abbey 38, Killamarsh 48, Westthorpe 47, canal, Chesterfield 37, Brailsford 24, Marston-on-Dove 22.

LORANTHACEAE

VISCUM L.

V. album L.   *Mistletoe*

Native, but also deliberately introduced. A hemi-parasite on a great variety of deciduous trees, most commonly apple; rarely on evergreen trees. Occasional.

E.   Rectory garden, Morley, and Morley Lane 3940, Stanley 4140.

S.   Clifton 1744 (now gone), Darley Moor and Cubley Common 1639, Kirk Langley 2739, Hartshorne 3221.

Linton: Chatsworth 26/27, Cromford 25, Beighton 48, Matlock 35, Pinxton 45, Allestree 33.

CORNACEAE

THELYCRANIA (Dumort.) Fourr.

T. sanguinea (L.) Fourr. (*Cornus sanguinea* L.)   *Dogwood*

Native. Woods, scrub and hedgerows on base-rich soils. Distributed throughout the county.

Recorded for all squares except:

> **W.**   09, 19, 28.
> **E.**   38, 37, 36, 45.
> **S.**   21.

Linton: Pinxton 45.

CORNUS L.

C. mas L.   *Cornelian Cherry*

Introduced and rarely naturalized.

E.   Cordwell Valley 3170.

Linton: Not recorded.

ARALIACEAE

HEDERA L.

H. helix L.   *Ivy*

Native. Climbing trees and hedgerow shrubs, rocks, cliff-faces and walls and creeping in woods; tolerant of deep shade but avoiding very acid, very dry or waterlogged substrata. Common except at high altitudes.

Recorded for all squares except:

> **W.**   19.

Linton: 'Common everywhere, except at higher levels'.

UMBELLIFERAE

HYDROCOTYLE L.

H. vulgaris L. *Marsh Pennywort*

Native. In damp or wet places, usually on non-calcareous soil. Rather local.

W. Recorded for 08, Derwent 1889, also Grindsbrook, Shatton and Win Hill in 18, Callow Bank, Hathersage 2582, also Bamford Moor and Higger Tor Lane in 28, Salter Sitch, Owler Bar 2977, also Leash Fen, Abney Clough, Haywood, Froggatt and Umberley Brook in 27.

E. Linacre 3272, Whitwell Wood 5278, also Markland Grips in 57, Wingerworth Pond 3667, also Tansley Moor in 36, Hardwick 4563, Wessington 3757, Morley Moor 3941, also Morley Lane in 34.

S. Lodge Farm, Ashbourne 1745, Hulland Moss 2546, also Mugginton, Mercaston, Spinnyford Brook and Shirley in 24, Breadsall 3339, Dale Abbey 4338, Swarkestone 3628, also Repton Rocks in 32.

Linton: Hollingworth Clough 09, above Buxton 07, Chee Tor 17, South Normanton 45, Burton 22.

SANICULA L.

S. europaea L. *Wood Sanicle*

Native. Woods and shady lanes.

Recorded for all squares except:

> **W.** 09, 19, 06, 16.
> **E.** 45.
> **S.** 33.

Linton: Stirrup 99, South Normanton and Newton Wood 45, Mickleover, Mackworth and Rough Heanor 33.

ASTRANTIA L.

A. major L.

Introduced. An occasional garden escape.

W. Yeld Wood, Baslow 2672.

E. Beauchief, garden weed 3381.

S. Railway bank nr. Breadsall 3839.

Linton: Not recorded.

CHAEROPHYLLUM L.

C. temulentum L. ('*C. temulum* L.') *Rough Chervil*

Native. Hedgebanks, waysides, etc. Frequent on moist base-rich soils.

Recorded for all squares except:

> **W.** 09, 19, 08.
> **E.** 38, 34.
> **S.** 33.

Linton: 'Almost everywhere common'.

ANTHRISCUS Pers.

A. caucalis Bieb. ('*A. vulgaris* Bernh.')   *Bur Chervil*
Native. Waysides, dry open places. Rare.
E.   Markland Grips 5074, Spital 3871.
S.   Nr. the Spread Eagle, Willington 2929, Milton 3226.
Linton: Trent banks, east of confluence with the Derwent 43.

A. sylvestris (L.) Hoffm.   *Cow Parsley, Keck, Queen Anne's Lace, Daily Bread*
Native. Hedgebanks, wood-margins, roadsides, moist limestone cliff-ledges.
Common and widespread in lowland areas but becoming less conspicuous on
the Millstone Grit, especially at high altitudes.
Recorded for all squares.
Linton: 'Common. General and abundant'.

A. cerefolium (L.) Hoffm.
Introduced and casual. Rare.
No recent records.
Linton: Burbage 28, Buxton 07.

SCANDIX L.

S. pecten-veneris L.   *Shepherd's Needle, Venus' Comb*
?Native. Arable fields on fertile soils. Rather uncommon.
W.   Hathersage Church 2381, recorded for 07, Miller's Dale Railway Station
1373, Bar Brook 2571.
E.   Carter Hall 3981, Markland Grips 5074, Whitwell Wood 5278, Glapwell
4966, Stony Houghton 4967, West Hallam 4142, Morley 4140.
S.   Lodge Farm, Ashbourne 1745, Longford 2038, Long Eaton 43.
Linton: Ridge Hall, Chapel 08, Norton 38, Renishaw 47, Pinxton 45, Pleasley
56, Yeldersley 24, Cubley 13, Hilton 23, Bretby 32, Cauldwell 21.

MYRRHIS Miller

M. odorata (L.) Scop.   *Sweet Cicely*
Native. Streamsides, moist waysides, field borders, etc., especially nr. villages.
Locally abundant on the Carboniferous Limestone and Grit-shales.
W.   Recorded for all squares except 09, 19, 18.
E.   Beauchief 3381, Barlow Woods 3575, nr. Stretton 3760, Pleasley Vale
5265, Ambergate 3451, Belper 3348, West Hallam 4142.
S.   Railway banks, Ashbourne 1747, Clifton 1541, Atlow 2348, Darley
Abbey 3538.
Linton: Mellor 09, Derwent Dale 19, Eckington 47, Codnor 44, Norbury 13.

TORILIS Adanson

T. japonica (Houtt.) DC. (*Caucalis anthriscus* (L.) Hudson)
Upright Hedge Parsley
Native. Hedge-banks, roadsides, field borders, etc., on fertile soils. Common at low altitudes.
Recorded for all squares except:

> **W.** 09, 19.
> **E.** 38, 45.

Linton: 'Common. Generally distributed and abundant'.

T. arvensis (Hudson) Link   *Spreading Hedge-parsley*
?Native. Arable fields and gardens in the south of the county. Rare. No recent records.
Linton: Dovedale 15, Shipley 44, Calke 32, Repton 32, Drakelow and Brizlincote 21.

T. nodosa (L.) Gaertner (*Caucalis nodosa* (L.) Scop.)   *Knotted Hedge-parsley*
Native. Dry, open banks and arable fields. Rare.
W. Recorded for 17.
S. Clifton, once only, from canary seed 1545.
Linton: Dovedale 15, Markland Grips 57, Breadsall 33, Repton 32.

CAUCALIS L.

C. platycarpos L. ('*C. daucoides* L.')   *Small Bur-parsley*
Introduced. A casual of waste places. Very rare.
S. Stores Road, Derby 3537.
Linton: Milton 32.

TURGENIA Hoffm.

T. latifolia (L.) Hoffm. (*Caucalis latifolia* L.)   *Great Bur-parsley*
Introduced. A casual of waste places. Very rare.
W. Gib Yard allotments, Buxton 0672.
S. Stores Road, Derby 3537.
Linton: Not recorded.

SMYRNIUM L.

S. olusatrum L.   *Alexanders*
Introduced and naturalized. Very rare.
W. Recorded for 18.
Linton: Not recorded.

CONIUM L.

C. maculatum L.   *Hemlock*

Native. Streamsides, ditch-banks, damp open woods, moist hedge-banks.
Locally frequent, especially in the south of the county. Very poisonous.

W. Buxton 0572, Hay Dale 1772, Monsal Dale 1871, Miller's Dale 1373,
Middleton Dale 2175, nr. Ashford 1869, Lathkill Dale nr. Alport 2164, Fenny
Bentley 1850, High Tor, Matlock 2959, Via Gellia 2556.

E. Eckington 4480, New Whittington 3974, Bolsover 4972, Ambergate
3551, Pye Hill 4452, Morley 3940, Stanley 4140.

S. Recorded for all squares.

BUPLEURUM L.

B. rotundifolium L.   *Hare's Ear*

Introduced. A casual weed of arable crops, not persisting. Rare.

S. Hanging Bridge 1545, Little Chester 3537.

Linton: Not recorded.

B. lancifolium Hornem.

Introduced. A casual, perhaps from bird-seed. Very rare.

S. Ockbrook 4236.

Linton: Not recorded.

B. subovatum Link

Introduced. A casual, seen on only one occasion.

S. Cubley 1638.

Linton: Not recorded.

APIUM L.

A. graveolens L.   *Wild Celery*

?Introduced. Wet places. Very rare.

W. Dimple, Matlock 2960.

Linton: Pinxton 45.

A. nodiflorum (L.) Lag.   *Marshwort, Fool's Watercress*

Native. Ditches, stream-margins, shallow ponds, wet places. Common.
Received for all squares except:

              W. 09, 19, 28.
              E.   38, 45.

Linton: 'Common'.

A. inundatum (L.) Reichenb. fil.

Native. Ponds and ditches, canal-sides, etc. Rare.

W. Chapel Reservoir 0379, Emperor Lake, Chatsworth 2771.

E. Pebley Pond, Barlborough 4879, Great Pond, Wingerworth 3667, nr. Ault Hucknall 4665, Nutbrook Canal, Ilkeston 4442.

S. Pool nr. Trent rifle range 4931, ponds by Swarkestone Bridge 3628, Overseal 2916.

Linton: Morley Moor 34.

PETROSELINUM Hill

P. crispum (Miller) Airy-Shaw (*Carum petroselinum* (L.) Bentham)   *Parsley*

Introduced. A casual in grassy waste places and on walls and rocks. Very rare.

E. Crich Quarry 3454.

Linton: Horsley Castle 34.

SISON L.

S. amomum L.   *Hedge Stonewort*

Native. Hedge-banks, roadsides. Rare, and only in the south of the county.

S. Stanton-by-Dale 4637.

Linton: Spondon 43, Calke 32, Lullington 21.

CARUM L.

C. carvi L.   *Caraway*

Introduced and casual. Roadsides. Very rare.

S. Stores Road, Derby 3537.

Linton: Ashby Road, Calke 32.

CONOPODIUM Koch

C. majus (Gouan) Loret (*C. denudatum* Koch)   *Pignut, Earthnut*

Native. Grassland, waysides and woods; on all except waterlogged or strongly podsolized soils. Common and widespread.

Recorded for all squares.

Linton: 'Abundant everywhere'.

PIMPINELLA L.

P. saxifraga L.   *Burnet Saxifrage*

Native. Dry grassy places, especially over limestone. Frequent, especially in the limestone dales.

Recorded for all squares except:

> **W.** 09, 19, 28.
> **E.** 45.
> **S.** 23, 43.

P

Linton: Etherow and Charlesworth 09, Newton Wood 45, Kirk Langley and Longford 23, Elvaston 43.

P. major (L.) Hudson   *Greater Burnet Saxifrage*
Native. Grassy places, wood-margins, hedge-banks. Frequent, especially over limestone.
Recorded for all squares except:

> **W.** 09, 19, 08, 28.
> **E.** 37, 45.
> **S.** 22, 32, 21.

Linton: Canal nr. Chesterfield 37, Alfreton 45, Burton and Willington 22, Calke, and between Melbourne and Kings Newton 32.

AEGOPODIUM L.

A. podagraria L.   *Goutweed, Bishop's Weed, Ground Elder*
Probably introduced. A persistent weed of cultivated ground and especially of gardens; also in hedgebanks and waste places and occasionally in valley-bottom woods. Usually near houses and probably an escape from former cultivation as a pot-herb. Locally abundant.
Recorded for all squares except:

> **W.** 09, 19.
> **E.** 48.

BERULA Koch

B. erecta (Hudson) Coville (*Sium erectum* Hudson)
                                              Narrow-leaved Water-parsnip
Native. Ditches, canals, rivers. Locally frequent; commonly associated with *Ranunculus penicillatus* in rivers of the limestone dales.
W. Recorded for 17, Middleton Dale 2175, Lathkill Dale 1865, Bradford Dale 2063, Dovedale (in quantity) 1551, etc.
E. Killamarsh 4681, Barlborough Park 4777, Renishaw Canal 4477, Scarcliffe Park Wood 5270, Pleasley Vale 5165, canal, Ambergate 3552.
S. River Dove 1546, Breadsall 3739, Sawley 4732, Trent Lock 4932, canal at Findern 3029, Weston-on-Trent 4128.
Linton: Brailsford Brook 24, Ecclesbourne Brook 34, Sudbury 13, Egginton 22, Drakelow Park 21.

OENANTHE L.

O. fistulosa L.   *Water Dropwort*
Native. Marshy places and shallow water. Local.
E. Morley Moor 3941.
S. Sawley 4732, Trent Lock 4931, Twyford 3228, Swarkestone Bridge 3727.
Linton: Sterndale 06, Renishaw 47, Pinxton 45, Sudbury/Scropton 13, Alvaston 33, canal, Egginton 22.

O. crocata L.   *Hemlock Water Dropwort*
Native. Ponds, ditches and wet places, usually avoiding highly calcareous water. Very rare.
S.   Canal nr. Repton 32.

O. aquatica (L.) Poiret (*O. phellandrium* Lam.)   *Water Horsebane*
Native. Slow-floating or stagnant water of rivers, canals, ditches and marshes. Local, and only in the Trent valley.
S.   Pools nr. Trent Station 4931, Willington Bridge 2928, ponds, Swarkestone Bridge 3727.
Linton: Trent Station 43, Swarkestone Bridge 32, Willington Bridge 22, Burton 22.

AETHUSA L.

A. cynapium L.   *Fools' Parsley*
Native. A weed of cultivated ground. Frequent.
Recorded for all squares except:
> **W.**   09, 19, 08, 28, 26.
> **E.**   45.
Linton: Chapel and Malcoff 08, Bamford 28.

FOENICULUM Miller

F. vulgare Miller   *Fennel*
Probably introduced. A casual of waste places. Rare.
W.   Calver New Bridge 2475, Bakewell 26.
E.   Creswell Colliery 5273.
S.   Ashbourne 1746, nr. Breadsall 3639, St. Mark's Churchyard, Derby 3637.
Linton: Spital, Chesterfield 37, Wingerworth 36.

SILAUM Miller

S. silaus (L.) Schinz & Thell. (*Silaus flavescens* Bernh.)   *Pepper Saxifrage*
Native. Meadows, grassy banks and roadsides. Occasional.
W.   Fenny Bentley 1750, Thorpe Station 1650.
E.   Whitwell Wood 5278, Morley 3940, Breadsall 3740, Smalley 4541, Mapperley 4443.
S.   Between Hollington and Longford 2239, Sawley 4732, Trent 4932.
Linton: Nr. Chatsworth 27, Atlow Win 24, Beauchief Abbey 38, Ault Hucknall 46, South Normanton 45, Bolsover 47, Swarkestone Bridge 32, Calke 32, Repton 32, brook above Egginton 22.

ANGELICA L.

A. sylvestris L.   *Wild Angelica*

Native. Woods, fens, stream-banks, ditches and damp meadows. Abundant.
Recorded for all squares except:

W. 09

Linton: Nr. Glossop 09.

PEUCEDANUM L.

P. ostruthium (L.) Koch   *Masterwort*

Introduced and occasionally escaping. Stream-banks and damp meadows.
Very rare.

W. Staden nr. Buxton 0772.

Linton: Axe Edge and nr. Buxton 06/07.

PASTINACA L.

P. sativa L. (*Peucedanum sativum* (L.) Bentham ex Hooker fil.)   *Wild Parsnip*

Native. Dry banks, roadsides, field-borders, especially on calcareous soils.
Uncommon.

W. Railway bank, Ashwood Dale 0872, Lathkill Dale 1966, Hoptonwood
Quarries 2655, Matlock Bath 2958.

E. Ambergate 3551.

S. Dale 4338, also Stanton-by-Dale, Trent and Sawley Junction in 43.

Linton: Mickleover, nr. Chaddesden and Breadsall 33, Calke 32.

HERACLEUM L.

H. sphondylium L.   *Cow Parsnip, Hogweed*

Native. Roadsides, meadows, hedge-banks and in woods. Common. Forms
with deeply laciniate leaves (var. *angustifolium* Hudson) have been recorded
from many localities.

Recorded for all squares.

Linton: Common.

Var. **angustifolium** Hudson

W. Ashwood Dale 0772, also Grin Plantation, Buxton 0472, Ravensdale
1773, also Miller's Dale in 17, Middleton Dale 2175, also Chatsworth in 27,
Bradford Dale 2063, also Lathkill Dale, Bakewell and Birchover in 26, nr.
Pike Hall 2159, also Kniveton, and Via Gellia in 25.

E. Old Brampton 3371, also Barlow and Linacre in 37, Ashover Hay 3560,
also Walton Wood and Stretton, Holymoorside in 36, Whatstandwell 3354,
also Brackenfield in 35.

Linton: Malcoff 08, Castleton 18, Scarcliffe Park Wood and Langwith Wood 57,
Newton Wood 45, Shottle 34, between Okeover and Ashbourne 14,
Idridgehay, Shirley and Bradley 24, between Longford and Boylestone 13,
Etwall 23.

H. mantegazzianum Sommier & Levier *Giant Hogweed*
Introduced. Commonly grown in gardens and occasionally establishing itself.

W. Roadside bank between Ashford-in-the-Water and Ashford Dale 1869.
Linton: Not recorded.

DAUCUS L.

D. carota L. *Wild Carrot*
Native. Rough pastures, grassy banks and roadsides, especially on calcareous soils. Local.

W. Shatton 1982, railway embankment, Ashwood Dale 0772, Peter Dale 1275, Parwich 1854, nr. Bonsall 2858, Dean Hollow 2855.

E. Bradway 3380, Linacre 3372, Pleasley Vale 5265, Boyah Grange 3944, West Hallam 4341.

S. Birchwood Quarry 1541, Kniveton 2147, Scropton 1930, Dale Abbey 4338, Breaston Station 4433, Sawley Junction 4732, Ticknall 3524, Cadley Hill 2719.

Linton: Mellor 98, Middleton Dale 27, Ashford 16, Dovedale 15, Matlock Bath 25, Holmesfield 38, Eckington 47, Bolsover 57, Ashover 36, Wigwell 35, Boylestone 13, Breadsall 33.

CUCURBITACEAE

BRYONIA L.

B. dioica Jacq. *White Bryony*
Native. Hedges, copses, scrub and rabbit-damaged grassland. Locally frequent at low altitudes in the south of the county but uncommon on the Carboniferous Limestone and absent from the Millstone Grit. See also map (page 90).

W. Miller's Dale and Monk's Dale 17.

E. Whitwell Wood 5279, Pleasley Vale 5165, Hazelwood 3244.

S. Recorded for all squares except 14, 24, 13.

Linton: South Normanton 45.

ARISTOLOCHIACEAE

ASARUM L.

A. europaeum L. *Asarabacca*
Probably introduced. Formerly much grown as a medicinal plant and occasionally seen as a garden plant. Woods and wood-margins. Very rare.

E. Allestree Park Lake 1963.

Linton: Not recorded.

EUPHORBIACEAE

MERCURIALIS L.

M. perennis L.   *Dog's Mercury*

Native. Woods and hedgerows on good mild-humus soils and especially abundant on calcareous soils; persisting in grassland after wood-clearance and also found in the open amongst shaded limestone rocks.

Recorded for all squares except:

W. 19, 08, 09.

Linton: 'Generally distributed'.

M. annua L.   *Annual Mercury*

Probably introduced. A casual in gardens and waste places. Very rare.

W. Hayes Leigh garden, West Road, Buxton 0473.

S. Clay Mills 2026.

Linton: Not recorded.

EUPHORBIA L.

E. lathyrus L.   *Caper Spurge*

Probably introduced. A casual of gardens and waste places. Very rare.

E. Chesterfield 37.

S. Ashbourne 1746, Clifton 1545, Mickleover 3234, Derby 3337.

E. helioscopia L.   *Sun Spurge*

Native. Cultivated ground, roadsides and waste places. Common.

W. Rowsley 2565, Tissington 1752, Mill Dale 1354, Brassington 2454.

E. Beauchief Abbey 3381, Brierley Wood 3775, Eckington 4379, Astwith 4464, Pleasley Vale 5265, Crich 3454, Little Eaton 3641, nr. Stanley 4040.

S. Ashbourne 1746, Mugginton 2943, Scropton 1930, Darley Abbey 3538, Repton 3123, Bretby 2722, Winshill 2623, Overseal 2918.

Linton: 'General throughout the County'.

E. peplus L.   *Petty Spurge*

Native. Cultivated ground, roadsides and waste places. Common and widespread.

Recorded for all squares except:

W. 09, 19, 18, 28, 16.

Linton: 'General and abundant'.

E. exigua L.   *Dwarf Spurge*

Native. Cultivated and waste ground. Uncommon, no recent records from the north.

E.   Millthorpe 3276, Brockwell 3772, Pleasley Vale 5265, railway nr. Black Rocks 3056.

S.   Ashbourne 1746, Brailsford 2544, Mickleover 3234, Trent 4931.

Linton: 'Common . . . In all the districts'.

E. esula L.

Introduced. Naturalized in waste places. Very rare.

This is the plant that has been named *E. uralensis* and is now thought to be a hybrid between *E. esula* ssp. *esula* and ssp. *virgata* (Waldst . & Kit.) A.R. Sm.

E.   Little Eaton 3641.

Linton: Not recorded.

E. amygdaloides L.   *Wood Spurge*

Native. Damp woods and thickets on light rich soils; hedgebanks. Very rare. No recent records.

Linton: Nr. Cromford 25, Chaddesden Common 33, Calke Park 32.

POLYGONACEAE

POLYGONUM L.

P. aviculare agg.   *Knotgrass*

Native. Cultivated ground, waste places. Common and widespread.

Two of the species of this aggregate certainly occur within the county, but their detailed distribution is not yet known:

P. aviculare L. (*P. heterophyllum* Lindm.) is a robust plant with leaves on the main stem much larger than those on flowering branches and with fruits having three more or less equal concave sides. It is characteristically a weed of arable fields and especially of cereal crops but occurs also on roadsides and waste places.

P. arenastrum Boreau (*P. aequale* Lindm.) usually forms a dense prostrate mat and has the leaves of main stem and flowering branches more or less equal in size. The fruits have two convex sides and one narrowly concave. It is a common weed of arable land but is less common in cereal crops than *P. aviculare* L. and is specially characteristic of trampled habitats such as roadsides, pavements, footpaths and waste places.

The aggregate has been recorded for all squares.

Linton: 'Abundant and generally distributed'.

P. bistorta L.   *Snake-root, Easter-ledges, Bistort*

Native. Damp meadows and pastures; streamsides. Chiefly on acid soils and in the northern moorlands, commonly near farms, presumably from being planted formerly as a pot-herb. Frequent.

Recorded for all squares except:
> **W.** 18, 28, 16.
> **E.** 45.
> **S.** 23, 22, 21.

Linton: South Normanton 45.

P. amplexicaule D. Don

Introduced. Occasionally naturalized. Very rare.

S.   A decreasing relic of the war-time aerodrome, on a hut site nr. Tinker's Inn, Ashbourne 1844.

Linton: Not recorded.

P. amphibium L.   *Water Persicaria*

Native. Pools, canals and slow-flowing rivers; the terrestrial form on banks by water. Frequent.

W.   Recorded for 08, Combs Reservoir 0379, Ashford 1969, nr. Rowsley 2565, Thorpe 1550, Bonsall Moor 2559.

E.   Recorded for all squares except 45, 38, 48.

S.   Recorded for all squares except 43, 32, 21.

Linton: Dinting 09, Bradwell and Hope 18, Pinxton 45, Ockbrook and Little Eaton 43, Thrumpton Ferry 53.

P. persicaria L.   *Red Hen, Persicaria*

Native. Cultivated land, waste places and by ponds. Common and widespread, especially as a field weed.

Recorded for all squares except:
> **W** 19.

Linton: 'Common. Generally distributed'.

P. lapathifolium L. (incl. *P. nodosum* Pers.)   *Pale Persicaria*

Native. Cultivated and disturbed ground, waste places, rubbish heaps, etc. Frequent, especially as a field weed.

Very variable in hairiness of leaves, presence or absence of red spots on the stem and of yellow glands beneath the leaves, and in flower colour.

W.   Recorded for 08, Yorkshire Bridge 1985, nr. King Sterndale 0871, Eyam 2077, recorded for 26, Kniveton 2050.

E.   Beauchief Abbey 3381, Linacre top reservoir 3272, Brierley Wood, Sheepbridge 3775, Wingerworth Great Pond 3667, Pleasley Vale 5265, recorded for 35, Longwood Hall, Pinxton 4454, Mapperley 4344.

S.   Recorded for all squares except 43.

P. nodosum Pers., recorded for 'Waste ground, Siddals Road, Derby 3536', falls within the normal range of variation of this extremely variable species and cannot be maintained as a separate species.

Linton: 'Common. Generally distributed'.

P. hydropiper L. *Water Pepper*

Native. In shallow water in ponds and ditches and in wet places. Frequent, except on the limestone.

W. Watford Bridge 0086, Yorkshire Bridge 1985, Combs Reservoir 0479, Calver New Bridge 2475, Bradbourne 1952, Bonsall Moor 2559.

E. Recorded for all squares except 45.

S. Recorded for all squares except 43.

Linton: 'Common. In all the districts'.

P. mite Schrank

Native. In ditches and by ponds and rivers. Rare.

E. Canal nr. Chesterfield 3871.

S. Nr. Anchor Church 3327, by the River Mease 2813 and 2711.

Linton: Not recorded.

P. minus Hudson

Native. By rivers and ponds and in damp places; avoiding calcareous substrata. Very rare.

S. Nr. Repton 3026.

Linton: Burton and Drakelow 22, Walton 21.

P. convolvulus L. *Black Bindweed*

Native. A climbing or scrambling weed of cultivated ground and waste places. Common and widespread.

W. Hathersage 2282, Calver New Bridge 2475, Hartington 1260, Bakewell 2168, Kniveton 2050.

E. Beauchief 3381, Holmesfield 3376, Mosbrough 4379, recorded for 36, nr. Ault Hucknall 4464, Fritchley 3552, Pinxton 4454, Morley 4040.

S. Recorded for all squares.

Linton: 'Common. Generally distributed'.

Var. subalatum Lejeune & Court.

E. Beauchief 3381.

S. Ashbourne 1847.

Linton: Between the Knob and Hulland 24.

P. cuspidatum Sieber & Zucc.

Introduced. Roads, railway banks, gardens and waste places; a persistent escape from cultivation. Frequent.

W. Near Charlesworth 0194, Hathersage 2380, Grindleford 2578.

E. Ladies' Spring Wood, Beauchief 3281, canalside, Eckington 4480, Belper 3447.

S. Roadside, Hulland 2446, very common in and around Derby 3537, Recorded for 32.

Linton: Not recorded.

FAGOPYRUM Miller

F. esculentum Moench   *Buckwheat*

Introduced. A casual of waste ground; an escape from cultivation. Occasional.

W. Coombs Dale 2373.

S. Roadside, Ashbourne, one plant 1846, Bretby 2922.

Linton: Buxton 07, Chesterfield 37, Breadsall Moor 34.

RUMEX L.

R. acetosella agg.   *Sheep's Sorrel*

Native. Bare places in short grassland, heaths, banks, cultivated ground and waste places; on all but the poorest soils but infrequent on calcareous soil. Common and widespread.

The only species of the aggregate certainly recorded for Derbyshire is *R. acetosella* L.

Recorded for all squares except:

W. 19.
E. 45, 56.

Linton: 'General through the County'.

R. acetosa L.   *Sorrel*

Native. Meadows, pastures and waysides, and in open places in woods. Common. The Derbyshire plant is the widespread ssp. *acetosa*. Recorded for all squares.

Linton: 'Common. Generally distributed'.

R. hydrolapathum Hudson   *Great Water Dock*

Native. In shallow water of streams, canals and ponds and in marshes. Locally frequent at low altitudes.

E. Renishaw Park 4378, also by the Chesterfield canal at Staveley, and along the canal from Chesterfield to Killamarsh in 47.

S. Bradley Dam 2245, Osmaston Ponds 2042, Shirley Millpond 2140, canal, Derby to Borrowash 3735, Sawley 4731, Trent Lock 4831, canal, Barrow to Stenson 3429, and Swarkestone 3728, by the River Mease, Netherseal 2812, River Trent 2017.

Linton: Burton and Willington 22.

R. alpinus L. *Monk's Rhubarb*

Introduced, Roadsides, streamsides and waste ground, especially near farms; a persistent relic of former cultivation. Occasional, chiefly in hilly areas.

W. Lane from Dore to Blacka Moor 2980, Brand Top 0568, nr. Flash 0267, Kirk Dale, Sheldon 1868, nr. Birchover 2361, Slaley 2757.

E. Freebirch 3072, nr. Spitewinter 3466, Kelstedge 3363, Gladwyns Mark 3066, recorded for 35.

S. Cockshead Lane, Snelston 1641.

Linton: Ludworth 98, Chinley 08, One Ash Grange 16, Dore 38.

R. longifolius DC.

Native. By streams, in ditches and in damp grassy places. Formerly occasional but not seen in recent years.

W. Birchover 2362, Winster 2460.

E. Cordwell Valley 3176, nr. Milltown 3561, Moorwood Moor 3656.

The above are all E. Drabble's records from the Journal of Botany and none is later than 1913.

Linton: Not recorded.

R. crispus L. *Curly Dock*

Native. Roadsides, waste places, cultivated ground. Common and widespread. Recorded for all squares except:

W. 19.

Linton: 'Common. Abundant everywhere'.

R. obtusifolius L. *Broad-leaved Dock*

Native. Roadsides, waste ground, hedgerows and field borders, especially where the ground has been disturbed. Common and widespread.

Recorded for all squares.

Linton: 'Common. Abundant and generally distributed'.

R. pulcher L. *Fiddle Dock*

Native. Dry waysides and waste ground. Very rare.

S. Once only, as a garden weed, Clifton 1644, Little Eaton Road, Derby 3538.

Linton: Burton 22, Repton, towards the Ferry 32.

R. sanguineus L. *Red-veined Dock, Bloody Dock*

Native. Waysides, waste places, hedgerows and woods. Common and widespread.

Var. *viridis* Sibth. is the common form in the county. Var. *sanguineus* seems comparatively rare.

Recorded for all squares except:

W. 09, 08, 07.
E. 57, 45, 34.
S. 23, 24.

Linton: 'Common and generally distributed'.
Var. *sanguineus*: Via Gellia 25, Newhall 22.

R. conglomeratus Murray   *Sharp Dock*
Native. Damp grassy places, open woods and waste ground. Frequent, and
probably under-recorded.
W. Dovedale 15, Kniveton 24.
E. Recorded for all squares except 37, 36, 34.
S. Recorded for all squares except 22.

URTICACEAE
PARIETARIA L.

P. judaica L. (*P. officinalis* L.)   *Wall Pellitory*
Recorded for: **W.** 17, 27, 26, 25.
                    E. all squares except 47, 45, 44.
                    S. 33, 43, 32, 31.

URTICA L.
U. urens L.   *Small Nettle*
Native. Cultivated ground and waste places. Uncommon in the north of
the county, more frequent in the south.
W. Ladmanlow Tip 0471, recorded for 27, Haddon Hall garden 2366,
Carsington 2553.
E. Cowley Chapel 3377, The Walls, Whitwell 5078, Birdholme 3369,
Stony Houghton 4966, Stanley 4141.
S. Recorded for all squares except 24, 43.

Linton: Mellor 98, Bradwell 18, Miller's Dale 17, Monyash 16, Renishaw 47,
Markland Grips 57, Belper 34, Shirley 24, Dale 43.

U. dioica L.   *Stinging Nettle*
Native. Woods, hedgebanks, grassy places, fens and near buildings, especially
where the ground is covered with litter or rubble; very sensitive to phosphate
deficiency and often marking the former site of farm-buildings, sheep-pens,
etc. Common and widespread.
Recorded for all squares.
Linton: 'Common. General'.

CANNABACEAE
HUMULUS L.
H. lupulus L.   *Hop*
Native. Hedges and thickets and clearings or margins of valley-bottom
woods; perhaps sometimes an escape from cultivation. Not uncommon.

W. Gritstone quarry, Manchester Road, Buxton 0573, nr. Chee Dale 1375, Clod Hall 2873, Fenny Bentley 1750, Bradbourne 2152.

E. Recorded for all squares except 35, 45.

S. Recorded for all squares except 22.

Linton: Pinxton 45, Bretby 22.

CANNABIS L.

C. sativa L.   *Hemp*

Introduced. A casual in waste places and rubbish-tips. Occasional.

E. Stonegravels 3872, Rowthorn 4764.

S. Garden weed, from bird-seed, Clifton, in 1941, 1645, rubbish dump, Stores Road, Derby 3537.

Linton: Corbar Wood, Buxton 17.

ULMACEAE

ULMUS L.

U. glabra Hudson (*U. montana* Stokes)   *Wych Elm*

Native. Moist woods on fertile soils, hedgerows and by streams; a characteristic tree of woods on both Magnesian and Carboniferous Limestones, but frequent also on non-calcareous soils. Frequent and locally abundant.

Recorded for all squares except:

W. 09.

Linton: 'Frequent'.

U. × elegantissima Horwood (*U. glabra* × *U. carpinifolia* var. *plotii* (Druce) Tutin).

Native, but often planted. Rare.

S. Ashbourne 1646, 1745, 1845, Offcote Grange 2047.

U. procera Salisb. (*U. surculosa* Stokes)   *English Elm*

Native, but often planted. In hedges and by roads. Infrequent in the north of the county, occasional in the south.

W. Recorded for 08, 07, nr. Friden 1660, Bubnell 2472, Eyam 2176.

E. Recorded for 48, New Whittington 3974, nr. Bolsover 4972, Ashover 3562, Hardwick Ponds 4563, Stony Houghton 5066, Morley 3940, Smalley Common 4042.

S. Spondon 3935, Littleover 3233, Locko 4138, Repton 3228, Drakelow 2319.

Linton: 'Frequent. General, mostly planted'.

U. × hollandica Miller var. (?*U. glabra* × *U. carpinifolia*)
W. Carsington 2453.
E. Barrow Hill 4175.

U. carpinifolia × **plotii** × **glabra**   *Elm*
S. Ashbourne 1745.

U. carpinifolia Gleditsch var. **plotii** (Druce) Tutin   *Plot's Elm*
Native, but often planted.
W. Dales of the Wye Valley 17.
E. Ford 4080, Cutthorpe 3474.
Linton: Not recorded.

JUGLANDACEAE

JUGLANS L.

J. regia L.   *Walnut*
Introduced. Planted for its fruit and occasional in hedgerows.
E. Upper Langwith 5169, Renishaw 4378.
S. Bretby 2923, Stapenhill 2522.
Linton: Not recorded.

BETULACEAE

BETULA L.

B. pendula Roth (*B. verrucosa* Ehrh.)   *Silver Birch*
Native. Woods and heaths, especially on light dry soils but rare on limestone;
a rapid colonizer of woodland clearings and of grassland and heath, except
where rabbits are very abundant. Common and widespread.
W. Recorded for all squares except:
> **W.** o8.
> **E.** 45.
Linton: 'Common'.

B. pubescens Ehrh.   *Birch*
Native. In similar situations to *B. pendula* but more tolerant of waterlogged
soils and of high altitudes. Less frequent than *B. pendula* in the south of the
county but increasing north-westwards and becoming the characteristic birch
of the northern moorlands.
Very variable in the pubescence and wartiness of the young shoots. The

common subspecies, with conspicuously pubescent shoots more or less devoid of warts, is ssp. *pubescens*. In upland areas of the north of the county forms are to be found that agree with descriptions of ssp. *carpatica* (Willd.) Ascherson & Graebner (ssp. *odorata* (Bechst.) E. F. Warburg), with young shoots scarcely or not pubescent and covered with brown resinous warts.

W.   Recorded for all squares except 15.

E.   Recorded for all squares except 45, 34.

S.   Bradley Wood 2046, Longford 2238, around Derby 33, Grange Wood, Lullington 2714, recorded also for 43, 22, 32.

Linton: 'Rather rare'.

ALNUS Miller

A. glutinosa (L.) Gaertner   *Alder*

Native. Streamsides, fen-woods, damp hedgerows. Common and widespread; forming pure woods locally in valley-bottoms.

Recorded for all squares.

Linton: 'Common. Generally distributed'.

CORYLACEAE

[CARPINUS L.

C. betulus L.   *Hornbeam*

Introduced. Occasionally planted in woods, copses and hedgerows.

W.   Buxton Gardens 0673, Hassop 2372, churchyard, Eyam 2176, riverside, Chatsworth 2570.

E.   Apperknowle 3778, Cutthorpe 3474, Holme Brook 3472, Uppertown 3264.

S.   The Green, Ashbourne 1947, Peathayes 2139, Dalbury Churchyard 2634, Netherseal 2913, Bretby 2923.

Linton: Mellor 98, between Mugginton and Cross o' th' Hands 24, Mickleover and Breadsall 33, Repton 32.

CORYLUS L.

C. avellana L.   *Hazel, Cob-nut*

Native. An important component of the shrub-layer in woods on a wide range of soil-types, though avoiding the poorest soils; also in hedges and persisting for some time on the site of former woodland, as locally on the sides of the limestone dales. Common and widespread.

Recorded for all squares.

Linton: 'General throughout the county'.

FAGACEAE
FAGUS L.
F. sylvatica L.   *Beech*

Introduced, though perhaps formerly native on the Magnesian Limestone.
Frequently planted in woods, copses and gardens and especially in wind-
breaks at high altitudes.

Recorded for all squares except:

> **W.** 09, 19.
> **E.** 45.

CASTANEA Miller

C. sativa Miller   *Sweet Chestnut, Spanish Chestnut*

Introduced. Much planted for its timber and for split-chestnut fencing;
woods and plantations. Frequent, and locally common.

W. Chatsworth Park 2670, Ashford 16, Robin Hood's Stride 2262, Fenny
Bentley 1650, Callow 2751, Bonsall Moor 2559.

E. Recorded for all squares except 36, 46, 45.

S. Recorded for all squares.

Linton: Ashopton 28.

QUERCUS L.

Q. cerris L.   *Turkey Oak*

Introduced. Frequently planted as an ornamental tree and occasionally
establishing itself from seed. Uncommon.

W. Pin Dale 1582, Upper Padley 2479.

E. Stubbing Wood 3977, Wingerworth 3667, Astwith 4464, Hardwick
4564, Little Eaton 3642.

Linton: Not recorded.

Q. ilex L.   *Holm Oak*

Introduced. Frequently planted as an ornamental tree; not naturalized in
Derbyshire.

W. Cromford 2957.

E. Barlow 3474, Renishaw Hall 4378, Barlborough Park 4778.

Linton: Not recorded.

Q. robur L. ('*Q. robur* var. *pedunculata* (Ehrh.)') *Common Oak, English Oak,*
                                                            Pedunculate Oak

Native. Woods and hedges, especially at low altitudes and on the heavier
base-rich or calcareous soils. Largely replaced by Q. *petraea* on the lighter and
more acid soils and at high altitudes. Locally common.

Recorded for all squares except 09, 19.

Linton: 'Common'.

Q. petraea (Mattuschka) Liebl. ('*Q. robur* var. *sessiliflora* (Salisb.)')

Durmast Oak, Sessile Oak

Native. The common oak of the gritstones, sandstones and shales of the southern Pennines.

Intermediates between Q. *robur* and Q. *petraea* are not uncommon.

W. Chew Wood 9992, nr. Glossop 09, Derwent Reservoir 1690, Napkin Piece 9987, nr. Whaley Bridge 0080, Ladybower 1986, Hope 1683, Hathersage 2281, Vale of Goyt 0174, Calver 2475, High Tor 2958.

E. Recorded for all squares except 44.

S. Recorded for all squares except 14, 13, 42.

Linton: Snake Inn 19.

SALICACEAE

POPULUS L.

P. alba L. *White Poplar, Abele*

Introduced. Frequently planted in woods and hedges but not naturalized.

W. Recorded for 08, Ashwood Dale 0772, Chee Tor 1373, Miller's Dale 1473, Froggatt 2476, Lathkill Dale 1966.

E. Coal Aston 3679, Cordwell 3176, Creswell Crags 57, Wingerworth Great Pond 3667.

S. Herdsman's Close, Offcote 1948, Osmaston Park, planted 2043, recorded for 23 and 43, Egginton 2628.

Linton: Mellor 98, Renishaw 47, Mickleover 33, Calke & Repton 32.

P. canescens (Aiton) Sm. *Grey Poplar*

Probably native. Damp woods and hedges. Rare.

W. Recorded for 07, Bakewell Bridge 2168.

E. Barlow Woodseats, Cordwell Valley 3175, Wingerworth Great Pond 3667.

Linton: Mellor 98, Burton and Bretby 22.

P. tremula L. *Aspen*

Native. Damp places in woods, by streams and flushes and in hedges; tolerant of a wide range of soil-types and forming dense thickets by suckering. Frequent.

W. Recorded for 08, Cressbrook Dale 1774, Miller's Dale 1473, Haywood, Grindleford 2477, Froggatt 2476, recorded for 16, by the Doveholes in Dovedale 1453, nr. Grange Mill 2458.

E. Twentywell, Totley 3280, Fox Lane, Cordwell 3076, Monk Wood 3576, Whitwell Wood 5278, Clay Cross 3963, Ault Hucknall 4765, Breadsall Moor 3742, Morley Moor 3841.

S. Recorded for all squares except 13, 32.

Q

Linton: Stirrup 99, Edale and Shatton 18, Hathersage 28, Goyts Clough 07, Renishaw 47, South Normanton 45, Cubley Brook 13, Calke and Repton 32.

SALIX L.

S. pentandra L.   *Bay-leaved Willow*

Native. Stream-sides, fens, marshes, wet woods. Local.

Derbyshire is close to the southern limit of this handsome willow. It is particularly attractive in late May and early June when the large and fragrant male catkins are in flower.

W. Alport Dale 1489, Buxton 0473, Combs Reservoir 0479, Miller's Dale 1773, Birchill Bank 2271, nr. Baslow 2875, Scow Brook 2552, Bradbourne 2053.

E. Nr. Stubley 3478, Great Ponds, Wingerworth 3667.

S. Brailsford 2544, nr. Hulland 2246, Cubley 1531, Anchor Church 3327.

Linton: Glossop 09, Barlboro' Common 47, Egginton 22.

S. alba L.   *White Willow*

Native. By streams and ponds and in marshes and wet woods. Frequent. The native form is ssp. *alba*.

W. Recorded for 08, Ashwood Dale 0772, Wye Valley 17, Calver New Bridge 2475, recorded for 06, Lathkill Dale 1966, Ashford 2069, Thorpe 1650, Bradbourne 2053.

E. Scarcliffe Park Wood 5170, West Hallam 4144, Ambergate 3551.

S. Recorded for all squares.

Linton: 'General through the County, often planted'.

Ssp. **vitellina** (L.) Arcangeli   *Golden Willow*

Introduced. Frequently planted as an osier or for ornamental purposes. Occasional.

S. Nr. the Henmore, Ashbourne 1746, ponds at Bradley 2245, Yeldersley 2144, Osmaston 2042.

Linton: Monsal Dale 17, Cromford 25.

Ssp. **caerulea** (Sm.) Rech. fil., the *Cricket-bat Willow* is often planted.

S. × **rubens** Schrank (*S. fragilis* × *S. alba*)

Native. By streams and ponds. Rare.

E. Bradway 3280, Barlow Woodseats 3175.

S. Ashbourne 1745, Hulland Carr 2645.

Linton: Pebley Pond 47.

S. fragilis L.   *Crack Willow*

Native. By streams and ponds and in hedges. Common and widespread.
Recorded for all squares except:

W. 19.
E. 34.

Linton: 'General; in all the districts, often probably planted'.

S. × meyerana Rostk. ex Willd. (*S. fragilis × S. pentandra*)
Possibly native. Very rare.
S. Derby 33.
Linton: Not recorded.

S. triandra L.   *Almond Willow*

Native, but often planted. Sides of streams and ponds, marshes and osier-beds.
Not common.
S. Clifton Mill 1644, Osmaston Ponds 2043, Chellaston 3728.
Linton: Chapel Reservoir 07, Miller's Dale 17, Longford 24, Cubley 13, Repton 32.

S. × mollissima Ehrh. (*S. triandra × S. viminalis*)
Probably native but grown as an osier. Uncommon.
S. Bentley Brook 1646, Yeldersley Pond 2144.
Linton: Miller's Dale 17, Dovedale 15, Egginton 22, Chellaston 32, Drakelow 21.

S. × speciosa Host (*S. triandra × S. fragilis*)
Native. By streams and ponds etc. Rare.
S. Bentley Brook 1646, Sturston Mill 2047.
Linton: Bentley Brook 14.

S. purpurea L.   *Purple Osier, Bitter Willow*

Native. Streamsides, marshes, hedges, wet woods. Occasional; sometimes
planted as an osier.
W. Miller's Dale 1573, Fishponds, Ashford 1769, Tissington 1752, Cromford 2957.
E. Barlow 3474, Holymoor 3469, recorded for 56.
S. Ashbourne Green 1847, Spinnyford Brook 2444, Corley 2147, Ednaston 2242, Osmaston 2042, Eaton 0936, Barton Blount 2135, Findern 3030, Scropton 1829, Egginton 2727, nr. Repton 3026, recorded for 21.
Linton: Mellor 98, Ashwood Dale 07, Hemsworth 38, Harlesthorpe 47.

S. × rubra Hudson (*S. purpurea* × *S. viminalis*)
Native. but often grown as an osier. Occasional.
W. Wye Valley 1072, Ashford 2065, Lode Mill 1453, Cromford 2957.
S. Doveridge 1134.
Linton: Mellor 98, River Derwent nr. Great Wilne 43, Trent Lock 53.

S. daphnoides Vill. ssp. **daphnoides**
Introduced. Sometimes planted in wet places.
W. Buxton 0573.
S. Bradley, by Lady's Pond 2245.
Linton: Not recorded.

S. viminalis L.   *Common Osier*
Native, but often planted. By streams and pools, in hedges and osier-beds.
Frequent.
Recorded for all squares except:
> **W.** 19, 28, 17.
> **E.** 35, 34, 44.
Linton: Mellor 98, Hemsworth 38.

S. caprea L. ssp. **caprea**   *Great Sallow, Goat Willow*
Native. Woods, scrub, hedges. Frequent.
Recorded for all squares except:
> **W.** 09.
> **S.** 21, 33.

S. × sericans Tausch ex A. Kerner (*S. caprea* × *S. viminalis*)
Native. Hedges, copses, streamsides. Occasional.
E. Pratthall 3273, Barlow, Engine Hollow 3474, Ashover 3463, Mapperley
4343, Shipley 4443.
Linton: Bradley, Edlaston Coppy and Overburrows, Brailsford 24, canal side
nr. Chellaston 33, Dale 43.

S. cinerea L. ssp. **atrocinerea** (Brot.) P. Silva & G. Sobr.   *Common Sallow*
Native. Woods and heaths, marshes and fens and by streams and ponds.
Common and widespread.
Recorded for all squares except:
> **W.** 09.
> **S.** 33, 21.
Linton: 'General throughout the County'.

S. × **smithiana** Willd. (*S. viminalis* × *S. cinerea*)

Native. Hedges, copses, streamsides. Frequent.

W. Big Moor 27, Dovedale 1550.

E. Holme Brook 37, Barlow, railway 3474, Smalley 4044.

Linton: Mellor 98, Ashwood Dale 07, by the Wye 17, Beauchief 38, Ashover 36, Mosborough 48, Shirley 24, Hulland Ward 24, nr. Brook Farm, Cubley 13, Marston-on-Dove 22, Repton 32.

S. aurita L. *Eared Sallow*

Native. Damp woods, heaths, by moorland streams and flushes and on rocks by streams; chiefly on acid substrata. Frequent.

W. Recorded for all squares except 09, 19.

E. Barlow 3472, Cathole 3267, Ambergate 3551, Morley 3940, Smalley 4044.

S. Recorded for all squares except 33, 43, 32.

Linton: Snake Inn 19.

S. repens L. ssp. **repens** *Creeping Willow*

Native. Damp places on heaths, moors and waysides; chiefly lowland. Rare.

W. Bradwell Moor 1480, Grin Plantation 0571, Chee Tor 1273, Middleton Dale 2075.

E. Hardwick Ponds 4563/4.

S. Spread Eagle Inn 2929.

Linton: Markland Grips 57.

ERICACEAE

LEDUM L.

L. groenlandicum Oeder

Introduced. A rare escape whose establishment on high and remote moorland in the north of the county is of considerable interest.

W. Nr. Barrow Stones, Bleaklow 1396, Old nurseries, Whitesprings, nr. Rowsley 2865.

Linton: Not recorded.

RHODODENDRON L.

R. ponticum L.

Introduced. Often planted in woods and elsewhere and spreading freely on suitably moist acid soils, both under shade and in the open. Locally abundant, and a fine sight when in flower.

W. Derwent to Upper Dams 19, Taxal 0080, recorded for 18, Blacka Moor 2880, Buxton 07, Grindleford 2778, Rowsley 2865, Fenny Bentley 1750.

E. Beauchief 3381, Cordwell, Unthank 3076, Wingerworth 3767, nr. Glapwell 4766, Ogston 3759, Shipley Wood 4543

S. Mapleton 1648, Shirley Wood 2042, Bentley Carr 1838, Culland Peat Moss 2439, also in 43, Bretby 2922, Melbourne and Repton 32.

Linton: No records.

ANDROMEDA L.

A. polifolia L.   *Bog Rosemary*

Native. Bogs and boggy moorland on the Millstone Grit. Very rare.

W. White Path Moss, Stanage Edge 2583, Axe Edge 0370.

Linton: Axe Edge, Combes Moss and Goyt's Moss 07.

ARCTOSTAPHYLOS Adanson

A. uva-ursi (L.) Sprengel   *Bearberry*

Native. Gritstone rocks and banks on the northern moorlands. Very rare.

W. Stainery Clough and a heathery bank above the River Derwent 1696, north side of Kinder Scout 0889, above Ladybower 1986, Ladybower end of Derwent Edge 2086.

Linton: Nr. Glossop 09.

PERNETTYA Gaudich.

P. mucronata (L.f.) Gaudich. ex Spreng.

Introduced. Often planted in gardens and occasionally escaping.

W. Dore 2980, Blacka Moor 2880, Nether Padley 2578, persisting in old nurseries at Whitesprings nr. Rowsley 2865, and at Farley Moor 2961.

Linton: Not recorded.

CALLUNA Salisb.

C. vulgaris (L.) Hull (*C. erica* DC.)   *Heather, Ling*

Native. Heaths, moors, bogs and open woods on acid soil or acid peat; a characteristic plant of podsolized soils and tending to promote podsolization. Common and widespread; locally dominant, especially on heather moors managed for grouse.

The dominance of heather is favoured by periodic and controlled burning on an 8-15-year rotation. It is readily suppressed or exterminated by heavy sheep-grazing and is therefore inconspicuous or absent over extensive tracts of moorland where there have been large flocks of sheep for a long time. It is also absent from most of the deeply hagged peat of the summit of Kinder Scout.

Recorded for all squares except:

E. 44.
S. 13, 43, 21.

Linton: Linton and Newhall 21.

ERICA L.

E. tetralix L. *Cross-leaved Heath*

Native. Bogs, wet heaths and moors. Locally frequent on the gritstone moorlands and in wet places on acid substrata elsewhere.

The white-flowered form is not uncommon.

W. Recorded for all squares except 19, 16, 15, 25.

E. Hipper Sick 3069.

Linton: 'General on moorlands'.

E. cinerea L. *Bell-heather, Fine-leaved Heath*

Native. Heaths, waysides, well-drained banks and rocky slopes on the gritstone moorland and on Edale and Coal Measure shales; absent from calcareous soils. Locally frequent, and often with *Ulex gallii* and *Calluna*.

W. Above the Snake Inn 19, Chinley 0383, Jaggers Clough 1487, Ringinglow 2983, Burbage Edge 0372, Axe Edge 0370, Totley Moss 2779, Cordwell Valley 2977, Washgate Valley 0468, Beeley Moor 2967, Tansley Moor 2962.

E. Bilberry Knoll 3058, Holy Moor 3368, Breadsall Moor 3742.

Linton: Barlow Lees 37, Shirley Wood 24, Repton Rocks 32.

E. vagans L. *Cornish Heath*

Introduced. A rare garden escape.

E. Matlock Moor Farm 3162.

Linton: Not recorded.

VACCINIUM L.

V. vitis-idaea L. *Cowberry, Red Whortleberry*

Native. Moors and open woods on acid soil or drained acid peat and especially on well-drained rocky slopes, cliff-ledges and screes on the higher gritstone moorlands; more shade-tolerant than heather. Locally abundant, and often with heather and bilberry. See also maps (page 92).

W. Alport Castles 1491, recorded for 08, Ladybower 1986, Ringinglow 2983, also Derwent Edge, Bamford Moor and Blacka Moor, Dore, in 28, Goyt's Moss 0172, Corbar Hill 0574, under Raven Tor 1573, Sir William Hill 2177, also Longshaw, Salter Sitch and Brampton East Moor in 27, Tansley Moor 2962, Whitesprings 2865.

E. Tansley Moor 3360, 3259, also Slagmill Plantation and Matlock Moor in 36, Cromford Moor 3055.

Linton: Glossop Moors 09, beyond Mam Tor and Win Hill 18, Wirksworth 25.

V. myrtillus L. *Bilberry, Whortleberry*

Native. Woods, heaths and moorlands on acid and especially on more or less strongly podsolized soils; also on drained acid peat and on gritstone rock-ledges and scree. More shade-tolerant than heather and dominating the the ground-layer of woods on podsolized soils if there is little or no grazing; but represented only by shoots a few inches high or completely eliminated where sheep-grazing is persistently heavy, as in both woods and open rough pastures on many of the high moorlands. Abundant, especially on the Millstone Grit and Edale Shales, but also very locally on acid loams of the Carboniferous Limestone plateau.

W. Recorded for all squares.

E. Ecclesall Wood 3282, Unthank 3075, Pratthall 3170, nr. Troway 3979, Tansley Moor 3360, Uppertown 3265, above Holloway 3356, nr. Wessington 3757, Breadsall Moor 3742.

S. Bradley Wood 1946, 2046, Hulland Moss 2546.

Linton: Nr. Ticknall and Repton Rocks 32.

V. × intermedium Ruthe (*V. vitis-idaea × V. myrtillus*)

Native. Heaths, moors and open woods, chiefly below 1,300 feet. Occasional, usually with both parents.

Recent records of this interesting hybrid are largely from the southern Pennines and many of them from Derbyshire. Studies by J. C. Ritchie and others show it to be intermediate between the parents in many features, including the degree of ridging of the stem; the colour, shape, thickness, toothing and persistence of the leaves; the number, shape and colour of the flowers and the colour of the fruit. It shows some hybrid vigour, patches often spreading at the expense of one or both parents. Its habitats seem almost always to have been recently disturbed by man: banks of cut peat, edges of ditches or drains, old cart tracks or moorland paths, old gun-sites etc. Such disturbance may well favour the close association of the parent species and so increase the chances of hybridization as well as providing bare ground for the establishment of hybrid seedlings.

W. Blacka Moor, Dore 2880, nr. Fox House 2780, Houndkirk Moor 2882, Big Moor 2676, nr. Nelson Monument on Birchen Edge 2773, nr. Darley 2964.

E. Slagmill Plantation, Matlock Bank 3060.

Linton: Not recorded.

V. oxycoccos L. (*Schollera oxycoccos* (L.) Roth)   *Cranberry*

Native. Wet bogs, moorland flushes and streamsides, often with *Sphagnum.* Local, and confined to the gritstone moors.

W. Nr. Ringinglow 2883, Stanley Moor 0471, also Burbage Edge and Combs Moss in 07, Salter Sitch 2878, Leash Fen 2973, Axe Edge 0369, Beeley Moor 2967.

Linton: Charlesworth 98, Hayfield and Kinder Scout 08, Dethick Common 35.

PYROLACEAE
PYROLA L.

P. minor L.   *Common Wintergreen*
Native. Woods and plantations, heaths. Very rare.
W. Heathfield nr. Buxton 0770, junction of Cowdale and King Sterndale
roads 0871 (believed extinct at both localities owing to tree-felling). Miller's
Dale 1372.
Linton: Nr. Miller's Dale 17.

P. rotundifolia L.   *Larger Wintergreen*
The two old records were almost certainly erroneous.
No recent records.
Linton: Ladywash Mine, Eyam 27.

MONOTROPACEAE
MONOTROPA L.

M. hypopitys L.   *Yellow Bird's-nest*
Native. Woods and plantations, especially of beech. Very rare.
The Derbyshire records seem all to be of ssp. *hypophegea* (Wallr.) So6.
W. Grin Wood, Buxton (last seen *c.* 1947) 0571, Via Gellia 2756.
Linton: Heights of Abraham and Black Rocks 25, nr. Ashover 36.

EMPETRACEAE
EMPETRUM L.

E. nigrum L.   *Crowberry*
Native. Moors, bog-margins and drier spots on bogs (such as tussocks of
cotton-grass); also gritstone scree and outcrops. Locally abundant, but
uncommon at altitudes below 1,000 feet. See also map (page 92).
W. Charlesworth Coombs 0092, Glossop 0393, Alport Dale 1390, Napkin
Piece 9987, Kinder Scout 0988, Edale 1285, Derwent Edge 2088, Burbage
Moss 2782, Withinleach Moss 9976, Combs Moss 0576, Coombs Edge 0475,
Salter Sitch 2878, Umberley Brook 2871, Totley Moor 2779, Axe Edge 0269,
0369, Beeley Moor 2867.
E. Slagmill Plantation 3068, Tansley Moor 3361.
Linton: Dethick Common 35.

E. hermaphroditum Hagerup
There is a specimen in the Herbarium of the British Museum said to have
been collected on Kinder Scout, but prolonged search for this northern species
has so far proved fruitless.
Linton: No records.

PRIMULACEAE

PRIMULA L.

P. veris L.   *Cowslip*

Native. Pastures, waysides, woods. Still widespread and locally abundant, but decreasing.

In grassland on loamy soils over limestone, and especially in the limestone dales, cowslips are often mixed with *Orchis mascula* and the two species flower at the same time to provide a most attractive sight.

Recorded for all squares except:

> **W.** 09, 19, 18, 28.
> **E.** 34, 46.
> **S.** 43.

Linton: 'Common. Generally distributed'.

P. veris x P. vulgaris   *'Oxlip', Common Oxlip*

Native. The hybrid between cowslip and primrose is intermediate between the parent species and is not uncommon where the parents are closely associated. The true oxlip is *P. elatior* (L.) Hill, with a restricted distribution in East Anglia.

W. Monk's Dale 1373, Cressbrook Dale 1773, Coombs Dale 2274, Middleton Dale 2175, Brassington Rocks 2154, Via Gellia 2857, Matlock Bath 2958.

E. Whitwell Wood 5279, Markland Grips 5074, Pleasley Vale 5165, Scarcliffe Park Wood 5270, Lea Mills 3156.

S. Woodcock Delph 1848, river bank, Mapleton, on the Staffs. side 1648.

Linton: Between Rowsley and Bakewell 26, Dovedale 15, Whatstandwell 35, East of Alkmonton 13, Peathayes 23, Mickleover and Breadsall 33, between Ockbrook and Dale Abbey 43, Calke, between Repton and Bretby and Repton Shrubs 32.

P. vulgaris Hudson (*P. acaulis* (L.) Hill)   *Primrose*

Native. Woods, hedgebanks and open grassy places on fertile soil. Linton describes it as 'common' and as 'general, though most abundant on clay or limestone'. There can be little doubt that it has greatly decreased since the early years of the century.

Recorded for all squares except:

> **W.** 09, 19.
> **E.** 45.
> **S.** 43.

Linton: 'Common. General, though most abundant on clay or limestone'.

HOTTONIA L.

H. palustris L.   *Water Violet*

Native. Ponds, ditches and slow streams. Local.

E. Morley Moor 3841.

S. Trent Lock 4831, Pond by Swarkestone Bridge 3628, nr. Repton 2927.
Linton: Osmaston-by-Derby, Breadsall, pond south of railway between Derby and Spondon 33.

LYSIMACHIA L.

L. nemorum L.   *Wood Pimpernel, Yellow Pimpernel*
Native. Damp places in woods, shady hedgebanks, flushes and stream-sides; characteristic of gleyed soils in valley-bottom woods, often with *Veronica montana*. Locally frequent.
Recorded for all squares except:
> **W.** 09.
> **E.** 45.
> **S.** 33.

Linton: Ripley and Newton Wood 45, Breadsall and Mackworth 33.

L. nummularia L.   *Money-wort, Creeping Jenny*
Native. Sides of ponds and ditches, damp meadows, moist hedgebanks and woods. Local, and chiefly in the south of the county.

W. Recorded for 08 and 18, by Derwent between Hathersage and Grindleford 28/27, Ashwood Dale 0872, Buxton 07, Combs Reservoir 0479, nr. Calver New Bridge 2475, Stoke Ford, nr. Abney 2179, Washgate Valley 0569, island in the Dove nr. Hollinsclough 0766, nr. Haddon Hall 2267, Woodeaves 1850, Grange Mill 2457.

E. Stubbing Top Pond 3567, Holy Moor 3369, Ault Hucknall 4665, Pleasley Vale 5265, Pleasley Forge 5164, Alfreton Brook 4056, Morley 4040.

S. Recorded for all squares except 22.

Linton: Barlborough, Renishaw and Walton 47, above Allestree Ford 34, Burton and Drakelow 22.

L. vulgaris L.   *Yellow Loosestrife*
Native. Stream-banks, fens, wet woods. Rare.

W. Nr. Hathersage Bridge 2280, Buxton College 0572, Miller's Dale 1373, nr. Calver New Bridge 2475, nr. Wye Farm, Rowsley 2465.

E. Stubbing Court 3567.

S. Railway bank, Clifton 1644, marsh below Brailsford Tip 2541, Grange Wood 2741.

Linton: Morley Moor 34, Markeaton Woods and Breadsall 33.

L. punctata L.
Introduced. Much grown in gardens and occasionally naturalized. Rare.
No recent records.
Linton: 'Brookside below Beeley, nr. the Derwent', 26.

TRIENTALIS L.

T. europaea L.   *Chickweed Wintergreen*

Probably native. Discovered recently nr. Sheffield in an extensive moorland flush where large numbers of plants grow with *Sphagnum recurvum* between tussocks of *Eriophorum vaginatum*. Very rare.

W. Houndkirk Moor 2882.

Linton: Not recorded.

ANAGALLIS L.

A. tenella (L.) L.   *Bog Pimpernel*

Native. In wet places on acid but moderately mineral-rich peat of bogs and fens. Occasional.

W. Recorded for 08, Win Hill 1984, Parkin Clough 1985, Shatton Edge 1982, Umberley Brook, Chatsworth 2870, Washgate Valley 0467.

E. Whitwell Wood 5278, Markland Grips 5074.

S. Hulland Moss 2546, by Bradley Brook 2244, Mugginton 2843.

Linton: Holymoorside 36, Pleasley 56, Bentley Brook, Matlock 35, Pinxton 45, Quarndon Common 34, Normanton 33, Melbourne Common, Foremark Park and Repton Rocks 32.

A. arvensis L.   *Scarlet Pimpernel*

Native. Cultivated land, waysides. Common and widespread.

There are two subspecies in the county:

Ssp. **arvensis,** with densely gland-fringed petals which are usually red, rarely blue. This is the common form.

W. Hathersage 2381, Grindleford 2477, Serpentine Walks, Buxton 0573, Miller's Dale 1373, Lathkill Dale 1966, Bonsall Moor 2559.

E. Recorded for all squares.

S. Recorded for all squares.

Linton: 'Common. Generally distributed'.

Ssp. **foemina** (Miller) Schinz & Thell., with non-overlapping sparsely fringed petals, always blue. This is an uncommon weed of cultivated land, rarely persistent. It is sometimes introduced with flower-seeds. Some of the older records may be of the blue-flowered form of ssp. *arvensis.*

W. In a garden, Manchester Road, Buxton 0673.

E. Holymoorside 3369, Aldercar nr. Loscoe 4248.

S. As a garden weed at Sandybrook Hall 1748, and Ashbourne 1746, Clifton goods yard 1644, Little Chester 3537.

Linton: Between Denby Common and Loscoe 44, Mickleover 33, Ockbrook 43.

A. minima (L.) E. H. L. Krause (*Centunculus minimus* L.)   *Chaffweed*
Native. Damp sandy places on heaths and cart-tracks. Very rare.
W. Recorded from 07.
Linton: Not recorded.

SAMOLUS L.
S. valerandi L.   *Brooklime*
Native. Wet places on mineral soil; marshy margins of streams and ditches.
Very rare.
No recent records.
Linton: Tansley Moor 36, Swarkestone Bridge 32.

OLEACEAE
FRAXINUS L.
F. excelsior L.   *Ash*
Native. Woods, scrub and hedges, especially on moist nutrient-rich soils; a
characteristic tree of valley-bottom alluvium and dominant on scree-slopes
in the limestone dales. Common and widespread.

The ash-woods of the limestone dales are often almost even-aged and appear to
have resulted from the colonization of the scree-slopes after a relaxation of
grazing pressure at various periods during the past two centuries or so. It
seems improbable that they represent an ancient stable type of woodland on
limestone.

Recorded for all squares.
Linton: 'Common in every district'.

SYRINGA L.
S. vulgaris L.   *Lilac*
Introduced. Much planted in gardens and hedges, but not often becoming
truly naturalized.
W. Taddington Dale (introduced) 1670/1.
S. In a hedgerow, Nether Sturston 1946, nr. Breaston 43.
Linton: Not recorded.

LIGUSTRUM L.
L. vulgare L.   *Privet*
Native. Hedges and scrub. especially on limestone. Locally frequent.
W. Chapel-en-le-Frith 0681, Cunning Dale 0872, Monk's Dale 1374, also
Monsal Dale, Miller's Dale and Cressbrook Dale in 17, Pilsley nr. Hassop 2370,
Washgate Valley 0571, Lathkill Dale 1966, Wolfscote Dale 1357, Via Gellia
2556.

E. Recorded for 48, Cutthorpe 3573, Pebley Pond 4878, Whitwell Wood 5278, also Creswell Crags and Markland Grips in 57, Morton 4464, Pleasley Vale 5265.

S. Recorded for all squares except 22.

Linton: Etherow district 09, Beauchief 38, Cromford Canal 35, Pinxton 45, between Duffield and Darley 34, Burton and Egginton 22.

APOCYNACEAE

VINCA L.

V. minor L.   *Lesser Periwinkle*
Probably introduced. Wood-margins, thickets, hedges. Occasional.

W. Leadmill Bridge 2380, King Sterndale Hall 0972, Sherriff Wood 2478, Lathkill Dale 1966.

E. Newbold 3573, Pleasley Vale 5265, nr. Quarndon 3441.

S. Roston 1340, Bradley 2245, Brailsford 2541, Cubley 1638, nr. Longford 2337, Dale Abbey 4338, Overseal 2915.

Linton: Matlock Bath 25, Lea Bridge, Cromford 35, Little Chester 33, Calke 32.

V. major L.   *Periwinkle*
Introduced. Open scrub, shady banks etc. Rare.

E. Pleasley Vale 5165.

S. Shirley 2141, Cubley 1637, Marston Montgomery 1337, Shardlow 4330, Repton 32, Lullington 2513.

Linton: Willersley 35.

GENTIANACEAE

CENTAURIUM Hill

C. erythraea Rafn (*Erythraea centaurium* (L.) Pers.)   *Centaury*
Native. Dry grassland, heaths, banks, wood margins and clearings, chiefly on light non-calcareous loams or sands. Locally abundant.

W. Calver New Bridge 2475, Bradford Dale 2063, Via Gellia 2556, Dovedale nr. stepping stones 1451.

E. Pebley Pond 4978, Scarcliffe Park Wood 5270, Tibshelf 4360, Pleasley 5168, Whatstandwell 3354.

S. Birchwood Quarry 1541, Bradley Brook 2244, Cubley 1539, Mackworth 3137, Dale Abbey 4338, Ticknall 3523.

Linton: New Mills 08, Whirlow Bridge 28, Wilday Green nr. Barlow 37, nr. Stubbing Court 36.

BLACKSTONIA Hudson

B. perfoliata (L.) Hudson   *Yellow Centaury, Yellow-wort*

Native. Dry limestone banks and limestone spoil-heaps. Rare; all recent records are from the Magnesian Limestone. See also map (page 92).

E. Nr. Pebley Pond 4878, Bolsover 4770, Whaley 5171 and 47, Steetley Quarry 5478, Pleasley Vale 5164.

Linton: Matlock 25/35, Ault Hucknall 46.

GENTIANELLA Moench

G. campestris (L.) Börner (*Gentiana campestris* L.)   *Field Gentian*

Native. Pastures, especially on non-calcareous soils; ascending to 1,500 feet. Local, and chiefly on the Millstone Grit and Edale Shales in the northern half of the county.

W. Corbar Hill 0574, Harpur Hill 0570, Dirty Rake, Longstone Edge 2075, Sir William Hill, Grindleford 2178 and 2757, Longdale 1860.

Linton: Nr. Glossop 09, Charlesworth 98, Chinley Hill 08, Axe Edge 07, nr. Goyt's Bridge 07, Taddington 17, Monsal Dale 17, Eyam 27, Osmaston-by-Ashbourne 24, Breadsall 34, Repton Shrubs 32.

G. germanica (Willd.) Börner

Doubtfully native. Recorded by E. Drabble for Castleton and Fallgate (J. Bot. 1911), but not seen since then.

Linton: Not recorded.

G. amarella (L.) Börner (*Gentiana amarella* L.)   *Felwort, Gentian*

Native. Dry pastures, especially on limestone; often abundant on spoil from old mineral workings on the Carboniferous Limestone. Local.

The Derbyshire plant is ssp. *amarella*.

W. Sparrowpit 0981, Eldon Hill 1181, nr. Dore 2881, Grin Wood 0571 and 0472, Diamond Hill 0570, Tansley Dale 1674, Ravensdale 1773, Hassop Mines 2172, Longstone Edge 2273, Dowel Dale 0767, Hartington 1260, Lathkill Dale 1765, Biggin Dale 1458, Hipley Toll Bar 2154, Dean Hollow 2855.

E. Owler Bar 3077, Pebley Pond 4878, Pleasley Vale 5265.

S. Ashbourne Green 1847, Atlow Rough 2047, Osmaston Pools 2042, Ticknall 3523.

Linton: Alport Dale 19, Bradwell 18, Pin Dale 18, Martinside 07, Heights of Abraham 25, Elmton 57, Creswell Crags 57, Snelston Common 14.

G. uliginosa (Willd.) Börner (*Gentiana uliginosa* Willd.)

There are no recent records and the record in Linton's Flora is almost certainly erroneous.

Linton: Miller's Dale 17.

MENYANTHACEAE

MENYANTHES L.

M. trifoliata L.   *Bogbean, Buckbean*

Native. Ponds, pools and the wetter parts of bogs and fens, in acid but moderately mineral-rich water. Local.

W. Hurst Brook, Hathersage, 2283, Moss Carr, Hollinsclough 0765, Leash Fen 2874, 2972, Big Moor 2774, under White Edge 2677, by the stream, Ramsley Moor 2975, Umberley Brook 2870.

E. Renishaw Lake 4478.

S. Nr. Snelston 1642, Bradley Brook 2242, Old Yeldersley 2144, Mugginton 2843.

Linton: Charlesworth Coombs 98, Tansley Common 36, South Normanton 45, Repton Rocks 32, Newhall 21.

POLEMONIACEAE

POLEMONIUM L.

P. caeruleum L.   *Jacob's Ladder*

Native. On steep slopes of the Carboniferous Limestone facing from somewhat west of north to north-east and therefore shaded from direct sunlight for most of the day even when in the open, but occasionally in woods and then in still deeper shade; always growing in moist black limestone soil or humus-rich limestone loam, often on scree or on broad rock-ledges, and usually associated with tall herbaceous species including *Arrhenatherum elatius, Festuca rubra, Filipendula ulmaria, Heracleum sphondylium, Mercurialis perennis, Silene dioica, Urtica dioica* and *Valeriana officinalis*; sometimes with the same associates by streams in the dales. Local. (See maps, page 95).

Jacob's Ladder is one of our most interesting and most beautiful species. Finds of pollen-grains in peat reveal that it was widespread in England and Wales during the last retreat of the ice and before the post-glacial forest period. Its present localities are the residual fragments of this much wider late-glacial distribution.

W. Parkhouse Hill, 0867, The Winnats 1382, Long Cliff, Castleton 1482, Nun's Brook, Fairfield 0674, Chee Dale 1272, Ashwood Dale 0772, Blackwell Mill 1172, Taddington Dale 1670, Monsal Dale 1771, Chrome Hill 0767, Ashford 1869, Lathkill Dale 1765, Bradford Dale 2063, Alport 2264, Rowsley 2465, Woodeaves, nr. Ashbourne 1850, Milldale 1455, Hipley Hill 2054.

E. Oaks Wood, South Wingfield 3775, Stanley 4140.

S. Fenny Bentley Brook 1748, Henmore Brook 1746, Dale 4338, in a ditch, Stanton-by-Bridge 3727.

Linton: Alfreton Brook 45, Cutler Brook, Kedleston 34, Breadsall, and below Derby 33, Drakelow 22.

BORAGINACEAE

CYNOGLOSSUM L.

C. officinale L. *Hound's-tongue*

Native. Woods, dry banks and dry waste ground on limestone. Rare.

W. Bradford Dale 2164, Dovedale (Staffs. side of river) 1451, Via Gellia 2656.

E. Markland Grips 5074.

Linton: Buxton 07, Cressbrook Dale 17, Creswell Crags 57, Ashover 36, Calke 32.

SYMPHYTUM L.

S. officinale L. *Comfrey*

Native. Sides of streams and ditches. Occasional.

True *S. officinale* has cream-coloured flowers. Its stems are clothed with long hairs not swollen at the base, the leaves are strongly decurrent and the anthers in front view are longer than the free part of the filaments. It seems more or less restricted to stream-banks and fens.

W. Recorded for all squares except 09, 19, 18, 28.

E. Canal side, Eckington 4480, Old Brampton 3371, Fallgate 3562, Pleasley 5164, Ambergate 3352, nr. Codnor 4250, Little Eaton 3641, Mapperley 4243.

S. Recorded for all squares.

Linton: Bradwell 18.

S. × uplandicum Nyman (incl. '*S. peregrinum* Ledeb.') *Hybrid Blue Comfrey*

Probably introduced. Hedgebanks, roadsides, wood-margins; rarely by water. Occasional.

Many of the older records for '*S. officinale*' as well as those for '*S. peregrinum*' are probably of the hybrid between native *S. officinale* and the introduced *S. asperum* Lepech., or of backcrosses of the hybrid with *S. officinale*. The name *S. × uplandicum* is given to the whole complex of forms intermediate between the two species. They usually have blue or purple-blue flowers but otherwise show various combinations of parental characters. A common type has clear blue flowers and non-decurrent leaves.

W. Calver New Bridge 2475, Lathkill Dale 1865, Haddon Hall 2366, Parwich 1954, Grange Mill 2457.

E. Ford, nr. Eckington 4081, Pleasley 5164.

S. Spend Lane 1748, Old Yeldersley 2245, Kirk Langley 2838.

Linton: By the river, Bradford and Youlgreave 16, Grange Mill 25.

S. tuberosum L. *Tuberous Comfrey*

Probably native. Damp woods and hedgebanks. Very rare; no recent records.

E. Crich (E. Drabble, J. Bot. 1913).

Linton: Not recorded.

R

S. orientale L.

Introduced. Occasionally naturalized in hedgebanks etc. Very rare.

E.   Bramley Vale 4566.

BORAGO L.

B. officinalis L.   *Borage*

Introduced. A rare garden-escape on waste ground near houses.

E.   Stanley 4141.

S.   Ashbourne 1746, Derby and district 33.

Linton: Elvaston 43, Mickleover 33.

PENTAGLOTTIS Tausch

P. sempervirens (L.) Tausch (*Anchusa sempervirens* L.)   *Alkanet*

Introduced. Occasionally naturalized in hedgerows and wood-margins near houses.

W.   Quarry, Manchester Road, Buxton 0474, by cottages on Parwich Hill 1854, Foufinside, Parwich 1855, Brassington tip 2355.

S.   Nr. Hanging Bridge 1545, Mugginton Lane End 2844, Etwall churchyard 2632, Overseal 2915.

Linton: Haddon 26, Etwall 23.

ANCHUSA L.

A. arvensis (L.) Bieb. (*Lycopsis arvensis* L.)   *Bugloss*

Probably introduced. Arable land on light soil, sandy heaths, waysides. Local, and chiefly in the south of the county.

W.   Gib Yard allotment, Buxton 0672.

E.   West Hallam Station 4240.

S.   Cornfields at Clifton 1743, and nr. Bradley Wood 2046, Dale Abbey 4338, Long Eaton 4831, Scropton 1829, nr. Church Gresley 2819, also Netherseal, Acresford and Cadley Hill in 21.

Linton: Breadsall Moor 33, Bretby 22, between Milton and Repton Rocks 32.

PULMONARIA L.

P. officinalis L.   *Lungwort*

Introduced. Occasionally naturalized in woods and hedgebanks.

W.   Sherbrook tip, opposite hospital, Buxton 0672, entrance to Chee Dale below Miller's Dale Station 1373.

E.   Norton 3682, Whitwell Wood 5278, Pye Hill Station 4452.

S.   Roadside, Kniveton (now gone) 1949.

Linton: Between Compstall Bridge and Marple Aqueduct 98, between Duffield and Kedleston 34.

MYOSOTIS L.

M. scorpioides L. (*M. palustris* (L.) Hill) *Water Forget-me-not*
Native. Streams, ditches, pools, wet places. Frequent, especially at low altitudes.
W. Recorded for all squares except 09, 07, 16.
E. Recorded for all squares except 45, 34, 35.
S. Recorded for all squares.
Linton: Nr. Buxton 07.

M. secunda A. Murray ('*M. repens* G. Don') *Creeping Forget-me-not*
Native. Boggy rills and flushes, boggy places by streams; avoiding calcareous substrata. Local, and more definitely an upland species than *M. scorpioides* and *M. caespitosa*.
W. Derwent 1889, Longshaw 2680, Salter Sitch 2877, below Sherriff Wood 2778, Elton 2161.
E. Cathole 3469.
S. Hulland Moss 2546, Repton Rocks 3221.
Linton: Snake Inn 19, Charlesworth Coombs 98, under Kinder Scout 08, Axe Edge 07, Killamarsh 48, Barlborough Low Common 47, Barton Blount 13, Markeaton 33.

M. caespitosa C. F. Schultz *Water Forget-me-not*
Native. Marshy places, ditches, wet trampled mud, sides of streams and ponds, wet places on moorland etc.; avoiding calcareous substrata. Locally frequent.
W. Watford Bridge 0086, Derwent 1889, about Buxton 0673, Topley Pike 1072, Chee Dale 1273, Baslow 2872, Ashford 1969, Alport Mill 2264, Thorpe Pastures 1551, Dovedale 1453, Via Gellia 2556, Bradbourne 2052.
E. Pebley Pond 4879, Barlow Brook 3474, Morley 3940, Mapperley 4343, Smalley 4144.
S. Recorded for all squares except 22.
Linton: Tarden 98, Beauchief 38, Glapwell and Ault Hucknall 46, Ambergate 35.

M. sylvatica Hoffm. *Wood Forget-me-not*
Native. Woods, hedges, banks of small streams. Frequent, and locally abundant, often with *Silene dioica*, in woods of the limestone dales.
Recorded for all squares except:

> **W.** 09, 19, 08, 18.
> **E.** 38, 48, 47, 45, 34.
> **S.** 33, 43, 23, 21.

Linton: Mellor 98, Ecclesbourne Brook 34, Mickleover 33, Burton 22.

M. arvensis (L.) Hill   *Field Forget-me-not*

Native. Arable land, waste places, wall-tops, banks, woods. Common and widespread.

W. Derwent Reservoir 1690, Bamford 2083, Peter Dale 1275, Chee Dale 1273, Sheriff Wood 2778, Chatsworth 2670, Conksbury Bridge 2165, Harthill Moor 2263, Tissington 1752, Dovedale 1452, Cromford 2855.

E. Beauchief 3381, Holmesfield 3277, Pebley Pond 4978, Ashover 3563, Astwith 4464, Pleasley Vale 5265, Crich 3454, Windley 3045, Shipley Hall 4444.

S. Recorded for all squares.

Linton: 'Abundant nearly everywhere'.

M. discolor Pers. (*M. versicolor* Sm.)   *Yellow-and-blue Forget-me-not*

Native. Open grassy places on light soils. Local.

W. Wolfscote Dale 1455, Minninglow gravel pit 2057, Ible 2557.

E. Tansley 3459, Stanley 4140, West Hallam 4341.

S. Above Bradley Wood 2046, Old Yeldersley 2244, Breadsall 33, Ockbrook 43, Drakelow 21.

Linton: Miller's Dale 17, Burbage Brook and above Padley Wood 27, Hilton Common 23, Bretby and Willington 22, Calke and Repton 32.

M. ramosissima Rochel ('*M. collina* Hoffm.')   *Early Forget-me-no*

Native. Bare ground on dry banks and slopes, especially on shallow limestone soils; also on limestone rocks and on walls. Locally common.

W. Recorded for all squares except 09, 19, 08, 28.

E. Birley Bottom 3173, Creswell Crags 5374.

Linton: Breadsall 33, Stanton-by-Bridge and Anchor Church 32.

AMSINCKIA Lehm.

A. lycopsoides Lehm.

Introduced. A rare casual.

S. Railway bank nr. Castle Donington 4428.

Linton: Miller's Dale 17, Hilton 23.

LITHOSPERMUM L.

L. officinale L.   *Gromwell*

Native. Woods and hedgebanks, chiefly on limestone. Local.

W. Via Gellia 2656.

E. Scarcliffe Park Wood 5170, Whitwell Wood 5278, Pleasley Vale 5265.

Linton: Matlock 25/35, Drakelow and Stapenhill 22, Calke and Repton 32.

L. arvense L.   *Corn Gromwell*
Native. Arable land, chiefly on light soils in the south of the county. Local.
E.   Stanley 4140, West Hallam 4341.
S.   Nr. Hanging Bridge, from bird-seed 1545, Dale 4338.
Linton: Charlesworth 09, Matlock 35, Mickleover, Allestree and Breadsall 33, Drakelow and Burton 22, Repton 32, Cauldwell 21.

ECHIUM L.
E. vulgare L.   *Viper's Bugloss*
Native. Banks, quarry spoil, waysides and fields on light and dry limestone or sandy soils, especially after disturbance. Rather rare.
W.   Peak Dale 0976, Harpur Hill 0570, Topley Pike 1172, Thorpe 1651, Harborough Brick Works 2554.
E.   Scarcliffe Park Wood 5270.
S.   Netherseal 2715, Burton Sidings 2524.
Linton: Clowne 47, Morley 34, Horsley Castle 34, Swarkestone Bridge 32, Drakelow 21, Caldwell 21.

CONVOLVULACEAE
CONVOLVULUS L.
C. arvensis L.   *Small Bindweed*
Native. Arable land, waysides, hedgebanks, waste places. Common and widespread.
Recorded for all squares except:
W.   09, 19, 17, 16.
Linton: 'Abundant throughout the county'.

CALYSTEGIA R. Br.

C. sepium (L.) R.Br. (*Volvulus sepium* (L.) Junger)   *Great Bindweed*
Native or introduced.

There are in Derbyshire two subspecies of *C. sepium*, one native and one introduced. Some authors take the view that they should be treated as distinct species despite the fact that they hybridize freely to give fertile hybrids.

Ssp. **sepium**, with white flowers and bracteoles neither appreciably inflated nor with overlapping margins. This is the native form and is common in hedges, fens and gardens.
W.   Nr. Charlesworth 0194, Buxworth 0282, Bamford 2083, nr. Middleton 1762, Bakewell 2169, Fenny Bentley 1750, Kirk Ireton 2650.
E.   Recorded for all squares.
S.   Recorded for all squares except 22.
Linton: 'Generally distributed'.

Ssp. **silvatica** (Kit.) Maire, with white flowers (rarely pink-striped outside) and with bracteoles strongly inflated and with overlapping margins. Introduced and naturalized in hedges and waste places in several localities.

W.   Calver 2475, Parwich 1854, Kniveton 2050.

E.   Killamarsh 4681, Whittington 3974, Clowne 4975, Pleasley Vale 5265, Shipley Colliery 4444.

S.   Ashbourne 1746, Rodsley 2040.

Linton: Not recorded.

Ssp. *pulchra* (Brummitt & Heywood) Tutin, with bright pink flowers and slightly inflated bracteoles, broader than long and with overlapping margins, is naturalized in neighbouring counties and should be looked for.

CUSCUTA L.

C. europaea L.   *Large Dodder*
Introduced.   A rare casual.
No recent records.
Linton: 'Casual.   E. G. Glover, 1829'.

C. epithymum (L.) L. (incl. *C. trifolii* Bab.)   *Common Dodder*
Perhaps native on *Ulex* and *Calluna*, but no recent records; when on clover probably introduced with the crop.
Linton: 'Very rare.   Glover 1829; Rept. F. & F., 1881'.

SOLANACEAE

LYCIUM L.

L. barbarum L. (*L. halimifolium* Miller)   *Tea-plant*
Introduced. Hedges, walls and waste places; an escape from gardens. Occasional.
W.   Stream flowing to Derwent Reservoir 19, Ashbourne Road, Buxton 0672, Baslow 2572, Middleton by Youlgreave 1963, Newhaven 1659, Tissington 1752, Hopton 2553.
E.   Spital 3871, also Wadshelf and Brampton in 37, Calow 47, Creswell Crags 57, Stretton 3961, Moorwood Park, Hardwick 46, West Hallam Station 4240.
S.   Recorded for all squares.
Linton: As *L. barbarum* and probably referring to the above sp., Buxton 07, Creswell Crags, by Nether Langwith 57, Mickleover 33, and 'in other parts of the county'.

ATROPA L.

A. bella-donna L.   *Deadly Nightshade, Dwaleberries*
?Native. Clearings in woods, open scrub, old quarries etc., especially on calcareous soils. Local and rare.

W.   On an allotment at Buxton for some years, originally sown 0672.

S.   Roadside nr. Melbourne Pool, originally sown or planted 3925.

Linton: Haddon Hall 26, nr. Wirksworth 25, Hardwick 46, Horsley Castle 34, Winshill 22, Repton Hall 32.

HYOSCYAMUS L.

H. niger L.   *Henbane*

Native. Waste land, roadsides, farmyards etc., on light soils and especially where the ground has been disturbed. Rare.

W.   Staden nr. Buxton 0772, one plant, Hindlow 0869, nr. Chelmorton Church 1169.

E.   Langer Lane, Chesterfield (one year) 3768.

S.   One plant, Ashbourne goods yard 1746, Ashbourne 1846.

Linton: Over Haddon 26, Fenny Bentley 15, Pinxton 45, Darley Abbey and Chellaston 33, Stapenhill 22, Calke 32.

SOLANUM L.

S. dulcamara L.   *Bittersweet, Woody Nightshade*

Native. Hedges, woods, fens, sides of ditches, waste ground. Common and widespread.

Recorded for all squares except 09, 19.

Linton: 'General throughout the county'.

S. nigrum L.   *Black Nightshade*

Probably native. Arable land, gardens. Rare.

W.   Ladmanlow 0471.

E.   Nr. Whittington 3974, Flash 3064.

S.   Ashbourne 1646, Derby 3435, Trent Lock 4831, Hartshorne 3221, Burton 2424.

Linton: Pinxton 45, Mickleover 33, Repton 32.

DATURA L.

D. stramonium L.   *Thorn-apple*

Introduced. A casual of waste places and cultivated land.

W.   Bradbourne 2052.

E.   Stanley 4180.

S.   Osmaston 2143, Cubley 1637, Kirk Langley 2839, Spondon 4036, Burton 2323.

Linton: Morley 34, Breadsall 33, Sudbury 13, Calke 32, Repton 32.

SCROPHULARIACEAE

VERBASCUM L.

V. thapsus L.   *Aaron's Rod, Great Mullein*
Native. Sunny banks, waysides, quarries and waste ground, usually on dry soils. Locally frequent, especially on limestone.

W. Dale Road, Buxton 0673, Priestcliffe Lees 1473, also Miller's Dale and Monsal Dale in 17, Longstone Edge 2073, Lathkill Dale 1965, Mill Close Mine 2562, Iron Tors 1456, Dovedale 1551, Brassington Rocks 2154, Via Gellia 2656.

E. Dronfield 3678, Bolsover 4972, Scarcliffe Park Wood 5270, Pleasley Vale 5164, Little Eaton 3640, Stanley 4141.

S. Birchwood Quarry 1541, roadsides about Brailsford and Bradley in 24, on soil dumped from quarries, not persisting, Dale Abbey 4338, Ticknall 3523, Grange Wood 2714.

Linton: Breadsall 33.

V. phlomoides L.
Introduced. A rare casual; no recent records.
Linton: Nr. Buxton 07.

V. lychnitis L.   *White Mullein*
Introduced. A rare casual or garden escape.
No recent records.
Linton: Haddon Hall 26.

V. nigrum L.   *Black Mullein*
Native. Waysides, fields, banks. Rather rare and chiefly on limestone.

W. Alport 2264, Stanton 2365, walls nr. Haddon Hall 2366, Bradford Dale 2163, Brassington Rocks 2154.

Linton: Alport and Stanton 26, west of Whatstandwell 35.

V. × semialbum Chaub. (*V. nigrum* × *V. thapsus*).
Native. Very rare.
No recent records.
Linton: Alport Mill 26, west of Whatstandwell 35.

V. blattaria L.   *Moth Mullein*
Probably introduced. Waste places, cultivated ground. Rare.
No recent records.
Linton: Haddon 26, Hardwick 46, Crich 35, Osmaston 24.

V. virgatum Stokes   *Twiggy Mullein*
Introduced. A rare casual.
W. Blacka Moor 2881.
Linton: Rowsley 26, Chellaston 33.

ANTIRRHINUM L.

A. orontium L. (*Misopates orontium* (L.) Raf.)   *Corn Snapdragon*
Probably introduced. Cultivated ground. Rare.
E.   Edge of cornfield, Beauchief 3381.
S.   Garden weed, Ashbourne 1746.
Linton: Matlock Bath 25, Pinxton 45, Breadsall 33.

A. majus L.   *Snapdragon*
Introduced. Naturalized on old walls. Rare.
W. On the rocks at Matlock 2957.
S.   Swarkestone Bridge 3826, walls at Ticknall 3424.
Linton: Cromford, Matlock 25/35, Rushall House and Burton Abbey walls.

LINARIA Miller

L. purpurea (L.) Miller   *Purple Toadflax*
Introduced. A rare garden escape.
E.   On high wall nr. a petrol station, Ashover 3463.
Linton: Not recorded.

L. repens (L.) Miller   *Pale Toadflax*
Probably introduced. An occasional casual.
W. Nr. Thornhill, Win Hill 1884, old railway, Bamford 2083, Monsal Dale
Station 1772, railway sidings, Parsley Hay 1463, walls, Fenny Bentley 1750.
S.   A garden escape in several spots at Ashbourne 1746, also Clifton 1644,
Derby 33.
Linton: Nr. Hill Top, Breadsall 33.

L. vulgaris Miller   *Yellow Toadflax*
Native. Banks, fields, waste places. Widespread and locally abundant.
Recorded for all squares except:

             W.   09, 19, 18.
             S.   23.

Linton: Thornhill 18.

CHAENORRHINUM (DC.) Reichenb.

C. minus (L.) Lange (*Linaria viscida* Moench)   *Small Toadflax*
Doubtfully native. Cultivated ground and waste places, and especially along
railways. Frequent.

W. Ashwood Dale, railway track 0872, nr. Peak Forest Station 0976, Harpur
Hill 0570, Miller's Dale, railway track 1373, Old High Peak railway, nr.
Friden 1660-1, Tissington Hall 1752, Hopton Incline 2654.

E. Bolsover 47, Markland Grips 5075, Stony Houghton 4966, Hale Wood,
Ashover 3463, Glapwell 4866, Pleasley Vale 5265, railway nr. Black Rocks
3056.

S. Ashbourne 1746, Clifton goods yard 1644, nr. Trent Station 4932,
Darley Abbey 3538, Breadsall Station 3639, Burton 2223.

Linton: Bamford Station 18, Millhouses 37, Etwall Station 23.

KICKXIA Dumort.

K. elatine (L.) Dumort. (*Linaria elatine* (L.) Miller)   *Sharp-pointed Fluellen*
Native. Arable land, chiefly on light soils. Very rare; decreasing like other
arable weeds.

W. Ladmanlow tip 0471.

E. Linacre 37 (E. Drabble, J. Bot. 1909).

Linton: Renishaw 47, Wingerworth 36, Fritchley 35, Pinxton 45, Boozwood
nr. Holbrook 34, Calke 32.

CYMBALARIA Hill

C. muralis P. Gaertner, B. Meyer & Scherb. (*Linaria cymbalaria* (L.) Miller)
                                                    Ivy-leaved Toadflax
Introduced. Old walls. Frequent.

W. Recorded for all squares except 09, 19, 08.

E. Ford, Eckington 4081, Barlow 3474, Brierley Wood 3775, Ashover
3462, Pleasley Vale 5265, Whatstandwell 3354, Milford 3545, Belper 3347,
Langley Mill 4448.

S. Recorded for all squares except 43, 21.

Linton: Glapwell 46, Sawley 43.

SCROPHULARIA L.

S. nodosa L.   *Figwort*
Native. Damp woods and hedgebanks, sides of ditches. Common.
Recorded for all squares except 09.

Linton: 'Generally distributed'.

S. auriculata L. (*S. aquatica* L.) *Water Betony*
Native. Streamsides, ditches, wet woods and meadows. Frequent.
Recorded for all squares except:
> **W.** 09, 19, 08, 18, 28.
> **E.** 38, 34, 44.

Linton: 'Frequent. Singularly absent from the valley of the Dove and its immediate surroundings'.

S. umbrosa Dumort.
Native. Shaded streamsides. Very rare.
E. Pleasley Vale 5265, Cromford Canal at High Peak Junction 3156.
Linton: Nr. Cromford Village 25, Burton 22.

MIMULUS L.

M. guttatus DC. *Monkey-flower*
Introduced. Streamsides and wet places. Locally frequent.
W. Recorded for all squares except 09, 19, 25.
E. Ford Valley 4080, Cordwell 3077, Upper Langwith 5169, Cromford Canal 3156.
S. Hanging Bridge 1644, Ashbourne Green 1947, Shirley Brook 2140, Foston 1831, Doveridge 1134, Sutton Brook 2333, Derwent Bank, Darley Abbey 3538, Willington 2928.

M. luteus L. *Blood-drop Emlets*
Introduced. Streamsides and wet places. Occasional; less frequent than *M. guttatus*.
W. Cavendish Golf Links 0774, Miller's Dale 1573, Monk's Dale 1374, Chee Dale 1273, Lathkill Dale 2165, River Wye at Bakewell 2268, Via Gellia 2656.
Linton: Stirrup Wood 98, Hathersage 27, Creswell Crags 57, Ashbourne 14, Morley 34, Knowle Hills 32.

M. guttatus and *M. luteus* have been confused in the past, and some of the records listed above may have been incorrectly ascribed. A further source of possible confusion is the existence of hybrids between the two species. These have been clearly recognized only recently, and they are likely to occur in more localities than the two given below.

M. guttatus × M. luteus
Arising by hybridization of the parents *in situ*. The hybrids have sterile pollen and may be recognized by the combination of large reddish spots on the corolla lobes together with numerous small dots in the throat. Occasional, but probably under-recorded.
W. Dovedale 1453.
S. Bradley 2245.

M. moschatus Douglas ex Lindley   *Musk*

Introduced. Commonly cultivated and occasionally naturalized in wet places in the north-west of the county.

W. Yorkshire Bridge 1985, Longshaw 2680.

E. The Ponds, Stubbing Court 3667, Tansley Moor 3261.

Linton: Not recorded.

LIMOSELLA L.

L. aquatica L.   *Mudwort*

Native. Margins of ponds and reservoirs, especially on mud bared during periods of exceptionally low water levels. Local and rare.

Seeds must remain viable but dormant for many years of high water levels and germinate in exceptional seasons.

W. Combs Reservoir 0379.

S. Pondside, Bradley Pastures 2246, pond now filled in.

Linton: Shipley Reservoir 44, Bradley Pastures 24, Willington, Repton and Foremark 32.

ERINUS L.

E. alpinus L.

Introduced. Naturalized on old walls and in rocky woods. Rare.

W. Longshaw House nr. Grindleford 2679, Manystones Lane. Longcliffe 2355.

Linton: Not recorded.

DIGITALIS L.

D. purpurea L.   *Foxglove*

Native. Open places in woods, open scrub, hedgebanks, heaths, waysides etc., on acid soils. Common except on limestone and locally abundant in cleared woodland.

Recorded for all squares.

Linton: 'Common except on limestone'.

VERONICA L.

V. beccabunga L.   *Brooklime*

Native. Ditches, streams, ponds, marshes and wet meadows. Common and widespread.

Recorded for all squares except:

W. 09.

E. 34.

Linton: 'Common in swamps'.

V. anagallis-aquatica L.   *Water Speedwell*

Native. Ditches, ponds, streams, wet meadows. Frequent.

W. Recorded for 08, Ashwood Dale 0872, Monk's Dale 1374, Chee Dale 1273, Middleton Dale 2175, recorded for 06, Hartington 1262, Bradford Dale 2164, Rowsley 2465, Dovedale 1451.

E. Scarcliffe Park Wood 5270.

S. River Dove, Mapleton 1648, recorded for 43, Scropton 1829, Willington 2927, Anchor Church 3327.

Linton: Haddon 26, Via Gellia 25, Cresswell 57, South Normanton 45, Derwent above Allestree 34, Barton Fields and Sudbury 13, Breadsall 33, Calke and Repton 32.

(*V. catenata* Pennell, with pink flowers and spreading fruit-stalks, is in neighbouring counties but has not been recorded for Derbyshire.)

V. scutellata L.   *Marsh Speedwell*

Native. Ponds, ditches, marshes, wet heaths. Local, and usually on acid substrata.

W. Recorded for 08, Swann's Canal 0470, Longshaw Estate 2679, Bar Brook and Blake Brook 2774.

E. Recorded for 57. Lydgate, Holmesfield 3077, Breadsall Priory 3841.

S. Bradley Brook 2244.

Linton: Nr. Snake Inn 19, Chee Dale 17, Markland Grips and Creswell Crags 57, Ault Hucknall 46, Clifton 14. nr. Derby 33, Sawley Junction 43, Egginton Heath and Burton 22, Calke 32.

V. officinalis L.   *Common Speedwell*

Native. Open woods, heaths and moors, banks, waysides, especially on dry non-calcareous soils. Frequent.

Recorded for all squares except:

> **W.** 09, 19.
> **E.** 47, 46, 35, 45, 34.

Linton: Renishaw 47, between Tibshelf and Hucknall 46, Ambergate 35, Pinxton 45, Morley 34.

V. montana L.   *Wood Speedwell*

Native. Damp woods and shady places by streams. Local.
Recorded for all squares except:

> **W.** 09, 08, 18, 16.
> **E.** 35, 44, 48.

Linton: 'Local'.

V. chamaedrys L.   *Germander Speedwell*

Native. Woods, hedgebanks, roadsides, grassland, on a wide range of soils, but avoiding strongly acid soils. Common and widespread.
Recorded for all squares except:

W. 09.

Linton: 'Common'.

V. repens Clarion ex DC.

Introduced. An occasional escape.

S. Appeared as a weed in the lawn of Alrewas House, Hanging Bridge, nr. Ashbourne 1545.

Linton: Not recorded.

V. serpyllifolia L.   *Thyme-leaved Speedwell*

Native. Pastures, arable land, waysides, gardens on moist soil. Common and widespread.

The Derbyshire plant is ssp. *serpyllifolia*.

W. Recorded for 08, Ladybower 1986, Ebbing and Flowing Well 0879, Woo Dale 0972, Chatsworth 2670, Lathkill Dale 1966, Rowsley 2465, Dovedale 1551, Brassington 2154.

E. Ford nr. Eckington 4081, Linacre 3372, Fallgate 3562, Stanley 4040.

S. Ashbourne 1746, Bradley 2245, Cubley Covert 1539, Locko 4138, Drakelow 2813, also Overseal and Grange Wood in 21.

Linton: 'Common'.

V. arvensis L.   *Wall Speedwell*

Native. Cultivated ground and bare places on dry soils in grassland and heath and on limestone rock-ledges. Common and widespread.

W. Ladybower 2086, Ladmanlow 0471, Chee Dale 1273, Taddington Dale 1770, Baslow 2672, Lathkill Dale 1966, Dovedale 1551, Carsington Pastures 2553.

E. Cordwell Valley 3176, Hate Wood 3463, Astwith 4464, Pye Hill 4352, Stanley 4040.

S. Snelston 1541, Ashbourne 1746, Yeldersley 2044, Weston-on-Trent 3927, Grange Wood, Lullington 2714, Smisby 3419.

Linton: 'Common. General and abundant'.

V. hederifolia L.   *Ivy-leaved Speedwell*

Native. Arable fields, gardens, hedgebanks, waste places. Common and widespread.

W. Cave Dale 1583, Cowdale 0872, Deepdale 0971, Taddington Dale 1671, Miller's Dale 1373, Coombs Dale 2274, Lathkill Dale 1966, 2165, Tissington 1752, Bonsall Moor 2559.

E.   Eckington 4081, Holmesfield 3277, The Butts, Ashover 3463, Stanley 4040.

S.   Ashbourne 1746, Sturston 2046, Rocester 1239, Kirk Langley 2838, Chester Green, Derby 3537, Bretby 2722, Smisby 3419.

Linton: 'Common and General'.

V. persica Poiret ('*V. tournefortii* C. Gmel.')   *Large Field Speedwell*

Introduced. Arable fields and gardens, waysides, waste ground. Common and widespread.

W.   Ladmanlow Tip 0471, Tissington 1752, Kniveton 2050, Bonsall Moor 2559.

E.   Beauchief 3381, Holmesfield 3277, Eckington 4379, Ashover 3462, Astwith 4464, Little Eaton 3641, Nutbrook, Ilkeston 4541.

S.   Recorded for all squares except 43.

Linton: 'Common. Generally distributed though scarcely plentiful'.

V. polita Fries   *Grey Speedwell*

Native. Cultivated ground. Common and widespread.

W.   Tissington 1752, Bonsall Moor 2559.

E.   Cutthorpe 3473, Stanley 4040.

S.   Ashbourne 1746, Osmaston 2042, Cubley 1638, Stanton-by-Dale 4638, common in cultivated ground in 22.

Linton: 'Common. In all the districts'.

V. agrestis L.   *Field Speedwell*

Native. Cultivated ground, waste places. Common and widespread.

W.   Bradwell 1680, Longstone Edge 1971, Hartington 1360, Tissington 1752, Bonsall Moor 2559.

E.   Pleasley Vale 5265, Stanley 4040.

S.   Ashbourne 1746, Rocester 1239, Longford 2137, Littleover 3233.

Linton: 'Common. Generally distributed'.

V. filiformis Sm.

Introduced. An escape from cultivation which is now extensively naturalized throughout the British Isles but is still uncommon in Derbyshire. It was first seen about twenty years ago as a weed of allotments near Buxton.

W.   Gib Yard allotments, Buxton 0672, Hartington 1260, laneside between Parwich and Pike Hall 1955, Thorpe 1550.

S.   In a meadow, Cubley 1638, Longford 2237, Bretby 2823.

Linton: Not recorded.

PEDICULARIS L.

P. palustris L.　　　　　　　　　　　　　　　*Marsh Lousewort*

Native. Marshes, fens and wet meadows. Rare.

W. Recorded for 07.

S. Hulland Moss 2546.

Linton: Woodseats nr. Castleton 18, Scarcliffe Park Wood and Whitwell 57, Tansley Moor 36, Quarndon Common 34, below Bradley Wood 14, Mackworth and Breadsall 33, Foremark and Repton 32.

P. sylvatica L.　　　　　　　　　　　　　　　　　*Lousewort*

Native. Damp heath and moorland, pastures and open woods on non-calcareous substrata. Frequent.

W. Recorded for 08, Derwent 1889, Jaggers Clough 1587, nr. Millstone Inn, Hathersage 2480, Burbage Edge 0373, Combs Moss 0576, Taddington Dale 1671, Longshaw 2578, under White Edge 2476, Washgate Valley 0567, Wye Farm 2565, Thorpe 1550, Newhaven 1758, Kniveton 2050.

E. Norton 3682, Linacre 3472, Totley 3079, Keltsedge 3363, by Birkinshaw Wood 3269, Wessington 3757, Whatstandwell 3354.

S. Ashbourne 1747, Hulland Moss 2546, Mansell Park 2644, Cubley 1639, Yeaveley 1839, Kirk Langley 2939, Repton Rocks 3221, recorded for 21.

Linton: Nr. the Snake and Alport Dale 19, nr. Renishaw 46, Pinxton 45, Morley 34, Mickleover, Breadsall and below Findern 33, Ockbrook 43, Winshill 22.

RHINANTHUS L.

R. minor L.　　　　　　　　　　　　　　　　　*Yellow-rattle*

Native. Grassland.

Two subspecies have been recorded for Derbyshire:

Ssp. **minor,** with suberect flowering branches from the middle and upper part of the stem, leaves parallel-sided for much of their length and usually no leaves between the uppermost branch and the bracts. Chiefly on dry nutrient-rich soils in meadows and pastures and on waysides. Common and widespread.

Recorded for all squares except:

　　　　　　　W. 09, 19.
　　　　　　　E. 47, 45, 38, 57.

Linton: 'Common, abundant everywhere'.

Ssp. **stenophyllus** (Schur) O. Schwarz, with ascending flowering branches from the lower and middle part of the stem, leaves tapering from the base and usually with one or two pairs of leaves between the uppermost branch and the bracts. Damp grassland. Occasional.

A plant from Shirley 2241 was named as 'probably this var.' by Dr. E. F. Warburg.

Linton: Owler Bar 27, Brailsford to Mercaston and Shirley, 'abundant' 24.

MELAMPYRUM L.

M. pratense L.   *Common Cow-wheat*

Native. Woods, open scrub, heaths; always on acid soils where woody plants are present. Local.

Derbyshire populations seem all to belong to ssp. *pratense;* the southern calcicole ssp. *commutatum* (Tausch ex Kerner) C. E. Britton has not been recorded.

W. Lower Small Clough, Upper Derwent 19, Hayfield 08, nr. Shatton 1982, Coombs 0478, Ramsley Reservoir 2876, Longshaw 2579, Grindleford 2477.

E. Twentywell 3280, Ecclesall Wood 3282, Unthank 3075, Barlborough 4678, Pumping Station, Holymoorside 3368, Cathole 3267, Whatstandwell 3155.

Linton: Snake Inn 19, Charlesworth 98, Jaggers Clough 18, Mytham Bridge 28, Mosboro' 48, Morley Moor 34, Shirley Park 24, Calke 32, Repton 32.

EUPHRASIA L.

E. officinalis L. *sensu lato*   *Eyebright*

Recorded for all squares except:

> **W.** 08, 18, 15.
> **E.**   37, 46.
> **S.**   13, 22, 21.

Linton: 'Frequent and rather local'.

E. nemorosa (Pers.) Wallr.   *Eyebright*

Native. Pastures, heaths, waysides. Locally frequent throughout the county.

W. Recorded for 08, Conies Dale 1280, Deep Dale 0971, Miller's Dale 1573, Longstone Edge 2173, Coombs Dale 2274, recorded for 06, Lathkill Dale 1765, Bradford Dale 2064, Dovedale 1453, Biggin Dale 1457, Via Gellia 2756, Minninglow Pit 2057.

E. Recorded for 47, 57, 36, Ogstone 3759.

S. Snelston 1541, Bradley 2046.

Linton: Matlock, Buxton 07, Ambergate 35, Millthorpe, Barlborough 47, Markland Grips, Scarcliffe Park and Whitwell 57, Yeldersley Rough nr. Ashbourne 24.

E. confusa Pugsley

Native. Limestone grassland. Locally frequent on limestone in the north-west of the county.

W. Cave Dale 18, Monk's Dale 1374, Taddington 1571, Longstone Edge 2073, Lathkill Dale 16, Gratton Dale 26, Dovedale 1453, Carsington Pastures 25, Minninglow Hill 2057.

S. Ashbourne Green 1847, Atlow Rough 2247.

Linton: Not recorded.

E. brevipila Burnat & Gremli

Native. Meadows and pastures. Local, and chiefly on limestone.

W. Monsal Dale 1771, Miller's Dale 1473, Buxton 1072, Middleton Dale 2175, Calver 2075, Lathkill Dale 1866, Over Haddon 26.

Linton: The Winnatts, Castleton 18, Wardlow and Chee Dale 17, Combes Moss 07, nr. the Snake Inn 19, Chatsworth 26/7, Bar Brook, above Baslow 27.

E. rostkoviana Hayne

Native. Moist meadows. Very local, and chiefly in the north-west of the county.

W. Kinder Scout 08, Edale 18.

Linton: Buxton 07, side of field nr. Wardlow Hay Cop 17, Chatsworth Park 26, 27, Whitwell 57, Upper Langwith 57.

E. anglica Pugsley

Native. Moist meadows and pastures. Local, and chiefly on limestone.

W. Recorded for 08, 07, Monsal Dale 1771, Miller's Dale 1373, Coombs Dale 2274, Calver Quarry 2374, Gib Hill 1563, Thorpe 1550, Parwich Moor 1757, Via Gellia 2656.

S. Clifton 1743, 1844, Atlow Rough 2247.

Linton: Not recorded.

ODONTITES Ludwig

O. verna (Bellardi) Dumort.   *Red Rattle*

Native. Arable land, waysides, waste places. Frequent.

Some Derbyshire material has been referred to ssp. *verna*, with branches coming off at an angle of less than 45 degrees and bracts longer than the flowers, and some to ssp. *serotina* Warb., with branches spreading at a wide angle and bracts equalling or shorter than the flowers. These subspecies are not always clearly distinguishable and most of our records are in any case for the species as a whole.

W. Miller's Dale 1373, 1473, Cressbrook Dale 1773, Ravensdale 1774, Calver New Bridge 2474, recorded for 26, Thorpe Station 1650, Ballidon 2054.

E. The Moss, Eckington 4080, Cordwell 3176, Stanfree 4873, Whitwell Wood 5278, Kelstedge 3363, Pleasley Vale 5265, Lea nr. Cromford 3257, Morley 3940, Stanley 4140.

S. Recorded for all squares except 13.

Linton: Bradwell 18, Pin Dale 18, Froggatt 27, Renishaw 47, Pinxton 45, Boylestone 13.

OROBANCHACEAE

LATHRAEA L.

L. squamaria L. *Toothwort*

Native. Woods, scrub and hedgerows. Local but not infrequent.

A parasite on the roots of various woody plants, especially elm and hazel.

W. Buxton 0572, Cressbrook 1774, also Ravensdale and Monsal Dale in 17, Shacklow Wood, Ashford 1769, Fenny Bentley 1950, Thorpe Rough 1450, High Tor, Matlock 2959, also Matlock Bath and Via Gellia in 25.

E. Ford, nr. Eckington 4081, Linacre 3372, Monk Wood 3576, Barlborough and Eckington in 47, Scarcliffe Park Wood 5170, Woods, Overton Hall 3462, Ashover 3464, Pleasley Park 5164, also Pleasley Vale and Langwith Wood in 56, Belper 3477.

S. Sturston 1946, river bank between Thorpe and Mapleton 1549, Corley nr. Bradley 2147, Darley Abbey 3438, Dale Abbey 4338.

Linton: Strines towards Marple & Mellor 98, Stirrup Wood 99, nr. Cromford Station 35, Calke 32.

OROBANCHE L.

O. ramosa L. *Branched Broomrape*

Introduced. There is a single old record and there must be some doubt about the identification.

No recent records.

Linton: Nr. Anchor Church, Ingleby 32.

O. purpurea Jacq. *Purple Broomrape*

Native. Parasitic on *Achillea millefolium* and a few other Compositae; near its northern limit in Derbyshire. Very rare.

E. Ilkeston Station 4642.

Linton: Not recorded.

O. rapum-genistae Thuill. *Greater Broomrape*

Native. Parasitic on broom and gorse. Very rare; no recent records.

Linton: Hag Tor, Matlock Bath 25/35, Stanton Moor 26, Little Eaton 34, Dale Hills 43, Repton Rocks 32.

O. minor Sm. *Lesser Broomrape*

Native. Parasitic chiefly on clover in fields and waste places. Very rare.

S. Repton 3026.

LENTIBULARIACEAE
PINGUICULA L.

P. vulgaris L.   *Common Butterwort*

Native. Fens and bogs, wet rocks, sides of upland streamlets; on calcareous to mildly acid substrata, and tolerant of some shade. Local.

W. Upper Alport Dale nr. Bleaklow 19, Lynch Clough nr. Howden Reservoir 1594, Oaken Clough 0786, Shatton 1982, Jaggers Clough 1687, Grindsbrook 1187, Conies Dale 1280, Toad's Mouth, Fox House 2681, Cupola, Hathersage 2481, Woo Dale and Ashwood Dale 0972, Grin Plantation 0572, Monk's Dale 1373/4, Brook Bottom, Tideswell 1476, Miller's Dale 1573, Topley Pike 1072, Peter Dale 1275, Coombs Dale 2174.

E. Whitwell Wood 5278/9.

S. Mugginton Bottoms 2843.

Linton: Hadfield 09, Froggatt and Umberley Brook 27.

UTRICULARIA L.

U. vulgaris L.   *Bladderwort*

Native. Pools and slow streams. Very rare.

No recent records.

Linton: Ponds nr. Swarkestone Bridge 32, Old Trent pools, Repton 32.

VERBENACEAE
VERBENA L.

V. officinalis L.   *Vervain*

Native. Waysides and waste places. Local and rare.

W. Nr. Monsal Dale Station 1772.

S. Cavendish Bridge, Shardlow 4429.

Linton: Winster and Via Gellia 25, Duffield, Quarndon and Morley 34, Mackworth and Little Chester 33, between Ockbrook and Borrowash 43, Winshill (not now in Derbyshire) 22, Calke and Melbourne 32, Linton 21.

LABIATAE
MENTHA L.

M. pulegium L.   *Penny-Royal*

Native. Wet heaths, pond-margins, especially on sandy soils. Very rare.

Linton (1903) reported that the species was extinct in most of its former stations and had not been seen for many years.

E. Roadside, Creswell Crags 5374.

Linton: Nr. Beighton 48, Pinxton 45, Radbourne and Langley Commons 23, Ockbrook Common 43.

M. arvensis L. *Corn Mint*

Native. Arable fields, woodland rides, damp places, ditches, ponds. Frequent and widespread.

Very variable, especially in length of internodes and size of leaves.

W. Derwent Reservoir 1691, Derwent 1889, Highlow Brook 2280, Monk's Dale 1374, Chatsworth 2771, Rowsley 2565, Thorpe 1550, Bonsall Moor 2559, Brassington 2454.

E. Recorded for all squares except 45.

S. Ashbourne 1745, Corley 2417, Cubley Covert 1539, recorded for 22, Swarkestone 3628, Netherseal 2913, Grove Farm, Drakelow 2318.

Linton: Glossop 19, Kingsterndale 07, Church Broughton and Etwall 23, Breadsall and Spondon 33, Locko 43.

M. × verticillata L. ('*M. sativa* L.'; *M. aquatica* × *M. arvensis*)

Native. Streamsides, ditches, pools, wet places. Frequent.

W. Chapel Milton 0581, Combs Reservoir 0379, Ladmanlow Tip 0471, Washgate 0567, Bradford Dale 2063, Rowsley 2465, Dovedale 1452, Carsington 2553.

E. Chesterfield canal 3974.

S. Ashbourne 1746, Cubley 1538, Kirk Langley 2939, Canal, Shardlow to Old Sawley 4530.

Linton: Nr. Castleton, Hope, Edale and Brough 18, Hathersage and Hood Brook 28, Monsal Dale and Miller's Dale 17, Abney Clough, Froggatt and Grindleford Bridge 27, Brackenfield and Cromford canal 35, Shirley and Atlow 24, Derby 33, Burton 22.

M. aquatica L. (incl. *M. hirsuta* Hudson) *Water Mint*

Native. Sides of rivers and ponds, canals, ditches, marshes, fens and wet woods. Common and widespread.

Very variable in hairiness, leaf-shape and form of inflorescence.

Recorded for all squares except:

W. 09, 28, 07.
E. 38.

Linton: 'Common'.

M. × gentilis L. (*M. arvensis* × *M. spicata*)

Native. Sides of ditches and streams, damp waste places. Occasional.

E. Press Reservoirs, Wingerworth 3565.

S. Yeldersley Pond and stream 2144.

Linton: Shatton 18, Bamford-Mytham Bridge, Burbage Brook and Fox House Inn 28, Miller's Dale and Chee Dale 17, Abney Clough 27, Thorpe 15.

M. × piperita L. (*M. aquatica* × *M. spicata*)   *Peppermint*

Native. Sides of streams and ditches, damp waysides etc. Frequent.

The Derbyshire plant is var. *piperita*.

W. Stoney Middleton 2275, Bubnell 2472, Highlow Brook 2279, Lathkill Dale 1765, Bradford Dale 2164, Dovedale 1453, Grange Mill 2457.

E. Staveley 4374, Press Reservoirs 3565, Cathole 3267, Stony Houghton 4966, Alfreton Brook 4056.

S. Ashbourne 1948, 1846, nr. The Grange, Cubley 1641, river-bank, Darley Abbey 3538, Trent 4831.

Linton: Chapel and Low Leighton 08, Hood Brook 28, Lea Bridge and south of Whatstandwell 35, nr. Belper 34, Barton Park 13, Anchor Church 32.

M. × smithiana R. A. Graham ('*M. rubra* Sm.'; *M. aquatica* × *M. arvensis* × *M. spicata*)

Native. Sides of rivers, ponds and ditches, damp hedgebanks and waste ground. Occasional.

W. Nr. Clarion Hut 2981, above Baslow 2572, Ashford 1869, Wolfscote Dale 1456, Dovedale 1452.

E. Roadside ditch nr. Barlow 3674, Press 3565, West Hallam 4240.

S. Fenny Bentley Brook 1646, ditch south of Hulland Moss 2445, Shirley Waters 2140.

Linton: Whaley Bridge 08, Beeley 26, Shottle 34, Eaton Wood 13, Ockbrook 43, Ticknall 32.

M. spicata L. ('*M. viridis* L.'; incl. '*M. longifolia* Hudson')   *Spear Mint*

Introduced. Naturalized on streambanks, damp roadsides and waste places. Occasional.

This is the mint most commonly cultivated as a pot-herb, and is an extremely variable species. Hairy forms have been confused in the past with *M. longifolia* (L.) Hudson, which is not a British plant.

W. Ladmanlow tip 0371, Deep Dale 1670, site of an old café, Taddington Dale 1770, recorded for 26, Via Gellia 2757 and 2050, Carsington 2452.

E. Buxton 0471, Holymoorside 3369, Northedge 3565, Ashover 3461, Pleasley Vale 5265, railway bank, West Hallam 4240.

S. River banks, Mapleton to Clifton 1545, nr. Clifton Station 1644, Old Derby Road, Ashbourne 1845, Brailsford Tip 2541, Eaton 0936.

Linton: River Lathkill below Alport, Youlgreave and Beeley 26, Dovedale 15, Pinxton 45, Aldercar 44, Marston Montgomery 13, Norbury 14, Duffield 34, Breadsall 33, Calke and Repton 32.

M. × niliaca Juss. ex Jacq. (incl. *M. alopecuroides* Hull; *M. spicata* × *M. suaveolens*)

Introduced. An occasional escape from cultivation; roadsides and waste places. The Derbyshire plant is var. *alopecuroides* (Hull) Briq., broad-leaved and often mistaken for *M. suaveolens* Ehrh. ('*M. rotundifolia* L.').

W. Longstone Edge 2037, roadside nr. Hartington 1260, Mill Dale 1454, Wolfscote Dale 1357, Beresford Dale 1259, Cromford 2857.

E. Nr. Scarcliffe Park Wood 5270, Kelstedge 3363, Walton Lees Farm, Ashover 3267.

S. Ashbourne 1845, Marston Montgomery 1440, Jingler's Lane, Bradley 2346, Biggin-by-Hulland 2643, Cubley Coppice Farm 1737, Sinfin 3431.

Linton: Kelstedge 36, Ambergate, Wessington and Brackenfield Green 35, Yelderlsey 24.

M. suaveolens Ehrh. ('*M. rotundifolia* Hudson')   *Apple-scented Mint*

Introduced. Streambanks and damp pastures. Very rare; no recent records.

It is likely that some of the older records are really of broad-leaved forms of *M. × niliaca*.

Linton: Miller's Dale 17, Mill Dale 15, Kedleston 34, Derby 33, Repton 32.

LYCOPUS L.

L. europaeus L.   *Gipsy-wort*

Native. Marshes and fens, ditches, streamsides. Frequent.

E. Recorded in all squares except 57, 56, 38.

S. Recorded in all squares except 14, 13, 23.

Linton: Creswell Crags 57, between Edlaston and Norbury 14, and Longford and Boylestone 13, Burnaston and Hatton 23.

ORIGANUM L.

O. vulgare L.   *Marjoram*

Native. Dry pastures, hedgerows and banks, chiefly on limestone and especially where the ground has been disturbed; common on partly stabilized limestone scree and on abandoned cultivated land over limestone. Locally abundant.
Marjoram is gynodioecious, with some plants hermaphrodite, some ovulate only.

W. Recorded for all squares except 19, 08.

E. Markland Grips 5074, Slagmill Plantation 3562, Pleasley Vale 5265, West Hallam 4142.

S. Snelston Limepits 1541, nr. Norbury Hall 1242, Locko 4138, Overseal 2915.

Linton: Conies Dale, Peak Forest 08, Bolsover 47, Crich 35, Breadsall 34.

THYMUS L.

T. pulegioides L.   *Larger Wild Thyme*

Native. Dry grassland, usually on calcareous soil. Very rare.

W. Miller's Dale 1573.

Linton: Not recorded, but for Linton 21 there is the comment 'approaching the form *Chamaedrys*'.

T. drucei Ronn. ('*T. serpyllum* Fr.')   *Wild Thyme*

Native. Dry pastures and rocky outcrops, especially in the limestone dales; also on heathy banks. Locally abundant.

W. Recorded for all squares except 09, 19, 08, 28.

E. Markland Grips 5074, recorded for 36, Pleasley Vale 5261, recorded for 35.

S. Birchwood Quarry, Snelston, on the limestone outlier 1541.

Linton: Nr. Hayfield 08, nr. Bolsover and Clowne 47, Turnditch 24, Radbourne 23, Littleover 33, Burton 22, Calke 32.

Also at Linton ('approaching the form *Chamaedrys*') 21.

CALAMINTHA Miller

C. ascendens Jordan ('*C. officinalis* Moench')   *Common Calamint*

Native. Dry banks and rocky outcrops, chiefly on limestone. Rare.

W. Via Gellia 2556.

E. Creswell Crags 5274, Fallgate, Ashover 3562.

Linton: Morley 34, Ockbrook 43, Repton Park 32, Linton 21.

C. nepeta (L.) Savi (*C. parviflora* Lam.)   *Lesser Calamint*

Doubtfully native. Dry banks. The two old records may be erroneous and cannot be confirmed because both sites have now been completely built over. There are no recent records.

Linton: Normanton 33, Ockbrook 43.

ACINOS Miller

A. arvensis (Lam.) Dandy (*Calamintha arvensis* Lam.)   *Basil Thyme*

Native. Dry sunny banks and south-facing rocky outcrops, chiefly in the limestone dales. Local.

W. Chrome Hill 9767 and 0767, Flagg Dale 1273, Tideswell Dale 1574, Cressbrook Dale 1773, Chee Dale 1273, Miller's Dale 1473, Tansley Dale 1774, Middleton Dale 2275, Dovedale 1454.

S. Birchwood Quarry, on the limestone outlier 1541.

Linton: Glapwell 46, Holbrook 34, Snelston 14, Foremark 32.

CLINOPODIUM L.

C. vulgare L. (*Calamintha clinopodium* Benth.)   *Wild Basil*
Native. Hedgebanks, wood-margins and scrub, disturbed ground etc., usually
on calcareous soils. Frequent.

W. Taddington Dale 1671, Coombs Dale 2275, Earl Sterndale 0966, nr.
Ashford 16, Wolfscote Dale 1351, Brassington Rocks 2254.

E. Pebley Pond 4878, Whitwell Wood 5278, Scarcliffe Park Wood 5270,
Bolsover 4869, Pleasley Vale 5265, Crich 3454, Morley 3940, Stanley 4140.

S. Mapleton 1647, Shirley 2140, Radbourne 2737, Breadsall 3739, Trent
Lock 4831, Sawley 4731, Ticknall 3523, Lullington 2513.

Linton: Ashwood Dale 07, Manners Wood, Bakewell 26, Middleton Dale 27.

MELISSA L.

M. officinalis L.   *Balm*
Introduced. Grown in gardens for its sweet scent and occasionally escaping.
Rare.

W. Froggatt village 2476.

E. Cartledge Lane 3277.

Linton: Not recorded.

SALVIA L.

S. pratensis L.   *Meadow Clary*
Introduced. A very rare casual in waste places; no recent records.

Linton: Crich 35, Swarkestone 32.

MELITTIS L.

M. melissophyllum L.   *Bastard Balm*
Introduced. Hedgebanks. Very rare; no recent records.

Linton: Between Sinfin and Barrow 33/32.

PRUNELLA L.

P. vulgaris L.   *Self-heal*
Native. Meadows and pastures, waysides, woodland clearings and rides,
waste places; common in limestone pastures on loamy soils, absent from
strongly acid soils. Common and widespread. White and rose-coloured
forms are not uncommon.

Recorded for all squares.

Linton: 'Common. Abundant everywhere'.

STACHYS L.

S. arvensis (L.) L.   *Field Woundwort*

Native. Arable and waste land on non-calcareous soils, especially sand or gravel. Rare.

E.   Nr. Beauchief Abbey 3281, nr. Pebley Pond 4880, Cartledge 3277, Troway 3879, Stubbing 3567, Tibshelf 4360, Milford 3445.

S.   Bradley Old Park 2344.

Linton: Elmton 57, Pinxton 45, Duffield 34, Breadsall 33, Long Eaton 43, Swarkestone 32, Egginton 22, Cauldwell 21.

S. palustris L.   *Marsh Woundwort*

Native. By streams, ditches and canals and in swamps, fens and wet woods. Fairly frequent.

W.   Recorded for 08, Buxton 0573, 0672, Kniveton 2050, Ashford 1769.

E.   Killamarsh Pond 4780, canal, Renishaw 4477, Whitwell Wood 5278, Wingerworth 3667, Whatstandwell 3354, canal, Ambergate 3452, Codnor Park 4350, Belper 3447, nr. Langley Mill 4545.

S.   Recorded for all squares except 13, 23.

Linton: Brough 18, Ashford 26, Sudbury 13, Thrumpton Ferry 53.

S. × ambigua Sm. (*S. palustris* × *S. sylvatica*)

Native. Damp hedgesides, ditches; growing with the two parent species. Rare.

E.   Nr. Holmesfield 37, Brierley Wood 3775, Cowley Bar 3278.

S.   Nr. Shirley 2143, nr. Mackworth 3038.

Linton: Mellor 98, Baslow and Hassop 27, Fritchley and Tansley 35, Mugginton 24, Morley 34, Willington and Stapenhill 22, Calke 32.

S. sylvatica L.   *Hedge Woundwort*

Native. Woods, hedgebanks, waysides, ditches and shady waste places; often with other tall herbs on rich moist soils in valley-bottom woods. Common and widespread.

Recorded for all squares.

Linton: 'everywhere abundant'

BETONICA L.

B. officinalis L. (*Stachys betonica* Bentham)   *Betony*

Native. Open woods, hedgebanks and pastures on nutrient-rich but usually non-calcareous soils and especially characteristic of shallow plateau-loams over limestone, as on the brows of the limestone dales where it is commonly associated with *Agrostis tenuis*, *Centaurea nigra* and *Lathyrus montanus*, and of soils derived from 'toadstone'; sometimes on poorly drained but only rarely and very locally on well-drained calcareous soil. Locally abundant.

Recorded for all squares except:

W. 19, 08.

S. 43.

Linton: 'In all districts'.

BALLOTA L.

B. nigra L. *Black Horehound*

Native. Roadsides, hedgebanks, waste ground. Frequent at low altitudes on the Magnesian Limestone and Trias, rare elsewhere.

The Derbyshire plant is ssp. *foetida* Hayek.

W. Bamford 2083, Lathkill Dale 1966.

E. Brockwell 3771, Creswell Crags 5374, Ashover 3463, Palterton 4768, Pleasley Vale 5165, Kirk Hallam 4540.

S. Scropton 1930, Burnaston 2930, Alvaston 3933, ballast heaps by Trent Station 4932, Swarkestone 3728, Lullington 2714.

Linton: Matlock 25/35, Staveley 47, Pinxton 45, Mugginton 24, Marston-on-Dove 22.

LAMIASTRUM Heister ex Fabr.

L. galeobdolon (L.) Ehrend. & Polatsch. (*Lamium galeobdolon* (L.) Crantz) *Yellow Archangel, Weasel-snout*

Native. Woods and hedgebanks on a wide range of soil-types, but absent from strongly acid soils. Frequent.

W. Chew Wood 9992, recorded for squares 08, 18, 07, 17, Chatsworth 2671, Longshaw 27, Lathkill Dale 1865, nr. Bradbourne 1951, Kniveton 2049.

E. Recorded for all squares except 45.

S. Recorded for all squares except 22.

Linton: Ashopton 19, Charlesworth 98, Pinxton 45.

LAMIUM L.

L. amplexicaule L. *Henbit*

Native. Arable land, waste places, especially on light soils; walls. Occasional in the south of the county. Formerly more frequent but now decreasing.

E. Pleasley Vale 5265, West Hallam 4341.

S. A persistent garden weed, Ashbourne 1746, Sawley Junction 4832, Willington 3028.

Linton: Osmaston and Shirley 24, nr. the Spread Eagle and Barton Fields 23, Breadsall Moor 33, Drakelow 22, Ingleby and Repton 32.

L. molucellifolium Fries (*L. intermedium* Fries)   *Intermediate Dead-nettle*
Probably introduced. A very rare casual of cultivated ground.
No recent records.
Linton: Drakelow 21.

L. hybridum Vill.   *Cut-leaved Dead-nettle*
Native. Cultivated and waste ground; roadsides. Rare.
No recent records.
Linton: Garden weed at Burbage, Buxton 07, Pleasley Station 56, Ticknall
Vicarage 32.

L. purpureum L.   *Red Dead-nettle*
Native. Cultivated and waste ground, hedgebanks, waysides. Common and
widespread.
Recorded   for all squares except:
>   **W.** 09,19.
>   **E.**   46.
Linton: 'Common. Generally distributed'.

L. album L.   *White Dead-nettle*
Probably native. Roadsides, hedgebanks, waste places. Common at low
altitudes but absent from some areas of the county.
Recorded for all squares except:
>   **W.** 09, 19, 08, 07.
Linton: 'Common. Generally distributed'.

L. maculatum L.   *Spotted Dead-nettle*
Introduced. Commonly cultivated in the form having a large whitish blotch
on the leaves and sometimes becoming naturalized on hedgebanks, roadsides and
waste places near houses. Not infrequent.
W. Alport Dale 19, Pin Dale 1682, riverside, Hope 1783, Ashopton 2086,
various localities in Buxton 0573, Wardlow Mires 1775, Newhaven 1659,
also Thorpe 1550, Fenny Bentley 1750.
E.   Milltown, Ashover 3561, Park Nook, Quarndon 3241.
S.   Fenny Bentley Brook 1646, Wyaston 1842, Kniveton Brook 2147,
roadsides, Overseal 2915.
Linton: Baslow 27, Darley Dale 26, Wirksworth, Hopton and woods by the
river, Matlock 25, Mickleover 33, Repton and Knowle Hills 32.

LEONURUS L.
L. cardiaca L.   *Motherwort*
Introduced. Hedgebanks, roadsides, waste places, banks. Rare.
W. Dovedale 15.

E. Barlow 3474.

S. The Mountain, Kirk Ireton 2649.

Linton: Via Gellia 25, by Derwent, Cromford 25/35, Spital, Chesterfield 37, Handley nr. Stretton 36, Mackworth 33.

GALEOPSIS L.

G. angustifolia Ehrh. ex Hoffm.   *Narrow-leaved Hemp-nettle*

Native. Arable land and gardens, waysides, waste ground. Rare.

W. Lathkill Dale 1765, Biggin Dale 1458.

E. Whitwell Woods 5278.

Linton: Peak Forest Station 07, Monsal Dale 17, Totley 38, Bolsover 47, Pinxton 45.

G. ladanum L.

Introduced. A very rare casual of cultivated land.

W. Biggin Dale 1458.

Linton: Nr. Elmton 57.

G. tetrahit L.   *Common Hemp-nettle*

Native. Arable land and gardens, hedgebanks, wood-margins, waste ground. Common and widespread as a field weed on a wide range of soil-types.

W. Derwent Reservoir 1691, Watford Bridge 0086, Bamford 2182, Miller's Dale 1573, Monsal Dale 1771, High Rake 1677, Calver New Bridge 2475, Hartington 1360, Monyash 1666, Dovedale 1453, Bradbourne 2052, Minninglow 2057.

E. Recorded for all squares.

S. Recorded for all squares.

Linton: 'Generally distributed'.

G. tetrahit L. var. *nigricans* Bréb., a 'large plant with blackish calyx and generally nigrescent stem and leaves' is recorded by E. Drabble (1929) for Cowley, Baslow. There are more recent records from:

W. Minninglow 25, Dovedale 15.

S. Mercaston 24, nr. Ashbourne 14, Spondon 3935.

Linton: Not recorded.

G. bifida Boenn.

Native. Arable land and waste ground. Often with *G. tetrahit* and escaping attention. Probably not infrequent but further observation needed.

Differs from *G. tetrahit* in having the red-tipped glandular hairs distributed all over the internodes (not confined to just below the nodes) and the middle lobe of the lower lip of the corolla deeply emarginate (not entire). The two species form hybrids.

E. Drabble (J. Bot. 1909) writes 'This is the common form round Chesterfield. We have not seen true *tetrahit* on the Coal Measures'.

E.   Tapton 37, Elmton 57, Hasland and Cat Hole 36.

Linton: Not recorded.

G. speciosa Miller (*G. versicolor* Curt.)   *Large-flowered Hemp-nettle*
Native. Arable land and gardens, wood-margins. Local, but not infrequent.

W.   Recorded for all squares except 09, 19, 17.

E.   Nr. Beauchief Abbey 3381, Freebirch 3072, Holymoorside 3369, Pleasley Vale 5265, Wessington Green 3757, nr. Langley Mill 4546.

S.   Recorded for all squares except 21.

Linton: Hayfield 09, Renishaw 47, Matlock Bath 25, Great Shacklow Wood 16, Shirley 24, Morley 34, Mickleover 33, Repton 32, Drakelow 21.

NEPETA L.

N. cataria L.   *Cat-mint*
Native. Hedgebanks and roadsides, dry stony or rocky places, especially on limestone. Rare.

W.   Ladmanlow tip, a few plants only, 0471.

E.   Beighton 48.

Linton: Creswell Crags 57, Matlock and Via Gellia 25, Ticknall 32, Calke 32, Linton 21.

GLECHOMA L.

G. hederacea L. (*Nepeta glechoma* Bentham)   *Ground Ivy*
Native. Woods, hedges, pastures, waste places; especially characteristic of woodland clearings and recently disturbed ground. Common and widespread. Gynodioecious, the flowers of hermaphrodite plants much larger than those of the 'female' plants lacking fertile stamens. White-flowered plants are found occasionally.

Recorded for all squares except 09, 19.

Linton: 'Common everywhere'.

MARRUBIUM L.

M. vulgare L.   *White Horehound*
Probably introduced. Waste ground, roadsides. Very rare; no recent records.

S.   Clifton tip 1644.

Linton: Middleton Dale 27, Repton 32.

SCUTELLARIA L.

S. galericulata L. *Common Skull-cap*

Native. Sides of streams, canals, ditches and ponds and in fens and wet meadows. Locally frequent.

W. Recorded for 08, 07, Monsal Dale 1771, Longshaw 2578, nr. Baslow 2774, recorded for 06.

E. Killamarsh 4481, Whittington 3974, Chesterfield canal 4074, Renishaw Lake 4478, Wingerworth 3766, Hardwick Hall 4563, Pleasley Vale 5265, Cromford Canal 3056, Alfreton Brook 4056, Langley Mill 4546.

S. Hulland 2446, canal, Derby 3537, Trent Lock 4831, Thrumpton Ferry 5031, Swarkestone 3728, canal nr. Weston 4027, Grange Wood 2714.

Linton: Edale 18, Allestree 34, Barton Blount 13, Marston-on-Dove 22.

S. minor L. *Lesser Skull-cap*

Native. Wet places on heaths and acid fens. Rare.

W. Blacka Moor, Dore 2980, Salter Sitch 2877, Greaves's Piece 2977.

E. Wessington 3757.

S. Hulland Moss 2446.

Linton: Pinxton 45, Calke and Repton Shrubs 32.

TEUCRIUM L.

T. scorodonia L. *Wood Sage*

Native. Woods, hedgebanks, heaths, cliff-ledges, partly stabilized limestone scree etc.; usually in habitats where the vegetation-cover is incomplete but on a wide range of soil-type, avoiding only very strongly acid, very heavy or waterlogged soils. Common and widespread.

Teucrium scorodonia has recently been shown to have races in Derbyshire which differ in their resistance to lime-chlorosis when grown on calcareous soils. Recorded for all squares except:

E. 57, 46, 45.

AJUGA L.

A. reptans L. *Bugle*

Native. Damp woods, hedgebanks and roadsides, damp meadows and pastures. Common and widespread.

Linton refers to a form without stolons found on the high ground between Matlock and Castleton.

Recorded for all squares except:

W. 09, 28.

Linton: 'Common. In all the districts'.

PLANTAGINACEAE

PLANTAGO L.

P. major L.   *Great Plantain*

Native. Roadsides, gateways, farmyards, waste ground; also in pastures where there is bared ground, as by trampling. Common and widespread.

Very tolerant of trampling and often accompanying *Poa annua, Polygonum arenastrum* and *Matricaria matricarioides* in heavily trampled places.

Recorded for all squares.

Linton: 'Common. General'.

P. media L.   *Hoary Plantain*

Native. Pastures, banks, waysides, especially on calcareous soils; characteristic of loamy limestone soils and usually avoiding shallow black limestone soils. Frequent and widespread.

Recorded for all squares except:

> **W.** 09, 19.
> **E.**   45, 38.
> **S.**   21.

Linton: Near The Snake Inn 19.

P. lanceolata L.   *Ribwort Plantain*

Native. Meadows, pastures and grassy waysides on basic or mildly acid soils. Common and widespread.

Recorded for all squares.

Linton: 'Common. Everywhere abundant'.

P. coronopus L.   *Buck's-horn Plantain*

Native. Dry open places on roadsides, pastures and waste ground, especially on sandy soil and avoiding limestone. Local and infrequent.

E.   West Hallam 4240.

S.   Trent 4931, Sawley 4732.

Linton: Little Eaton 34, Osmaston Manor Reservoir 24, Breadsall Moor 33, Melbourne, between Calke and Melbourne, and Willington Junction 32, Cauldwell 21.

LITTORELLA Bergius

L. uniflora (L.) Ascherson (*L. juncea* Bergius)   *Shore-weed*

Native. In shallow water at the sandy or gravelly margins of pools and reservoirs. Rare.

W. Combs Reservoir, Chapel-en-le-Frith 0379, Emperor Lake, Chatsworth 2670.

E.   Linacre 3272.

Linton: Chatsworth Park 27/26, Pebley Pond, Barlborough 47, Wingerworth 36.

CAMPANULACEAE

WAHLENBERGIA Schrader

W. hederacea (L.) Reichenb. *Ivy-leaved Bell-flower*

Native. Damp, acid, peaty places on moors and heaths or by clough streamlets. Very rare.

W. Mill Clough, Derwent 1889, River Derwent at Bamford 1984.

Linton: Jarvis Clough 18, Repton Rocks 32.

CAMPANULA L.

C. latifolia L. *Broad-leaved Bell-flower, Giant Throatwort*

Native. Woods and hedgebanks on nutrient-rich calcareous or mildly acid soils. Frequent.

Recorded for all squares except:

> **W.** 09, 19.
> **E.** 38, 47, 45.
> **S.** 22, 32.

Linton: Stirrup Wood 98, Holmesfield 38, Renishaw 47, Bretby 22, Repton Rocks 32.

C. trachelium L. *Nettle-leaved Bell-flower*

Native. Woods, scrub and hedgebanks, especially on limestone soils. Local. See also maps (page 93).

W. Buxton 0772, Chee Dale 1273, Monk's Dale 1374, Miller's Dale 1473, Monsal Dale 1871, Coombs Dale 2274, Calver 2374, Lathkill Dale 1966, Dovedale 1454, Matlock Bath 2959.

E. Scarcliffe Park Wood 5270, Pleasley Vale 5265, Shipley 4444.

S. Atlow 2249.

Linton: Cresswell 57, Langwith Wood 56, Spondon 43, between Duffield Bridge and Makeney 34, Breadsall 33.

C. rapunculoides L.

Introduced. Open woods, disturbed grassy places. Very rare.

E. Alfreton Brook 4056.

Linton: Not recorded.

C. persicifolia L.

Introduced. Established in open woods, hedgebanks etc. Rare.

W. Alport 2164.

Linton: Not recorded.

T

C. glomerata L.   *Clustered Bell-flower*

Native. Grassy places on limestone. Local and infrequent.

W. Buxton 0673, Cunning Dale 0872, Deepdale 0971, Chee Dale 1172, Topley Pike 1072, Long Dale 1362, Biggin Dale 1458.

Linton: Haddon Hall 26, Stanton-by-Dale 43, Spondon 43.

C. rotundifolia L.   *Harebell*

Native. Pastures, banks and waysides on a wide range of soils but avoiding the poorest and most strongly podsolized soils. Common and widespread.

Recorded for all squares.

Linton: 'Abundant everywhere'.

C. patula L.   *Spreading Bell-flower*

Probably native. Open woods, hedgebanks and cultivated fields. Very rare.
S. Melbourne, footpath to Castle Donington 4026.

Linton: Stapenhill 22.

Linton gives a few other stations (Morley, Osmaston and Breadsall) but states that they had become uncertain even in his day.

LEGOUSIA Durande

L. hybrida (L.) Delarbre (*Specularia hybrida* A.DC.)   *Venus' Looking-glass*

Probably native. A weed of arable land, especially of cornfields on a sandy or dry soil. Rare.

E. Stanley 4140.

S. Cornfield north-east of the Knob, Hulland 2546, Dale 4338.

Linton: Breadsall 33, Melbourne Common, Foremark Park Farm and Repton 32, Cauldwell 21.

JASIONE L.

J. montana L.   *Sheep's Bit*

Native. Dry banks and grassy places, especially on light, sandy or gravelly soils; avoiding calcareous soils. Local.

W. Recorded for 08, 18, Ladybower 2086, Barmoor Clough 0779, Abney 1979, Stoney Middleton 2375, Washgate 0567.

E. Railway track, Bramley Vale 4566, Whatstandwell 3354, Coxbench 3843, Morley 3940, Stanley 4140.

S. Dale 4338, Netherseal 2715.

Linton: New Mills 98, Bugsworth 08, Birchover 26, Wirksworth 25, Wingerworth 36, Pleasley 56, Tansley 35, Spinnyford Brook 24, Derby 33, Repton 32.

RUBIACEAE

SHERARDIA L.

S. arvensis L.   *Field Madder, Spur-wort*

Native. Arable land and waste places. Frequent, especially in the south of
the county.

W. Recorded for 07, Miller's Dale 1373, Chee Dale 1273, Monsal Dale
1771, Middleton Dale 2173, Hassop Quarries 2273, Lathkill Dale 1966,
Conksbury Bridge 2165, Dovedale 1551, Wolfscote Dale 1456.

E. Cordwell 3277, Elmton 5073, Glapwell 4866, nr. Whatstandwell 3255,
Stanley 4042.

S. Birchwood Quarry 1541, Ticknall 3523, Overseal 2915.

Linton: Youlgreave 16, Wensley 25, Norton 38, Renishaw 47, Markland
Grips 57, Whitwell 57, Tapton 36, Pleasley 56, Shirley 24, Quarndon Common
34, Alvaston 33, Ockbrook 43.

ASPERULA L.

A. cynanchica L.   *Squinancy-wort*

Native. Dry grassy slopes and banks on limestone. Rare and perhaps extinct.
No recent records.

Linton: The Warren, Ashford 16, Bonsall 25, Pinxton 45, Normanton 33.

A. arvensis L.

Introduced. A casual of arable land and waste places; not persisting. Rare.

E. Beauchief, on an allotment 3381.

S. Clifton, as a weed from canary seed 1545, waste ground nr. Darley Abbey
3538.

Linton: Not recorded.

GALIUM L.

G. odoratum (L.) Scop. (*Asperula odorata* L.)   *Sweet Woodruff*

Native. Woods on moist but well-drained substrata; common on limestone
and often on shaded scree-slopes but extending to quite acid mulls. Locally
abundant.

Recorded for all squares except:

        W. 09, 19, 18.
        E. 45, 34, 44.

S. Fenny Bentley 1849, Norbury 1342, Dale Abbey 4338, Grange Wood
2714.

Linton: Otterbrook, Edale 18, Newton Wood 45, Smalley 44, Burton 22.

G. cruciata (L.) Scop.   *Crosswort, Mugwort*
Native. Open woodland and scrub, hedgebanks, pastures and waysides,
especially in limestone areas. Abundant and widespread.
Recorded for all squares except:

> **W.** 09, 19, 08.
> **E.** 45.
> **S.** 23, 32.

Linton: 'Common'.

G. mollugo L.   *Great Hedge Bedstraw*
Native. A very variable species in which two subspecies are commonly
recognized.

Ssp. **mollugo**, the common form in Derbyshire, is a plant of open woodland,
hedgerows and roadsides.
W. Nr. Whaley 98, Miller's Dale 1373, Parwich Hill 1854, above Cromford
2857, Brassington 2355.
E. Spinkhill 4578, Pebley Pond 4878, nr. Bolsover 4972, Creswell Crags
5074, Pleasley Vale 5265, Bilberry Knoll 3057.
S. Above Bradley Wood 2046, Scropton 1930, Darley Moor, nr. Ashbourne
1741.
Linton: Barlborough 47, Winster 25, Dethick 35, Pinxton 45, Osmaston 24,
Mickleover 33, nr. Melbourne, Repton, Bretby and nr. Stanton-by-Bridge 32.

Ssp. **erectum** Syme (*G. erectum* Hudson) is a very rare plant of dry roadsides
and hedgebanks.
W. Calver Sough 27, Taddington by-pass 1471.
Linton: Yeldersley 24, Mackworth 34, Heanor 44.

G. × pomeranicum Retz. (*G. mollugo* x *G. verum*)   *Hybrid Bedstraw*
Native. Roadsides and hedgebanks. Rare.
E. Roadside, nr. Whitwell 5277.
Linton: Not recorded.

G. verum L.   *Yellow Bedstraw, Lady's Bedstraw*
Native. Grassland, hedgebanks and waysides on all but the most acid soils
and abundant on loamy soils over limestone. Frequent and widespread.
Recorded for all squares except 09, 19, 28, 45.
Linton: Hayfield 09, Snake Inn 19.

G. saxatile L.   *Heath Bedstraw*
Native. Heaths, moors, grassland and woods on acid soils, including surface-
leached loamy soils over limestone. Very common and widespread.
Recorded for all squares except 43, 22.
Linton: 'Frequent but somewhat local'.

G. sterneri Ehrendorf. ssp. **sterneri** ('G. *silvestre* Poll.')   *Slender Bedstraw*
Native. Cliff-ledges, rock outcrops and open grassy slopes of the Carboniferous and Magnesian Limestones; replaced by *G. saxatile* where the soil is acid at the surface and therefore more characteristic of dry south-facing or very steep slopes than of gentler north-facing slopes. Locally frequent.

W. Cave Dale 1583, Ashwood Dale 0972, Deepdale 0971, Chee Dale 1273, Miller's Dale 1473, Ravensdale 1774, Middleton Dale 2175, Longstone Edge 2073, Earl Sterndale 0966, Lathkill Dale 1765, Conksbury Bridge 2165, Dovedale 1454, Wolfscote Dale 1455, Brassington Rocks 2154, Haven Hill Dale 2152.

E. Markland Grips 5074, Whitwell Wood 5278.

G. palustre L.   *Marsh Bedstraw*
Native. Marshes, fens, flushes, ditches, streamsides and reed-swamp. Common and widespread.

Recorded for all squares except 09, 45.

Linton: 'Generally distributed'.

This very variable species is represented in the county by the two main subspecies:

Ssp. **elongatum** (C. Presl) Lange (incl. 'var *lanceolatum* Uechtr.'), the common form of reed-swamps, stout-stemmed but weak and supported by surrounding vegetation, with flowers 4·5 mm. in diameter and mericarps 1·6 mm. in diameter, the fruit-stalks spreading but not reflexed. Occasional, but rarely recorded as a distinct subspecies.

E. Wingerworth Great Pond 3667, Cromford Canal, High Peak Junction 3156 ('G. *palustre* var. *lanceolatum*').

Linton: Below Snake Inn 19, Harlesthorpe nr. Clowne 47, Hilton 23.

Ssp. **palustre** (incl. var. *witheringii* Sm.), with slender more or less erect stems, flowers only 3 mm. and mericarps 1·2 mm. in diameter, and with fruit-stalks strongly reflexed. Common in marshy or peaty areas where water stands chiefly in winter.

No recent records for the subspecies separately.

Linton: Axe Edge 07, Brierley Wood 37, Morley 34, between Locko and Spondon, Chaddesden and Ockbrook 43, Willington 22, Repton 32.

G. tricornutum Dandy ('G. *tricorne* Stokes')   *Rough Corn Bedstraw*
Doubtfully native. Cornfields, especially on calcareous soil. Very rare.

S. 'Once only (in 1941), as a garden weed from canary-seed; not persisting', Clifton 1545, Stores Road, Derby 3537.

Linton: Bolsover 47.

G. aparine L.   *Goose-grass, Cleavers*

Native. Hedges, stream-banks, limestone scree, margins and clearings of
woods on moist fertile well-drained soils, waste places. Abundant and
widespread.

Recorded for all squares except 19.

Linton: 'Abundant and generally distributed'.

G. spurium L.   *False Cleavers*

Probably introduced. Arable fields. Very rare.

S.   Longford Hall 2138.

Linton: Not recorded.

G. uliginosum L.   *Bog Bedstraw*

Native. Fens. Much more restricted than *G. palustre* but scattered throughout
the county.

W.   Upper Derwent 1696, Roych 0783, Derwent Edge 1989, Buxton 0473,
0875, Combs Reservoir 0379, Baslow 2871, Salter Sitch 2977, Bar Brook nr.
Curbar Road 2774, Umberley Brook 2871, Umberley Sick 2970, Brand Top
0468, Woodeaves 1850.

E.   Donkey Racecourse, Chesterfield 3572.

S.   Nr. Tinkers' Inn 1744, Hulland Moss 2546, between Ednaston Lodge
and Bradley 2244, canal-side, Trent 4931, Cauldwell 2517.

Linton: Hathersage 28, Totley Moss 28, Mosborough 48, Scarcliffe Park
Wood 57, Pleasley 56, Coxbench 34, Wyaston 13, Findern 32.

CAPRIFOLIACEAE

SAMBUCUS L.

S. ebulus L.   *Dwarf Elder, Danewort*

Doubtfully native. Roadsides and damp grassy places, especially in limestone
areas. Local and rather rare; perhaps always a relic of cultivation.

W.   Monsal Dale 1871, Stoney Middleton 2275, Ashford 1969.

E.   Frechville 3984.

S.   Nr. Longford 2136.

Linton: Buxton 07, Alport 26, Wirksworth 25, Dethick 35, South Normanton
45, Brailsford 24, Boulton 33.

S. nigra L.   *Elder, Bourtree*

Native. Woods, scrub, hedges, roadsides and waste places; especially
characteristic of disturbed nutrient-rich soils; avoided by rabbits and often
prominent where they are abundant. Common and widespread.

Recorded for all squares except 19.

Linton: 'Generally distributed'.

Var. *laciniata* L.

W. Knockadown, Carsington 2551.

Linton: 'In hedge of main road, south-west of Carsington' 25.

S. canadensis L.   *Canadian Elder*

Introduced. This shrub differs from *S. nigra* in being stoloniferous and clump-forming and in having smaller purplish-black berries.

W. Alsop-en-le-Dale 1555, Alsop Station 1554.

Linton: Not recorded.

S. racemosa L.   *Red-berried Elder*

Introduced. An escape from cultivation in a few localities both on gritstone and limestone.

W. Buxton 0573, Ashwood Dale 0772, Darley 2964.

Linton: Not recorded.

VIBURNUM L.

V. lantana L.   *Wayfaring Tree, Mealy Guelder Rose*

Probably native in a few localities, introduced elsewhere. Hedges, scrub and woods. Rare and local; absent from the Carboniferous Limestone.

W. Grindleford Station 2578.

E. Roadside, Dronfield to Norton 3479.

Linton: Glapwell 46, Langwith 56, Repton Shrubs 32, Measham (not now in Derbyshire) 31.

V. opulus L.   *Guelder Rose*

Native. Woods, scrub and hedges especially on damp soils. Frequent. Recorded for all squares except:

W. 19.
E. 45, 44.

Linton: Newton Wood 45.

SYMPHORICARPOS Duhamel

S. rivularis Suksdorf   *Snowberry*

Introduced. Commonly planted and naturalized in a few places.

W. Derwent 1889, Hathersage 2282, Grindleford Station 2673, Picory Corner 2365, Newhaven 1659, Kniveton 2050.

E. Beauchief Abbey 3381, Horsley Gate 3177, Newbold 3672, Langwith Wood 4976, Wingerworth 3667, Astwith 4464, Milford 3545, nr. Langley Mill 4546.

S. Recorded for all squares.

Linton: Not recorded.

LONICERA L.

L. xylosteum L. *Fly Honeysuckle*
Introduced. Woods and hedges. Rare.
W. Shrubbery, Hathersage 2381, Beresford Dale, with other alien shrubs 1258, Tissington, planted 1752.
S. Ticknall Quarry 3523.
Linton: Wormhill 17, Calke 32.

L. periclymenum L. *Honeysuckle, Woodbine*
Native. Woods, scrub and hedges on all but waterlogged or strongly pod-solized soils. Common and widespread.
Recorded for all squares except:
W. 09, 19, 17.
E. 45.
Linton: 'Common. Generally distributed and abundant.'

ADOXACEAE

ADOXA L.

A. moschatellina L. *Moschatel, Townhall Clock*
Native. Woods, hedgebanks, shaded limestone rocks; on a wide range of shaded, well-drained, nutrient-rich substrata. Frequent and widespread.
Recorded for all squares except:
W. 09, 19.
E. 46, 45.

VALERIANACEAE

VALERIANELLA Miller

V. locusta (L.) Betcke (*V. olitaria* Poll.) *Corn Salad, Lamb's Lettuce*
Native. Cultivated land, hedgebanks, limestone rocks etc. Frequent, and the commonest species in the county.
W. Deepdale 1072, Monk's Dale 1374, also Chee Dale, Cressbrook Dale, in 17, Middleton Dale 2275, Lathkill Dale 1865, Alport 2264, Dovedale 1453, also Milldale and nr. Alsop-en-le-Dale in 15.
E. On a wall nr. Whatstandwell 3353, Stanley 4141.
S. Clifton Station 1644, between Trent and Thrumpton 5031, nr. Findern 3129, Bretby 2923.
Linton: Killamarsh 48, Tapton 37, Renishaw 47, Morley 34, Ednaston 24, Cauldwell 21.

V. carinata Loisel. *Keeled Corn Salad*

Native. Rock outcrops and cliff-ledges of the Carboniferous Limestone. Local.

W. Recorded for 07, Tideswell Dale 1573, Chee Dale and Monk's Dale 1373, Parkhouse Hill 0967, recorded for 26, Havenhill Dale 2152.

Linton: Castleton 18, Dovedale 15.

V. rimosa Bast.

Doubtfully native. Cornfields. Probably extinct.

No recent records.

Linton: Monsal Dale 17, Dovedale 15, Locko Park 43, Breadsall 33.

V. eriocarpa Desv.

Introduced. A casual of arable land and waste places. Very rare.

W. Between Miller's Dale and Litton Mill 1673.

Linton: Dovedale 15

V. dentata (L.) Poll.

Native. Cornfields. Formerly frequent, especially in the south of the county; now rare.

W. Deepdale 0971.

E. Stanley 4143.

Linton: Bamford 28, Miller's Dale 17, Killamarsh 48, Barlow Lees, Tapton and Linacre 37, Elmton and Whitwell 57, Hardwick Wood 46, Matlock and Fritchley 35, Codnor Park Station 45, Morley 34, Breadsall 33, Ockbrook and Risley 43, various localities about Calke and Repton 32, Newhall 22.

VALERIANA L.

V. officinalis L. (incl. *V. mikanii* Syme, *V. sambucifolia* Willd.)
Cat's Valerian, Allheal

Native. Rough pastures and scrub, usually on moist soils and often in wet places and by streams. Frequent.

Very variable. Some authors distinguish *V. sambucifolia* Mikan f., with epigeal stolons, 3 to 4 pairs of leaflets and glabrous fruit, from *V. mikanii* Syme (?*V. collina* Wallr.), with stolons short and hypogeal or 0, 4 to 6 pairs of leaflets and hairy fruit. It has been claimed that the former has 56 chromosomes and grows in wet places, the latter 28 chromosomes and is the typical form of drier habitats in the limestone dales; but the distinction is far from clear-cut and more research is needed.

Recorded for all squares except:

      W. 09.
      E. 46.

Linton: Hardwick 46.

V. pyrenaica L.

Introduced. Established in a few localities, presumably as a garden escape.
W. Naturalized in wood below Hopton Hall 2552.
Linton: Not recorded.

V. dioica L. *Marsh Valerian*

Native. Marshy meadows and fens. Frequent.

W. Recorded for 08, Shatton 1982, Back Dale 0970, Ashwood Dale 0772, Monsal Dale 1770, Taddington Dale 1670, White Edge 2677, Umberley Brook 2970, recorded for 06, Hartington Station 1461, Robin Hood's Stride 2261, Umberley Sick 2969, nr. Thorpe Station 1650, Middleton Wood 2756.

E. Cordwell 3176, Whitwell Wood 5279, Scarcliffe Park Wood 5270, Kelstedge 3364, Coxbench 3743.

S. Tinker's Inn 1744, Hulland Moss 2546, Ticknall 3523, Repton Rocks 3221.

Linton: Charlesworth 09, Ault Hucknall 46, Newton Wood 45, Siddalls and Breadsall 33, Drakelow 22.

CENTRANTHUS DC.

C. ruber (L.) DC. *Red Valerian*

Introduced. Frequently cultivated and well naturalized on old walls, cliffs, banks etc.

W. Ashford 1969, Carsington 2453, Quarry, Kniveton 2150.

E. Whatstandwell 3354, West Hallam 4240, Breadsall 3739, Chellaston 3830.

S. Littleover 3334, Dale Abbey 4338.

Linton: Litton Mill 17.

DIPSACACEAE

DIPSACUS L.

D. fullonum L. (*D. sylvestris* Huds.) *Teasel*

Native. Roadsides, field borders, stream-banks, wood-margins, especially where heavy soil has been disturbed. Locally frequent in the south of the county but occasional throughout.

W. Tideswell 1575, Kniveton 2050.

E. Dronfield 3678, Renishaw 4378, Scarcliffe Park Wood 5270, Ault Hucknall 4665, Pleasley Vale 5265, Milford 3645, Little Eaton 3641, Stanley 4140.

S. Scropton 1930, Derby 3434, Sawley 4731, Egginton 2628, Ticknall 3523, Netherseal 2913.

Linton: Radbourne 23.

D. pilosus L. *Small Teasel*

Native. Woods and hedgebanks, chiefly on the Carboniferous Limestone. Local and rare; near its northern limit and probably marking areas of surviving ancient woodland.

W. Dimmins Dale 1670, Coombs Dale 2274, Clough Wood 2561, Dovedale 1551, Via Gellia 2756, 2556.

E. Whatstandwell 3354.

S. Mapleton 1648, nr. Thorpe 1549, Hollington 2339.

Linton: Matlock, High Tor 25, Eaton Woods 34, Repton Shrubs 32, Bretby Mill 22.

KNAUTIA L.

K. arvensis (L.) Coulter (*Scabiosa arvensis* L.) *Field Scabious*

Native. Grassy banks, field borders, roadsides, waste places; also on ledges and at the foot of limestone crags. Frequent and widespread.

Recorded for all squares except:

> **W.** 09, 19.
> **E.** 45.
> **S.** 23, 22.

Linton: 'Generally distributed though nowhere very plentiful'.

SCABIOSA L.

S. columbaria L. *Small Field Scabious*

Native. Cliff-ledges, rocky outcrops and grassy slopes, chiefly on limestone. Locally abundant.

W. Recorded for all squares except 09, 19, 28.

E. Markland Grips 5074, Fallgate Quarries 3562, Whitwell Wood 5278, Pleasley Vale 5265.

Linton: Duffield–Allestree and Horsley Castle 34, Willington Junction 22.

SUCCISA Haller

S. pratensis Moench (*Scabiosa succisa* L.) *Devil's Bit Scabious*

Native. Marshes, damp grassland, open woodland and waysides; common on moist loamy soils over limestone. Frequent and widespread.

Recorded for all squares except:

> **S.** 33.

Linton: Breadsall 33.

COMPOSITAE

RUDBECKIA L.

R. hirta L.

Introduced. Grown in gardens and sometimes escaping. Very rare.

S. A few plants, once only, on the site of an aerodrome above Bradley Wood 2046.

Linton: Not recorded.

BIDENS L.

B. cernua L.   *Nodding Bur-Marigold*

Native. By ponds, streams and canals and especially in wet places where water stands in winter but not during the growing season; absent from very acid substrata. Locally common in the south of the county, infrequent in the north.

W. Pond at Heathcote 1460.

E. Hardwick Ponds 4563, Sutton Scarsdale 46, Allestree Park 3440, Little Eaton 3541.

S. Recorded for all squares except 22, 21.

Linton: Renishaw Canal 47.

B. tripartita L.   *Bur-Marigold*

Native. Sides of pools, streams and canals and round water-filled gravel- and sand-pits. Locally common.

W. Watford Bridge 0086, Chapel Reservoir 0379.

E. Beauchief 3381, Norwood 4681, Chesterfield Canal 3871, 4074, Eckington 4379, Pebley Pond 4878, Wingerworth 3687, Hardwick Ponds 4563, Cromford 3354, Butterley Reservoirs 4052, Little Eaton 3541.

S. Mammerton 2136, canal banks, Chaddesden to Spondon 3736, and Stenson to Willington 3230, Darley Abbey 3538, Twyford 3228, Melbourne and Repton 32.

Linton: Barton Blount 13, Sandiacre 43.

GALINSOGA Ruiz & Pavon

G. parviflora Cav.   *Gallant Soldier*

Introduced. A weed of cultivated and waste ground. Infrequent, but spreading.

E. Chesterfield 3771, 3772, Duffield 3447, Woodhouse Mill, 4385.

Linton: Not recorded.

G. ciliata (Rafin.) Blake

Introduced. A casual of waste places. Very rare.

E. Tibshelf 4360.

S. Derby, several stations 33.

Linton: Not recorded.

SENECIO L.

S. jacobaea L.   *Ragwort*

Native. Pastures, roadsides and waste places; extending to cliff-ledges and rock outcrops, especially of the Carboniferous limestone. Common and widespread.

Recorded for all squares.

Linton: 'Common. Generally distributed'.

S. aquaticus Hill

Marsh Ragwort

Native. Marshes, wet meadows, stream-banks. Common and widespread.
Recorded for all squares except:

W. 19, 09.
E. 44.
S. 33.

Linton: 'Generally distributed'.

S. erucifolius L.

Hoary Ragwort

Native. Roadsides, banks, wood-margins. Frequent but scattered and showing a preference for calcareous and heavy soils.

W. Bamford 2083, Chee Dale 1273, Bubnell 2472, nr. Matlock 2960, nr. Tissington 1650.

E. Troway 3980, Killamarsh 4481, Barlow Lees 3476, Pebley Pond 4878, nr. Eckington 4478, Markland Grips 5074, Glapwell Wood 4766.

S. Ashbourne 1847, Agnes Meadow 2147, Cubley Common 1639, Trent 4932.

Linton: Chapel-en-le-Frith 08, Castleton 18, Buxton 07, Old Brampton 37, Eckington 47, Ashover 36, Pleasley 56, Kirk Langley 23, Mickleover 33, Ticknall 32.

S. squalidus L.

Oxford Ragwort

Introduced. Roadsides, waste places, railway banks and tracks, walls, etc. First recorded for the county on tips of the Sheepbridge Iron Works, just north of Chesterfield, by Percy Biggin in 1941. Now abundant and spreading rapidly.

W. Chew Wood 9992, Hathersage 2281, Harpur Hill 0570, Blake Brook 2874, Matlock 2960, Kniveton 2050.

E. Recorded for all squares except 46, 56, 34.

S. Recorded for all squares.

Linton: Not recorded.

S. sylvaticus L.

Wood Groundsel

Native. Disturbed ground on light-textured non-calcareous soils, chiefly in the south of the county. Locally frequent.

W. Bamford 2083, Cressbrook Dale 1772, Eyam 2276, Rowland 2172, Rowsley Bar 2766.

E. Recorded for all squares except 56, 35, 45.

S. Recorded for all squares except 13, 21.

Linton: Chapel-en-le-Frith 08, Robin Hood's Stride 26, Cubley Common 13.

S. viscosus L.   *Stinking Groundsel*
Doubtfully native. Waste ground, railway banks and tracks etc. Occasional.
W. Buxton 0673, Monsal Dale Station 1771, Calver 2475, Bakewell 2269,
Parwich 1855, Longcliffe 2554.
E. Recorded for all squares.
S. Recorded for all squares.
Linton: 'Rare'. Dronfield 37, Cotmanhay 44, West Hallam Colliery 43.

S. vulgaris L.   *Groundsel*
Native. Cultivated and waste ground. Very common and widespread.
Recorded for all squares except: **W** 09.
Linton: 'An abundant weed'.

Var. **hibernicus** Syme (see D. E. Allen, Watsonia 6 (1967), 280–282), with
ray-florets, has been recorded from several localities.
W. Bradwell 1680, Pavilion Gardens, Buxton 0573, Curbar 2474, Bakewell
2169, Beeley 2767, Darley Bridge 2762, Matlock 2959.
E. Beauchief 3281, Spitewinter 3466.
S. Ashbourne 1746, 1846, Spondon 3935, Breadsall 3639.
Linton: Not recorded.

S. fluviatilis Wallr. ('*S. saracenicus* L.')   *Broad-leaved Ragwort*
Introduced. Streamsides, fens and damp woods. Very rare and probably a
garden escape.
W. Moscar Lodge 2387.
Linton: Nr. Chatsworth 26.

DORONICUM L.
D. pardalianches L.   *Leopard's Bane*
Introduced. Naturalized locally in woods and plantations; occasionally on
old walls. Rare.
W. Grin Plantation 0572, Ashwood Dale 0872, Haddon Hall 2366, Rowsley
2266, Slaley 2657, Alsop Moor 1556.
S. Clifton 1644, Biggin 2548, Hulland Grange 2447.
Linton: Wormhill 17, Matlock Bath 25, nr. Ashover 36.

TUSSILAGO L.
T. farfara L.   *Coltsfoot*
Native. Cultivated and waste ground, banks, quarry spoil-heaps etc.; especially
abundant on heavy soils. Abundant and widespread.
Recorded for all squares:
Linton: 'Common. Abundant everywhere'.

PETASITES Miller

P. hybridus (L.) P. Gaertner, B. Meyer & Scherb. (*P. officinalis* Moench)

Butterbur

Native. Streamsides and damp shady places. Locally abundant. 'Female' plants are common in some of the limestone dales, by the Derwent near Froggatt, round Ashbourne and elsewhere.

Recorded for all squares except:

> **E.** 46, 45, 44.
> **S.** 43.

Linton: 'Frequent'.

P. fragrans (Vill.) C. Presl   *Winter Heliotrope*

Introduced. A very rare casual.

W. Lightwood Road, Buxton 0574, rubbish heap, Parwich 1854, Springfield, Darley Dale 2763.

E. Ryecroft Glen 3281, Hazelwood Road, Duffield 3344.

S. Roadside nr. Wyaston Grove 1942, Winshill 2523/4, Lullington Hall grounds 2413.

Linton: Nr. Burton 22.

INULA L.

I. helenium L.   *Elecampane*

Introduced. A rare escape from former cultivation.

W. Roadside nr. Comb's Chapel 0478, Dovedale 1453, by River Derwent at Hathersage (E. Drabble) 28.

E. Ilkeston 4642.

Linton: Not recorded.

I. conyza DC.   *Ploughman's Spikenard*

Native. Wood-margins, rock outcrops, disturbed ground and roadsides on limestone and especially on shallow black limestone soil; avoided by rabbits. Local.

W. Hayfield 08, Cressbrook Dale 1772/3, Monsal Dale 1772, Hay Dale 1277, Miller's Dale 1373, Longstone Edge 2073, Greave's Piece 2877, Hassop Quarries 27, Calver New Bridge 2475, Baslow 2872, Coombs Dale 2274, Middleton Dale 2175, recorded for 16 and 26, Via Gellia 2556.

E. Works tip, Dronfield 3678, Markland Grips 5074, Creswell Crags 5374, Whitwell Wood 5278, Langwith Wood 5068, Pleasley Vale 5265, rail track, Pleasley 5164.

Linton: Taddington Dale 17, Matlock Tor 25, Steetley 57.

PULICARIA Gaertner

P. dysenterica (L.) Bernh.   *Fleabane, Lesser Fleabane*
Native. Marshes, wet pastures and damp roadsides. Frequent.
W. Chinley 0482, Combs Lane, Chapel 0478/9, Ashwood Dale 0772, Monk's Dale 1374, recorded for 27.
E. Whirlow Bridge 3182, nr. Holmesfield 3176, Cowley 3477, Pebley Pond, Barlborough 4878, Coldwell, Ashover 3764, nr. Ault Hucknall 4464, Hardwick Ponds 4564, Pleasley Vale 5265, Holloway 3256, also nr. Riber and Whatstandwell in 35, nr. Langley Mill 4647.
S. Ashbourne Green 1948, Marston Montgomery 1439, Cubley 1639, Kirk Langley 2838, Trent Lock 4931, canal side, Sawley 4831, Sandiacre 4737, Twyford 3228, Cauldwell 2517.
Linton: Derwent Dale 19, Edale 18, nr. Rowsley 26, Thorpe, Tissington and Dovedale 15, Matlock Bath 25, Markland Grips, Elmton and Harlesthorpe 57, Pinxton 45, Holbrook and Denby 34, Nether Biggin 24, about Derby, Mickleover and Mackworth 33.

FILAGO L.

F. germanica (L.) L.   *Cudweed*
Native. Heaths, dry pastures, arable land and roadsides, chiefly on acid sandy soils. Local and rare.
W. Alsop 1653.
E. South of Little Eaton 3640, West Hallam 4240, Stanley 4140.
S. Birchwood Quarry 1541, Dale 4338, Trent 4931, nr. Scropton 1829, Marston-on-Dove 2329, Weston-on-Trent 4027.
Linton: Tideswell 17, Renishaw 47, Markland Grips 57, Langwith Junction 56, Shirley 24, Derby 33, Calke 32.

F. minima (Sm.) Pers.   *Slender Cudweed*
Native. Sandy and gravelly heaths and fields; calcifuge. Local and rare.
No recent records.

Linton: Findern 33, nr. Repton Rocks 32.

GNAPHALIUM L.

G. sylvaticum L.   *Heath Cudweed, Wood Cudweed*
Native. Dry open woods, heaths, dry pastures and waysides on acid soils. Local and infrequent.
W. Below Fallinge Farm, Rowsley (last seen 1937) 2666.
E. Breadsall Moor 3742.

S. Edlaston Coppice (last seen 1937, now probably extinct) 1743, Breward's Car 2843, Bretby 2923.

Linton: Charlesworth 98, Combs Moss 07, Abney 27, Whitwell Wood 57, Brackenfield Green 35, Bradley Wood 24, Cross-o'-the-Hands 24, Repton 32.

G. uliginosum L. *Marsh Cudweed*

Native. Damp roadsides and waste ground and poorly drained arable land. Locally frequent.

W. Dale Bottom, Hathersage 2481, Dennis Knoll 2283, Combs Reservoir 0479, Calver New Bridge 2475, Bonsall Moor 2559.

E. Beauchief Abbey 3381, Ford Valley 3880, Linacre 3272, Brockwell 3771, Wingerworth 3766, Stubbin Ponds 3567, nr. Wigwell 3254, Pinxton 4454, Breadsall Moor 3742.

S. Ashbourne 1847, Shirley Wood 2042, Cubley 1537, Barton Blount 2034, Littleover 3233, Dale Abbey 4338, Scropton 1829, Bretby Woods 3023, Robin Wood 3525, Bretby 2923.

Linton: Rowarth 08, Killamarsh 48, Renishaw 47, Whitwell Wood 57. Egginton Common 22.

ANAPHALIS DC.

A. margaritacea (L.) Bentham *Pearly Everlasting*

Introduced. Banks, walls and waste places. Very rare.

W. Cunning Dale 0872, Harpur Hill 0570, railway bank, Friden 1760.

Linton: Not recorded.

ANTENNARIA Gaertner

A. dioica (L.) Gaertner *Mountain Cudweed, Catsfoot*

Native. In short and more or less open turf of surface-leached loamy soils especially on south-facing limestone slopes but also on nutrient-rich but acid moorland. Local and rare.

W. Above Grin Plantation, Buxton 0571, Back Dale 0870, Diamond Valley, Buxton 0571, Brook Bottom 1477, Rolley Low nr. Wardlow 1873, Arbor Low 1663, upper reaches of Lathkill Dale 1765, Longdale 1860, between Brassington Rocks and Longcliff 2255.

Linton: Glossop 09, between Hayfield and Kinder Scout 08, Axe Edge 07, 06, Fin Hill, Brushfield 17.

SOLIDAGO L.

S. virgaurea L. *Golden-rod*

Native. Woods, hedgebanks and open grassland or moorland, especially on loose substrata (including limestone scree); tolerant of low levels of mineral nutrients but not found on strongly podsolized soils; often accompanying *Teucrium scorodonia*. Locally frequent.

U

W. Recorded for all squares except 15.

E. Beauchief Abbey 3381, Brierley Wood 3775, Marklands Grips 5074, Pleasley Vale 5164, Crich Quarries 3454, Little Eaton 3640, nr. Langley Mill 4547.

S. Mapleton Wood 1646, Dale Abbey colliery spoil-heaps 4338, Swadlincote 2819.

Linton: Repton 32.

ASTER L.
A. tripolium L.   *Sea Aster*

Introduced. On a mud-flat of colliery waste, with *Typha latifolia, T. angustifolia, Juncus inflexus, Phragmites* etc.

E. By the colliery, Killamarsh 4481.

Linton: Not recorded.

ERIGERON L.
E. acer L.   *Blue Fleabane*

Native. Dry open grassland, especially over limestone; also limestone spoil-heaps, railway banks and walls. Local.

W. Baslow 2872, Middleton Dale 2175, Kniveton Quarry 2150.

E. Upper Langwith 5169, Mapperley 4344.

S. Birchwood Quarry 1541, Bradley 2246.

Linton: Clowne 47, Creswell Crags 57, Ticknall and Calke 32.

E. strigosus Muhl.   *Daisy Fleabane*

Introduced. A casual weed of waste places. Very rare.

S. Bradley Wood 2046 ('a few plants in the aerodrome field, once only, in 1948').

Linton: Not recorded.

CONYZA Less.

C. canadensis (L.) Cronq. (*Erigeron canadensis* L.)   *Canadian Fleabane*

Introduced. A weed of waste and cultivated ground, waysides etc., especially on light soils. Rare.

S. 'Occasional about the Cattle Market, Ashbourne' 1847, Burton-on-Trent 2523.

Linton: Not recorded.

BELLIS L.
B. perennis L.   *Daisy*

Native. Abundant in short turf of pastures, waysides, lawns and playing fields and extending to at least 1,500 feet.

Recorded for all squares.

Linton: 'Very common'.

EUPATORIUM L.

E. cannabinum L. *Hemp Agrimony*
Native. Ditches, stream-banks, moist woods, wet roadsides. Frequent but rather local.

W. Recorded for all squares except 09, 19, 18, 28, 27.

E. Brierley Wood 3775, Eckington 47, Markland Grips 5074, Kelstedge 3363, Milltown 3561, Whatstandwell 3354, Milford 3545, Langley Mill 4643.

S. Darley Abbey 3538, Sawley 4731, Swarkestone 3728.

Linton: Bradley Brook and Mugginton 24, Boylestone 13, Burton-on-Trent 22.

ANTHEMIS L.

A. tinctoria L. *Yellow Chamomile*
Introduced. A rare casual.

No recent records.

Linton: Renishaw 47.

A. cotula L. *Stinking Mayweed*
Native. Not uncommon as a casual weed of cultivated ground, roadsides etc., but not usually persisting.

W. Great Rocks Dale 0974, High Rake 1677, nr. Hartington 1259.

E. Beauchief Drive 3381, Creswell 5274, Stoney Houghton 4966.

S. Nr. Ashbourne 1948, Yeldersley 2144, Overseal 2915.

Linton: Chapel 08, Baslow 27, Elmton 57, Pinxton 45, Holbrook 34, Mickleover and Breadsall 33, Calke 32.

A. arvensis L. *Corn Chamomile*
Native. A weed of arable land, not usually persisting. Rare, and apparently decreasing.

S. Eaton (one plant) 0936, field by the Rat Pond, Osmaston (one plant) 2042.

Linton: Miller's Dale 17, Renishaw 47, between Pleasley and Rowthorne 46, Breadsall 33, Newhall 22, Repton 32.

CHAMAEMELUM Miller

C. nobile (L.) All. (*Anthemis nobilis* L.) *Chamomile*
Native. Sandy commons, heaths and roadsides; formerly used medicinally and as a lawn plant. Local and rare.

E. Boythorpe nr. Chesterfield 3869.

Linton: Tansley Moor 36, Morley Moor 34, Ockbrook 43, Linton 21.

ACHILLEA L.

A. millefolium L.   *Yarrow, Milfoil*
Native. Pastures and roadsides on a wide range of substrata but avoiding
black limestone soils, well-developed podsols and wet places. Abundant and
generally distributed.
Recorded for all squares.
Linton: 'Common. Generally distributed'.

A. ptarmica L.   *Sneezewort*
Native. Marshes, stream-banks and moist roadsides. Frequent in suitable
habitats throughout the county.
Recorded for all squares except:
> **W.** 19.
> **E.** 56.
> **S.** 23, 33, 22.

Linton: Church Broughton, Barton Park and Radbourne Common 23,
Mickleover 33.

A. tanacetifolia All.
Introduced. Rare; presumably a garden escape.
No recent records.
Linton: Matlock and Cromford Moor 25/35.

TRIPLEUROSPERMUM Schultz Bip.

T. maritimum (L.) Koch ssp. **inodorum** (L.) Hyland. ex Vaarama (*Matricaria
inodora* L.)   *Scentless Mayweed*
Native. Arable land and disturbed ground of roadsides etc. Frequent,
especially on somewhat acid soils.
Recorded for all squares except:
> **W.** 09, 19, 26.
> **E.** 45, 44.

MATRICARIA L.

M. recutita L. ('*M. chamomilla* L.')   *Wild Chamomile*
Native. Cultivated and disturbed ground, roadsides etc., especially on light
loams and sands. Frequent.
W. Hathersage 2183, Cressbrook 1772, Miller's Dale 1573, Bubnell 2473, Alsop
1555, Dovedale 1452, Bonsall Moor 2559.
E. Beauchief Abbey 3381, High Moor 4680, Beighton 4483, Totley 3179,
Chesterfield 3871, Staveley 4174, Press 3565, Pleasley Vale 5265.
S. Eaton 0936, Lodge Farm 1745, Bradley Moor 2045, Church Broughton
1934, Marston-on-Dove 2329, Swarkestone 3628, Grove Farm 2318.
Linton: Renishaw 47, Whitwell 57, Ambergate 35, Kirk Hallam 43, Egginton
22.

M. matricarioides (Less.) Porter (*M. discoidea* DC.)   *Pineapple Weed,*
                                                                  Rayless Mayweed

Introduced. Trampled ground of roadsides, paths, gateways, waste ground near villages etc. Locally abundant and increasing.

Recorded for all squares.

CHRYSANTHEMUM L.

C. segetum L.   *Corn Marigold*

Probably native. Arable land on acid sands and gravels. Formerly locally abundant but now infrequent and chiefly in the south of the county.

E.   Beauchief 3381, Old Racecourse, Chesterfield 3672, Creswell 5274, Whatstandwell 3354, West Hallam 4240.

S.   Ashbourne 1746, Dale Abbey 4338, Swarkestone 3628, Overseal 2819, Stanton-by-Newhall 2620.

Linton: Beighton 48, Eckington 47, Horsley 43, Shirley 24, Breadsall 33.

C. leucanthemum L.   *Moon-daisy, Ox-eye Daisy, Marguerite*

Native. Grassland of pastures, meadows, waysides and banks on all but strongly podsolized or very wet soils. Abundant except on the moorlands of the northern part of the county.

Recorded for all squares except:

                          W. 19.

Linton: 'Common'.

C. parthenium (L.) Bernh.   *Feverfew*

Doubtfully native. Banks, walls, waste places. Local, and always close to habitations.

Recorded for all squares except:

                     W.   09, 19.
                     E.   47, 57.

Linton: Mellor 98, Duckmanton 47.

C. vulgare (L.) Bernh. (*Tanacetum vulgare* L.)   *Tansy*

Native. Roadsides, hedgerows, streambanks, waste places. Local, and chiefly at low altitudes.

W. Chapel 0579, Buxton 0474, Miller's Dale 1473, Lathkill Dale 1966, Cromford 2856.

E.   Recorded for all squares.

S.   Recorded for all squares except 14, 23.

Linton: Edale 18, Hathersage 27.

ARTEMISIA L.

A. vulgaris L.   *Mugwort*

Native. Roadsides, river banks and shingle, field borders and waste ground.
Frequent and widespread, but scarce on the Carboniferous Limestone; abundant
in the Trent Valley and round Derby.

Recorded for all squares except 09, 19, 28, 15.

Linton: Dinting 09, Hathersage 28.

A. absinthium L.   *Wormwood*

Probably native. Waste ground near houses, river banks and river shingle,
factory-sites and tip-fields etc. Common in the south and east of the county
but scarce on the Carboniferous Limestone; abundant on shingle of the
River Trent.

W. Recorded for 08, 07, Tideswell Dale 1573, Calver New Bridge 2475,
Lathkill Dale 1966, Kniveton 2150.

E. Recorded for all squares except 38, 46, 45, 34.

S. Recorded for all squares except 23.

Linton: Horsley 34.

CARLINA L.

C. vulgaris L.   *Carline Thistle*

Native. Dry grassy slopes, especially over limestone and especially where
ground has been disturbed when prospecting for lead. Local.

W. Recorded for all squares except 09, 19, 18.

E. Pebley Pond 4878, Creswell Crags 5374, Pleasley Vale 5265.

S. Ticknall Quarries 3523.

Linton: Whitwell 57, Old Covert, Mugginton 24.

ARCTIUM L.

A. lappa L. (*A. majus* Bernh.)   *Great Burdock*

Native. Waysides and waste places. Very rare.

E. About Quarndon 3341, Cumber Hills, Duffield 3342.

S. Breadsall 3639, Church Lane, Swarkestone 3327.

Linton: Mellor 98, Breadsall 33, Repton 32.

A. minus Bernh.   *Common Burdock*

Native. Roadsides, woodland margins and clearings, waste places. A common
but very variable species in Derbyshire, with all three of the commonly
recognized subspecies present as well as intermediates:

Ssp. **minus**, with heads small, sessile or short-stalked, in racemes, contracted at the top in fruit.

Recorded for all squares except 09, 08; but typical S. English ssp. *minus*, with heads often less than 2 cm. in overall diameter (including the spines), seems uncommon. The commonest Derbyshire forms seem intermediate between ssp. *minus* and ssp. *nemorosum* or ssp. *pubens*.

Ssp. **pubens** (Bab.) Arènes, like ssp. *minus* but heads rather larger and with stalks up to 15 cm. long, remaining open at the top in fruit. No localized records.

Ssp. **nemorosum** (Lejeune) Syme, with larger short-stalked or subsessile heads, the top 2–4 in a subcorymbose aggregate, widely open in fruit. Derbyshire forms are not clearly separable from *pubens* and *minus* but have been recorded from:

W. Roadside tip nr. Taddington 1470, Taddington village 1471.

E. Steetley 5478, Woolley 3660, Holymoorside 3369, Astwith 4414, Hardwick Ponds 4563, Holloway 3256.

S. Wyaston, on waste ground 1842, Bretby 2922.

Linton: Hope 18, Longstone Edge 27, Lathkill Dale 16/26, Shirley 24.

CARDUUS L.

C. tenuiflorus Curt. *Slender Thistle*

Native. Waysides and waste places. Rare, no recent records.

W. Via Gellia 25. (E. Drabble, Journ. Bot. 1909)

Linton: Castleton 18.

C. nutans L. *Nodding or Musk Thistle*

Native. Disturbed ground of pastures, waysides, hedge-banks, field-borders, especially over limestone. Locally frequent in the limestone areas, occasional elsewhere.

W. Recorded for all squares except 09, 19.

E. Whitwell Wood 5278, Markland Grips 5074, Scarcliffe Park Wood 5270, Harewood Grange 3067, Pleasley Vale 5265, Crich 3454, West Hallam 4242.

S. Birchwood Quarries 1541, Bradley Old Park 2345, Brailsford Common 2643, Chellaston 3830, Ticknall 3523.

Linton: Glossop 09, Sandiacre 43, Thrumpton Ferry 53, Drakelow 21.

C. acanthoides L. ('*C. crispus* L.') *Welted Thistle*

Native. Moist waysides, streambanks, hedges and waste ground; commonly with *Urtica dioica*, *Galium aparine* and *Poa trivialis*. Frequent but local.

Recorded for all squares except:

> **W.** 09, 19.
> **E.** 38, 37, 36, 46.

Linton: Bolsover 47, Sutton Scarsdale 46.

Linton's records under 'var. *acanthoides* L.' were almost certainly of large solitary-headed forms of *C. acanthoides* (which he named *C. crispus*), not of the continental species now called *C. crispus* L.

C. × orthocephalus Wallr. (*C. nutans* × *C. acanthoides*)
Native. No recent records.
Linton: Cressbrook Dale, east side 17, Via Gellia 25.

CIRSIUM Miller

C. eriophorum (L.) Scop. (*Cnicus eriophorus* (L.) Roth)
Woolly-headed Thistle
Native. Pastures, hedgebanks, open scrub and waysides, chiefly on limestone. Local and infrequent.
W. Winnatts 1382, Litton Mills 1573, Coombs Dale 2274, Hassop Mines 2273, Middleton Dale 2175, Via Gellia 2857.
Linton: Buxton 07, Stanton Moor 26, Dovedale 15, Calke 32.

C. vulgare (Savi) Ten. (*Cnicus lanceolatus* (L.) Willd.)   *Spear Thistle*
Native. Pastures, waysides, waste ground etc. Common.
Recorded for all squares.
Linton: 'Abundant everywhere'.

C. palustre (L.) Scop. (*Cnicus palustris* (L.) Willd.)   *Marsh Thistle*
Native. Marshes and moist pastures, rides and clearings in damp woods, roadsides etc. Common.
The white-flowered form is locally common.
Recorded for all squares.
Linton: 'Generally distributed and plentiful'.

C. arvense (L.) Scop. (*Cnicus arvensis* (L.) Roth)   *Field Thistle*
Native. Fields, roadsides, waste ground. Common.
Recorded for all squares.
Linton: 'Very common'

C. acaulon (L.) Scop. (*Cnicus acaulos* (L.) Willd.)   *Stemless Thistle*
Native. In short turf of south-facing slopes over limestone. Rather rare.
See also maps (page 94).
W. Roadside bank above Dowel Dale 0668, recorded for 18, Ladmanlow Tip 0471, Brook Bottom 1575, Longstone Edge 2073, Long Dale 1860, Gratton Dale 2060, Wolfscote Hill 1358, Longcliffe 2255.
E. Markland Grips 5074.
S. Calke Park 3623.

C. heterophyllum (L.) Hill (*Cnicus heterophyllus* (L.) Roth)
Melancholy Thistle
Native. Moist shaded slopes and valley-bottoms of the limestone dales, rarely elsewhere. Locally frequent. See also maps (page 94).

W. Malcoff 0782, Cunning Dale 0873, Harpur Hill 0670, Grin Wood 0572, Ashwood Dale 0772, Burbage 0473, Miller's Dale 1573, Monk's Dale 1374, Cressbrook Dale 1774, Priestcliffe 1472, Monsal Dale 1771, Parsley Hay 1363, Lathkill Dale 1865, Hartington 1260, Beresford Dale 1258, Beresford Hill 1358, Via Gellia 2556.

E. Beauchief Abbey 3381.

S. Ednaston 2442.

C. dissectum (L.) Hill (*Cnicus pratensis* (Hudson) Willd.)
Marsh Plume-thistle
Native. Fens. Local and very rare.

S. Willington 2928.

Linton: Buxton 07, Ashbourne 14, Radbourne 23, Findern 33.

SILYBUM Adanson

S. marianum (L.) Gaertner (*Mariana lactea* Hill) *Milk Thistle*
Introduced. Naturalized on waste ground and roadsides. Very rare.

S. Trent Lock 4831.

ONOPORDUM L.

O. acanthium L. *Scottish Thistle, Cotton Thistle*
Introduced. Naturalized on waste ground and roadsides. Rare.

W. Staden 0772, recorded for 17, Rowsley 2566.

E. Creswell Colliery sidings 5273, Duffield 3443.

S. Clifton Tip 1644, Willington 2928.

Linton: Bentley Hall 13, Burton Road, Repton 32.

CENTAUREA L.

C. scabiosa L. *Greater Knapweed*
Native. Rock outcrops, cliff-ledges and sunny slopes of the limestone areas and on banks, roadsides and field borders elsewhere. Locally abundant but chiefly on limestone.

W. Recorded for all squares except 09, 19, 08, 18, 28.

E. Clowne 4975, Whitwell Wood 5278, Markland Grips 5075, Bolsover 4869, Pleasley 5164, West Hallam 4243.

S. Scropton 1930, river nr. Scropton 1829, Church Gresley 2819.

Linton: Hulland 24, Morley 34, Ticknall Quarry and Repton 32.

C. montana L.

Introduced. A garden escape. Very rare.

W. Between Hartington Station and the main road 1561.

Linton: Not recorded.

C. cyanus L.   *Cornflower, Bluebottle*

Native. Cornfields and waste ground. Formerly not infrequent but now very uncommon.

E. Pebley Pond 4880, Lea 3357.

S. Cornfield, Ashbourne 1646, by river at Allestree 3539, Trent 4932, between Twyford and Willington 3128, Bretby 2722.

Linton: 'Not common'. Bamford 18, Grange Mill 25, Renishaw 47, Wyaston 14, Breadsall 33, Morley and Spondon 43, Calke and Repton 32.

C. paniculata L.   *Panicled Knapweed*

Introduced. A rare casual.

W. Buxton 07.

Linton: Not recorded.

C. nigra L.   *Lesser Knapweed, Hardheads*

Native. Grassland, waysides, banks, rock outcrops and cliff-ledges; often with *Arrhenatherum elatius*. Common and widespread.

Recorded for all squares.

Two subspecies, more or less distinct morphologically, ecologically and geographically but with some intermediates, are commonly recognized and both occur in Derbyshire:

Ssp. **nigra**, with stout hispid stems conspicuously swollen beneath the heads and with broadly lanceolate toothed or pinnatifid leaves, is the commonest form in the county;

Ssp. **nemoralis** (Jord.) Gugl., with more slender and pubescent stems not much swollen beneath the heads and with narrowly lanceolate almost entire softly hairy leaves, is much more local. Intermediates also occur.

Linton: 'Abundant everywhere'.

C. calcitrapa L.   *Star Thistle*

Probably introduced. A casual of roadsides and waste places. Very rare.

S. 'Near a fowl-pen at Hanging Bridge in 1944, persisting for several years but sometimes with aborted heads' 1545.

Linton: Not recorded.

SERRATULA L.

S. tinctoria L.   *Saw-wort*

Native. Open woods and copses, little-grazed grassland, waysides, railway banks etc.; usually on good loamy soil and avoiding both black limestone soil (unless periodically waterlogged) and podsols; common on surface-leached soils over limestone and on soils derived from igneous and volcanic rocks. Locally frequent.

W. Ashwood Dale 0772, Cunning Dale 0872, Chee Dale 1273, Monk's Dale 1374, Cressbrook Dale 1274, Bretton Clough 2079, Long Dale 1959, Biggin Dale 1458, Brassington Rocks 2154, Scow Brook 2450, Grange Mill 2457.

E. Bramley Mill 4080, Cordwell Valley 3176, Markland Grips 5074, Somersal Lane 3569, nr. Cromford 3057, nr. Ambergate 3451.

S. Nr. Kniveton 2048, Bradley Brook 2244, nr. Sudbury 1532, Trent 4932.

Linton: Norton Lees 38, Harlesthorpe 47, Mickleover 33, Spread Eagle 32.

CARTHAMUS L.

C. tinctorius L.   *Safflower*

Introduced. A rare casual.

W. Dimple Tip, Matlock 2656.

Linton: Not recorded.

CICHORIUM L.

C. intybus L.   *Chicory, Wild Succory*

Doubtfully native. Roadsides, fields, waste places. Local and not common; chiefly in the south of the county.

W. Buxton 0573, Longstone 2072, Hindlow 0869, Friden 1760, Hopton 2753.

E. Wigley 3172, Lea 3357, Allestree 3540, Shipley 4444.

S. Ashbourne 1744, Bradley Old Park 2345, Ballast tip, nr. Trent 4831, Anchor Church 3327, Netherseal 2813, Burton 2523.

Linton: Clowne 47, Cauldwell 21.

LAPSANA L.

L. communis L.   *Nipplewort*

Native. Hedge-banks, waysides, waste ground. Common.

Recorded for all squares except:

W. 09, 19.

Linton: 'Common. Generally distributed'.

HYPOCHAERIS L.

H. radicata L.   *Cats-ear*

Native. Meadows and pastures, grassy banks, roadsides and waste places on
a wide range of well-drained non-calcareous soils but avoiding strongly
podsolized or waterlogged soils. Very common.

Recorded for all squares.

Linton: 'Common. Abundant throughout the county'.

LEONTODON L.

L. hispidus L.   *Rough Hawkbit*

Native. Meadows, pastures, grassy slopes, waysides etc., especially on
calcareous soils but also on well-drained fertile non-calcareous soils. Common.

Recorded for all squares except:

> **W.** 09, 19, 08.
> **E.** 46.
> **S.** 21.

Linton: 'Common. General'.

L. autumnalis L.   *Autumnal Hawkbit*

Native. Meadows, pastures, waysides. Common.

Recorded for all squares except:

> **E.** 34.

Linton: 'Common. More or less generally distributed'.

L. taraxacoides (Vill.) Mérat ('*L. hirtus* L.')   *Hairy Hawkbit*

Native. In short turf of dry slopes, waysides and waste ground. Occasional,
but uncommon on the Carboniferous Limestone.

W. Derwent Reservoir 1690, Back Dale and Woo Dale 0972, Burbage
Edge 17, nr. Churn Hole 1071, Bar Brook, Baslow 2672, Leam Lane, Hathersage 2379, Dean Hollow, Wirksworth 2855.

E. Ford nr. Eckington 4081, Pebley Quarry 4978, Wilday Green 3274,
nr. Troway and nr. Holmesfield 37, nr. Pleasley 5164.

S. Mugginton 2843, nr. Blackwall 2649, Findern 3030, nr. Willington 2928,
Ticknall Quarries 3523, Weston-on-Trent 3828.

Linton: Charlesworth 09, Greenhill to Bradway, and Norton 38, Elmton and
Whitwell 57, Wingerworth 36, Tibshelf to Hucknall and Scarcliffe 46,
Crich Common 35, Codnor 44, Cubley Common 13.

PICRIS L.

P. echioides L.   *Bristly Ox-tongue*

Native. Roadsides, hedge-banks, field borders and waste places, especially
where stiff or calcareous soil has been disturbed. Local and rare.

W. Miller's Dale 1473, Kniveton Churchyard, one plant 2050.

E. Between Holmesfield and Millthorpe, one plant 3276.

S. Ashbourne, one plant 1746.

Linton: Nr. Winster 26, Chellaston and Breadsall 33, between Stapenhill and Cauldwell 22, Repton 32.

P. hieracioides L. *Hawkweed Ox-tongue*

Native. Waysides, hedge-banks and waste ground, especially on limestone; stabilized limestone scree. Local.

W. Deep Dale 0971, Ashwood Dale 0872, Cressbrook Dale 1774, also Chee Dale, Tansley Dale and Burfoot in 17, Bubnell 2472, also Middleton Dale and nr. Bretton in 27, Middleton-by-Youlgreave 1963, Lathkill Dale 16/26, Dovedale 1453, Wolfscote Dale 1456, Wirksworth 2854, also Middleton-by-Wirksworth and Via Gellia in 25.

E. Fallgate, Ashover 3662, Ashover Hay 3560, Pleasley Vale 5265, Crich 3454, nr. Stanley 4240.

S. Littleover 3334, Chellaston 3830.

Linton: Bolsover 47, Palterton 46, between Sandybrook and Thorpe 14/15, Ockbrook and Sandiacre 43, Stapenhill and Newton Solney 22, Calke and Repton 32.

TRAGOPOGON L.

T. pratensis L. *Goat's-Beard, Jack-go-to-bed-at-noon*

Native. Meadows, pastures, banks, field borders, waysides.

The common form in Derbyshire is:

Ssp. **minor** (Miller) Wahlenb.

Recorded for all squares except:

W. 09, 19.

Ssp. **pratensis,** with ray-florets about equalling the involucral bracts, is very doubtfully native and far less common.

W. Recorded for 06.

E. Pleasley Vale 5265, Whitwell Wood 5278.

Linton: Not recorded.

T. porrifolius L. *Salsify*

Introduced. An occasional escape from cultivation.

E. West Hallam 4240.

S. Dale 4338.

Linton: Between Matlock Bath and Bonsall 25, by the water works, Breadsall 34.

LACTUCA L.

L. virosa L.   *Wild Lettuce*

Probably introduced. An infrequent casual.

E.   Common on rubbish dumps, Whitwell 57.

S.   Widely scattered on waste ground, old race-course, Derby 3637.

Linton: Matlock 35, Radbourne 23, between Derby and Borrowash 33/43, nr. Stanton-by- Bridge 32.

MYCELIS Cass.

M. muralis (L.) Dumort. (*Lactuca muralis* (L.) Gaertner)   *Wall Lettuce*

Native. Walls and rocks, in shade or in the open; also on hedgebanks and on the ground in woods. Locally common, especially on limestone.

Recorded for all squares except:

> **W.**   09, 08, 07.
> **E.**   45.
> **S.**   22, 21.

Linton: Mellor 98, Glossop 09.

SONCHUS L.

S. arvensis L.   *Corn Sow-Thistle, Field Milk-Thistle*

Native. Cultivated and waste ground, roadsides, banks. Common and widespread.

Recorded for all squares except:

> **W.**   09, 19, 17.
> **E.**   45.

Linton: 'Generally distributed'.

S. oleraceus L.   *Sow-Thistle, Milk-Thistle*

Native. Cultivated ground, waysides, waste places, walls etc. Common and widespread.

Recorded for all squares except:

> **W.**   09, 19.
> **E.**   45.

Linton: 'Generally distributed'.

S. asper (L.) Hill   *Spiny Sow-Thistle, Spiny Milk-Thistle*

Native. Cultivated ground, waysides, waste places. Common and widespread.

Recorded for all squares except:

> **W.**   09, 19.
> **E.**   45.

Linton: 'Common'.

CICERBITA Wallr.

C. macrophylla (Willd.) Wallr.   *Large-leaved Blue Sow Thistle*
Introduced. A garden escape, established in a few places.
W. Roadside at Bradwell 1781, Hathersage 2381, Foolow 1976.
Linton: Not recorded.

HIERACIUM L.

Derbyshire is rich in hawkweeds and on that account has attracted the attention of many experts in what is certainly one of the most difficult genera in the British flora. Linton was himself deeply interested in hawkweeds and several microspecies were first recognized by him; but since the publication of the Flora of Derbyshire there have appeared H. W. Pugsley's Prodromus of the British Hieracia (1948) and the acount by P. D. Sell and C. West in the Critical Supplement to the Atlas of the British Flora (1968). There was therefore a serious need for a revision of Derbyshire Hieracia, and it is a pleasure to acknowledge the great help given by P. D. Sell and Professor J. N. Mills in drawing up the following account. It will be seen that there is still much to be done, particularly in the treatment of the *exotericum* aggregate and in placing forms that have been named *H. holophyllum* and *H. caledonicum*.

Where a microspecies recorded by Linton is clearly identifiable one locality is given for it from each 10 km. square additional to those covered by recent records. Additional records are sometimes available from old herbarium specimens which have been named by Sell and West and used in maps for the Critical Supplement to the Atlas of the British Flora.

SUBGENUS **HIERACIUM**

OREADEA Zahn

H. saxorum (F. J. Hanb.) P. D. Sell and C. West
Native. Limestone cliffs and outcrops. Rare.
Only known in the county by old specimens from Tideswell 17 and Hay Dale 07.

H. stenopholidium (Dahlst.) Omang
Native. Very rare.
The only Derbyshire locality is on the gritstone at Alport Castles.
W. Alport Castles 1491.

H. britannicum F. J. Hanb.
Native. An abundant and characteristic hawkweed of the limestone dales, chiefly on cliff-ledges and rock outcrops but also on cliff-foot debris.
W. Middleton Dale 2175, 2275, Bradwell Dale 1780, Miller's Dale 1473, Litton Mill 1672, 1673, Great Rocks Dale 1172, Parkhouse Hill 0767, Ashwood Dale 0772, Hitter Hill 0866, Topley Pike 1172, Tansley Dale 1774, Cressbrook

Dale 1775, Tideswell Dale 1573, Peak Forest 0976, Peter Dale 1276, Chee Dale
1273, Horseshoe Dale 0970, Monk's Dale 1373, 1374, 1375, Lathkill Dale
1765, nr. Middleton Wood 2656. Dovedale 1452.

Linton: Parkhouse Hill 06.

H. dicella P. D. Sell and C. West

Native. Limestone dales, often in small clumps with *H. britannicum* and then
distinguishable by the more elliptic leaves, narrowing towards the base.
Frequent.

W. Cowdale 0872, Deep Dale 0971, 0970, Woo Dale 0972, Hitter Hill 0867,
Pictor 0872, Peak Dale 0877, Bradwell Dale 1780, Litton Mill 1672, Great
Rocks Dale 1172, 1272, Miller's Dale 1473, 1372, Topley Pike 1172, Tideswell
Dale 1573, Cressbrook Dale 1774.

H. naviense J. N. Mills.

Native. Cracks and ledges on limestone cliffs. The Derbyshire locality is
the only one for this newly recognized species.

W. Winnats Pass 1382.

H. subplanifolium Pugsl.

Native. Limestone cliffs and outcrops. Locally common.

W. Tideswell 1574, Wye Dale 1072, Monk's Dale 1374.

There is also an old specimen of *H. subplanifolium* from Deep Dale 07.

H. decolor (W. R. Linton) A. Ley

Native. Cliffs, outcrops and rubble in the limestone dales. Uncommon.

W. Chee Dale 1172.

There are also old specimens of *H. decolor* from King Sterndale 07 and Deep
Dale 07.

H. holophyllum W. R. Linton

Native. Limestone outcrops in the dales. There are no certain recent records
and some of the older records may in fact have been of *H. caledonicum.* The
following are correct:

W. Deep Dale 17, Great Rocks Dale 17, Dovedale 15.

H. caledonicum F. J. Hanb.

Native. Cliff-ledges and rock outcrops in the limestone dales. Frequent.

W. Tansley Dale 1774, Deep Dale 0971, Back Dale 0870, Dove Holes 0778,
Cressbrook Dale 1774, Peter Dale 1276, Lathkill Dale 1666, nr. Hartington
1360.

There are also old specimens of *H. caledonicum* from: Cave Dale 18, nr. Lode
Mill 15, Dovedale 15.

VULGATA F. N. Williams

H. exotericum agg.

Native or introduced. The microspecies of this aggregate are still imperfectly known and more work is needed before a satisfactory account can be given. They are specially characteristic of disturbed habitats in the limestone dales, as on quarry waste, spoil-heaps of old mineral workings; some are abundant in old toadstone quarries. This type of distribution suggests that they may not be native to the county, a view supported by evidence from other areas.

The records below are of forms that cannot be assigned to any described microspecies.

W. Deep Dale 1072, 1071, 0971, Chee Dale 1273, Bradwell Dale 1780, Topley Pike 1072, Great Rocks Dale 1272, Middleton Dale 2275, Miller's Dale 1473, Ashwood Dale 0772, 0872, Taddington Dale 1571, Woo Dale 0972, Back Dale 0970, Tideswell Dale 1574, 1573, Peak Dale 0877, Peak Forest 0976, 0975, Lathkill Dale 1765, Chapel-en-le-Frith 0682, Glossop 09, Dove Holes 0878, Smalldale 0977.

H. grandidens, H. severiceps and *H. 'sublepistoides'* belong to the *exotericum* aggregate.

H. grandidens Dahlst.

Native or introduced. This handsome hawkweed occurs on cliff-ledges as well as in disturbed habitats. Local.

W. Deep Dale 1071, 0971, Ashwood Dale 1172, 0772, Woo Dale 0972, Lathkill Dale 1666, 1765, Chee Dale 1273, Millstone Edge 2480.

H. severiceps Wiinst.

Probably introduced. Very rare.

Only known in Derbyshire by an old specimen from Ashwood Dale 37.

'H. sublepistoides (Zahn) Druce'

Probably introduced. The commonest Derbyshire member of the *exotericum* aggregate and usually in disturbed habitats; not confined to limestone, being abundant in the toadstone (dolerite) quarry at Tideswell. Our plant is not quite identical with *H. sublepistoides* and is meanwhile called 'dark-headed sublepistoides' by P. Sell (*in litt.*).

W. Pictor 0872, Chapel-en-le-Frith 0682, Tideswell Quarry 1573, Peter Dale 1275, Deep Dale 0971, Chee Dale 1273.

The hawkweed named '*H. silvaticum* Gouan var. *lepistoides* Johanss.' in Linton's Flora probably belongs here.

H. pellucidum Laest.

Native. Cliff-ledges, limestone rubble, spoil-heaps of old mineral workings, etc., in the limestone dales. Frequent.

W. Ashwood Dale 0772, Miller's Dale 1473, 1672, Chee Dale 1172, Topley Pike 1072, Tansley Dale 1774, Wye Dale 17, Wirksworth 25.

There is also an old specimen of *H. pellucidum* from Matlock 26.

H. subprasinifolium Pugsl.

Native. Grassy limestone slopes in the dales. Occasional.

W. Chee Dale in square 17.

There is also an old specimen of this species from Dovedale 15.

H. cymbifolium Purchas

Native. Limestone dales. Frequent.

W. Ashwood Dale 0772, Hitter Hill 0866, Deep Dale 1072, 0971, Topley Pike 1072, Winnats Pass 1382, Pin Dale 1582, Chee Dale 1273, Cressbrook Dale 1773, Miller's Dale 1473, Litton Mill 1673, 1772, Tideswell 1574, 1573, Bradwell 1780, Tansley Dale 1774.

Linton: Dovedale 15.

H. maculatum Sm.

Introduced. Walls, quarries, pit-heaps, grassy slopes and waste ground. Locally frequent in neighbouring counties but only two recent records from Derbyshire.

E. Pleasley Vale 46.

S. Scropton, alabaster waste 1930.

H. diaphanum Fries

Probably native. Cliff-ledges, rock outcrops, roadsides, banks, waste ground etc., on a wide range of rocks and soils, from limestone to gritstone and from calcareous to moderately acid. Frequent.

W. Chapel-en-le-Frith 0580, 0579, 0582, 0682, Ashwood Dale 0772, Back Dale 0870, Peter Dale 1276, Tideswell 1573, Bradwell Dale 1780, Miller's Dale 1473, 1573, Winnats Pass 1382, Monk's Dale 1373, 1374, Froggat Edge 2577, Millstone Edge 2480, Hathersage 2381, Harborough Rocks 2355, Masson Hill 2859.

E. Whitwell Wood 57.

Linton: Alport Castles 19.

H. diaphanoides Lindeb.

Native. Cliff-ledges and rock outcrops, usually on limestone. Occasional.

W. Miller's Dale 1473, Litton Mill 1673, Pin Dale 1582, Chapel-en-le-Frith 0582.

Linton: Harborough Rocks 25, Dovedale 15, Owler Bar 27.

There are also old specimens of *H. diaphanoides* from Chee Dale 17, Castleton 18 and Monk's Dale 17.

H. strumosum (W. R. Linton) A. Ley (incl. *H. sciaphilum* Uechtr.)

Native. The commonest hawkweed in the county, found in a variety of habitats including cliff-ledges on limestone and gritstone, stabilized scree-slopes, woods, grassy banks, road-sides and waste places. It seems likely to be native in some of these habitats but to have spread recently to others.

H. strumosum as treated by P. D. Sell and C. West includes *H. lachenalii* of many authors (*H. vulgatum* var. *sciaphilum* Uechtr.), which cannot be clearly separated from it.

The records listed below include only one from each 10 km. square.

W. Glossop 09, Alport Dale 1292, Cave Dale 1482, Deep Dale 0971, Cressbrook Dale 1774, Gardham's Rocks nr. Baslow 2773, Lathkill Dale 1666, Dovedale 15, Via Gellia 25, recorded for 08.

E. Markland Grips 5174.

S. Ashbourne 1745, Osmaston 2042, Repton 32.

Linton: Hathersage 28, West Hallam 44, Coal Aston 37, Dale 43, Houghton 46, Langwith Wood 56, north of Cubley 13, Church Broughton 23, Egginton Station 22.

There are also old specimens of *H. strumosum* from Lathkill Dale 26, and Alton 36.

H. rubiginosum F. J. Hanb.

Native. Rocks and cliff-ledges, grassy slopes and banks, walls; always on limestone. Occasional.

W. Back Dale 0869, Deep Dale 0970, Topley Pike 1172, Chee Dale 1273, Middleton Dale 2175, Lathkill Dale 1666, 1765.

There are also old specimens of *H. rubiginosum* from Bradwell Dale 18, Cave Dale 18, nr. Buxton 07 and Hartington Dale 16.

H. vulgatum Fries

Native. A common hawkweed of rocky and grassy places, quarries, banks and waste ground on a wide variety of substrata; commonly on limestone but also on sandstone.

The records listed below include only one from each 10 km. square.

W. Nr. Glossop 0593, Ashwood Dale 0772, Cressbrook Dale 1774, nr. Stoney Middleton 2175, Parsley Hay 1563, also recorded from 08, 18 and 25.

E. Nr. Ripley 3752, Woodnook 3472.

Linton: Dovedale 15, Belper 34, Dale 43, Calow 47, Dalbury 23, Mickleover 33, Anchor Church 32.

PRENANTHOIDEA Koch

H. prenanthoides Vill.

Native. Rocks and grassy banks. Local.

W. Ashwood Dale 0772, Miller's Dale 1473, 1572, Tideswell Dale 1573.

TRIDENTATA F. N. Williams

H. scabrisetum (Zahn) Roffey

Only known in Derbyshire by an old specimen from Chapel-en-le-Frith 08.

H. eboracense Pugsl.

Native. Grassy banks. Very rare.

W. Eyam 2176.

FOLIOSA Pugsl.

H. subcrocatum (E. F. Linton) Roffey (*H. strictum* Fries var. *subcrocatum* Linton)

Native. Only known in the county from one locality.

W. Chapel-en-le-Frith 0579.

UMBELLATA F. N. Williams

H. umbellatum L.

Native. Roadsides, banks, heaths. Occasional.

W. Chinley 0582, Chapel-en-le-Frith 0579.

E. Smalley 4044, Shottle 3246, Barlow 3575.

S. Nr. Osmaston 1844.

Linton: Upper Dove valley 06, Charlesworth 09, Edale 18, Mickleover 33, Hollington 23, Shirley 24, West Broughton 13, nr. Ockbrook 43, Egginton 22, Repton 32.

SABAUDA F. N. Williams

H. perpropinquum (Zahn) Druce

Native. Open woodland and heath, roadsides, especially on acid soils. Frequent.

W. Chapel-en-le-Frith 0580, 0579, nr. Monyash 1766, Chee Dale 1273, Froggatt 2476, Eyam 2276.

E. Birchwood 4254.

There is also an old specimen of *H. perpropinquum* from Shirley 24.

Intermediates between *H. perpropinquum* and *H. vagum* are sometimes encountered.

H. vagum Jord.

Native. Rocky places and open woodland; also roadsides, grassy slopes and railway banks. Widespread and locally common.

The records listed below include only one from each 10 km. square.

W. Nr. Glossop 0593, Hayfield 0486, Hope 1783, nr. Whaley Bridge 0079, Thorpe 1550, also recorded for 98, 17 and 25.

E. Killamarsh 4481.

S. Ashbourne 1745, Brailsford 2542.

There is also an old specimen of *H. vagum* from Bamford 28.

SUBGENUS **PILOSELLA** (Hill) Tausch

PILOSELLA (Hill) Tausch

H. pilosella L. (*Pilosella officinarum* C. H. and F. W. Schultz)

Mouse-eared Hawkweed

Native. Grassy pastures and heaths, banks, rocks, walls, etc., on calcareous to moderately acid substrata. Common and widespread.

W. All squares except 09, 19.

E. All squares except 34.

S. All squares.

Linton: 'Very general through the County',

PRATENSINA Zahn

H. brunneocroceum Pugsl. (*Pilosella aurantiaca* ssp. *brunneocrocea* (Pugsl.) P. D. Sell and C. West)

Introduced. A garden escape now widely naturalized on roadsides, railway banks, churchyards, walls etc. Frequent.

W. Chapel-en-le-Frith 0682, nr. Woodeaves 1850, nr. Parwich Moor 1757, Wirksworth 2854, Harborough Rocks 2355.

E. Matlock 3062, Pleasley 5064, Little Eaton 3642, Derby 3343, Allestree 3540.

S. Nr. Ashbourne 1747, Ashbourne Churchyard 1746, Clifton 1644, Littleover 3334, Bretby 2923.

Linton: Not recorded.

H. aurantiacum L. (*Pilosella aurantiaca* (L.) C. H. and F. W. Schultz ssp. *aurantiaca*)

Introduced. Naturalized in gardens, churchyards etc., and on roadsides and grassy banks. Much less common than *H. brunneocroceum* and many of the records may in fact be of that species.

W. Fairfield Common 0774, Buxton 0672, nr. Great Longstone 1971, Baslow Churchyard 2572.

E. Clay Cross Cemetery 3962, Chesterfield Cemetery 3871, Dronfield Cemetery 3578, Tansley Churchyard 3260.

Linton: Cemetery at Staveley 47 (perhaps *H. brunneocroceum*).

CREPIS L.

C. vesicaria L. ssp. **taraxacifolia** (Thuill.) Thell. (*C. taraxacifolia* Thuill.)

Beaked Hawk's-beard

Introduced. Waysides, walls, railway banks, waste places. Rare.

E. Pleasley Vale 5265.

S. By River Trent 4932, nr. Repton 3027.

Linton: Matlock 35, Yeldersley Lane 24, Bradley Park 24, Shardlow 43.

C. biennis L.   *Rough Hawk's-beard*
Probably native. Pastures, waysides, fields. Rare.
E.   Markland Grips 5074.
S.   Nr. Shirley 2040.
Linton: Barmoor Clough 98, Buxton 07, Renishaw 47, Burnaston 23, Breadsall
33.

C. capillaris (L.) Wallr. (*C. virens* L.)   *Smooth Hawk's-beard*
Native. Grassland, fields, banks, waysides, waste places. Common and
widespread.
Recorded for all squares except:
                        W. 09, 19, 07.
Linton: 'Generally distributed'.

C. paludosa (L.) Moench   *Marsh Hawk's-beard*
Native. Streamsides, wet woods, flushes, marshy places. Local.
W. Recorded for 08, Jaggers Clough 1487, Derwent 1889, Yorkshire Bridge
1985, Burbage 0473, Grin Plantation 0472, White Edge 2677, Dalehead Farm
0469, Carsington 2552.
E.   Scarcliffe Park Wood 5270.
S.   Tinkers' Inn 1744, Bradley Brook 2244, Shirley Wood 2141.
Linton: Snake Inn 19, River Sett 08, Castleton 18, Dore 28, Coxbench 34,
Horsley Carr 34.

TARAXACUM Weber

T. officinale Weber, *sensu lato*   *Dandelion*
Native. Pastures, meadows, lawns, waysides, walls, waste places etc. Common
and widespread.
Recorded for all squares.
Linton: 'Very common . . . Everywhere abundant'.

T. palustre (Lyons) DC. ('*T. officinale* var. *palustre* (DC.)') *Marsh Dandelion*
Native. Marshes and fens and by streams. Occasional.
W. Marsh below Thorpe station 1650.
E.   Whitwell Wood 5278, Pleasley Vale 5265, Shottle 3147, Duffield 3443.
S.   Ashbourne Green 1847, nr. Orman's Close Farm 1848, Hulland Moss
2546, nr. Shirley Mill 2141.
Linton: Grindsbrook 18, Axe Edge 06/07, Brassington Rocks 25, Cubley and
Abbots Clownholme 13, Radbourne 23, Drakelow 22.

T. spectabile Dahlst.   *Broad-leaved Marsh Dandelion*
Native. Marshes, streamsides, boggy pastures etc. Local and uncommon.
W. Foxhouse 2781, Peter Dale 1275, Wirksworth 2854.
E. Totley Bents 3080, Eckington 4081, Pebley Quarry 4978.
S. Idridgehay 2849.
Linton: Not recorded.

T. laevigatum (Willd.) DC. (*'T. officinale* var. *laevigatum* (DC.)')
                                                         Lesser Dandelion
Native. Dry places on sandy or calcareous soils, heaths, walls etc. Locally
frequent, especially on the limestone.
W. Winnats 1382, Monsal Head 1871, Coombs Dale 2274, Lathkill Dale 1765,
2063, Parwich Hill 1854, Thorpe Pasture 1551, Carsington Pasture 2553,
Bonsall 2858.
E. Markland Grips 5074, Creswell Crags 5374, Duffield 3443.
S. Snelston Common 1541, Swarkestone 3728.
Linton: Normanton 33.

ANGIOSPERMAE
MONOCOTYLEDONES

ALISMATACEAE

BALDELLIA Parl.

B. ranunculoides (L.) Parl. (*Alisma ranunculoides* L.)   *Lesser Water-Plantain*
Native. By ponds and streams and in ditches. Very rare.
E. Pebley Pond 4878.
Linton: 'Betwixt Derby and Burton'. Repton, 'perhaps now extinct'.

ALISMA L.

A. plantago-aquatica L.   *Water Plantain*
Native. In shallow water and on mud by ponds, streams and canals. Common.
W. Watford Bridge 0086, Dovedale 1451, Cromford 2956.
E. Recorded for all squares.
S. Recorded for all squares.
Linton: 'Common. Generally distributed'.

A. lanceolatum With.

Native. In similar situations to *A. plantago-aquatica* but very rare or perhaps overlooked.

E.   Renishaw 4478, Whatstandwell 3354.

S.   Canal, Spondon 3935, pool behind Rifle Butts, Trent Lock 4931.

Linton: Dovedale 15.

SAGITTARIA L.

S. sagittifolia L.   *Arrow-head*

Native. On mud in shallow water of streams and canals. Frequent at low altitudes and especially in the south of the county.

E.   Canals at Killamarsh 4681, Staveley 3974, Renishaw 4477, Whatstandwell 3354, Kirk Hallam 4145, Sawley Canal 4732.

S.   Recorded for all squares except 14, 24, 13, 23.

Linton: 'Frequent at low levels'.

BUTOMACEAE

BUTOMUS L.

B. umbellatus L.   *Flowering Rush*

Native. Margins of rivers, ponds and canals and in ditches. Frequent at low altitudes and especially in the south of the county.

E.   Canal, Ambergate 3452, Butterley Reservoir 4052, Nutbrook Canal, Ilkeston 4442.

S.   Ednaston 2442, Findern to Chellaston 33, nr. Sawley 4731, Thrumpton Ferry 5031, nr. Melbourne 32.

Linton: Norbriggs nr. Staveley and Mastin Moor 47, canal, Egginton 22.

HYDROCHARITACEAE

ELODEA Michx.

E. canadensis Michx.   *Canadian Pondweed*

Introduced. Naturalized in streams, canals and ponds. Frequent.

W.   Recorded for all squares except 09, 19, 27.

E.   Beauchief Abbey 3381, Ford 4080, Pebley Pond 4878, Scarcliffe Park Wood 5070, Wingerworth 3768, Hardwick Ponds 4563, recorded for 56, Whatstandwell 3354, Shipley Gate 4843.

S.   Recorded for all squares except 23.

Linton: Mellor 98, Alfreton 45, Morley 34.

JUNCAGINACEAE

TRIGLOCHIN L.

T. palustris L.   *Marsh Arrow-grass*

Native. Marshy meadows and by ponds, canals and streams. Local.

W. Nr. Charlesworth 0194, Jaggers Clough 1687, Countess Cliff 0571, also Cavendish Golf-course and moors west of Buxton, Axe Edge and Brook Bottom in 07, Monk's Dale 1374, Salter Sitch 2577, also Umberley Brook, Leash Fen and Big Moor in 27, Washgate Valley 0567, Bradbourne 2052.

E. Bramley Vale, Chesterfield 4676, Whitwell Wood 5278, Wingerworth Great Pond 3667, recorded for 56.

S. Ashbourne 1846, Hulland Moss 2546, also Bradley Brook and Brailsford in 24, Alkmonton Bottoms 2037.

Linton: Stirrup 99, Charlesworth 09, nr. Snake Inn 19, Malcoff and nr. Rowarth 08, Morley 34, nr. Findern 33, Ockbrook and Sandiacre 43, Stapenhill and Burton 22, Calke and Repton Rocks 32, recorded for 21.

POTAMOGETONACEAE

POTAMOGETON L.

Mr. J. E. Dandy has kindly read the proofs of this account, and has given us much valuable information about Derbyshire pondweeds. All 'old specimens' to which reference is made have been authenticated by him.

P. natans L.   *Broad-leaved Pondweed*

Native. Ponds, rivers, canals and ditches. Frequent.

Recorded for all squares except:

W. 09, 19, 28.

Linton: Chisworth 09, Lee nr. Mellor 98, Creswell Crags 57, Kirk Hallam 44, Egginton and Bretby 22, Repton 32.

P. polygonifolius Pourret   *Bog Pondweed*

Native. Moorland streamlets, flushes, pools and ditches, in shallow acid water. Locally frequent.

W. Recorded for 08, Jagger's Clough 1687, also Derwent, Win Hill and above Yorkshire Bridge in 18, Swann's Canal, Turncliff 0470, nr. the Grouse Inn, Longshaw 2578, also Salter Sitch by Owler Bar, under Baslow Edge, Umberley Brook and Ramsley Moor in 27, recorded for 26.

Linton: Charlesworth Coombs 09, Westend, and nr. the Snake 19, Offerton Moor and Callow Bank nr. Hathersage 28, Tansley Moor 36, Dethick Common 35, Hulland Moss 24, Repton Rocks 32.

Also an old specimen (1881) from 'near Glossop'.

P. lucens L.   *Shining Pondweed*
Native. Streams, canals and pools on base-rich inorganic substrata. Rather rare.
S.   Sutton Brook 2234, New Stanton 4539, Anchor Church 3327.
Linton: Breadsall 33, Willington 22.
Also an old specimen (1849) from the R. Trent at Stapenhill (now in Staffs.).

P. alpinus Balb.   *Reddish Pondweed*
Native. Ponds and slow streams; chiefly in non-calcareous water and on richly organic substrata. Rare.
W.   Emperor Stream, Brampton Moor 2970. Umberley Brook 27.
Linton: Nr. Bar Brook Hall, Chatsworth 27, Archer's Pool and Calke 32.
Also an old specimen (1900) from 'millpond N. E. of Fritchley' 35.

P. praelongus Wulfen   *Long-stalked Pondweed*
Native. Canals and pools. Rare.
S.   Trent Lock 43, Willington 2928.
Linton: Canal, Osmaston–Chellaston and Chaddesden–Spondon 33.

P. perfoliatus L.   *Perfoliate Pondweed*
Native. Streams, canals, ponds; chiefly on moderately organic substrata. Frequent in the south of the the county.
S.   Derby 3537, Trent Lock 4831, Sawley 4731, Wilne canal 4530, Thrumpton 5031, Swarkestone 3728, Weston-on-Trent 4127, Shardlow 4530.
Linton: Killamarsh 48, Renishaw 47, Whatstandwell 35, Kirk Hallam and Shipley Gate 44, Brailsford Brook 24, Sudbury 13, Barton Fields 23, Burton, Willington and above Egginton 22.

P. friesii Rupr.   *Flat-stalked Pondweed*
Native. Canals and slow streams, especially on a muddy bottom. Occasional.
E.   Killamarsh canal 48, Canal nr. Ambergate 3552.
S.   Canal nr. Allenton 3732, canal nr. Spondon 3935, Weston-on-Trent 4127.
Linton: Renishaw canal 47, Kirk Hallam and Shipley Gate 44, canal nr. Egginton 22.
Also an old specimen (1897) from Sandiacre canal 43.

P. pusillus L. (*P. panormitanus* Biv.)   *Lesser Pondweed*
Native. Streams, canals and ponds, especially in highly calcareous water. Rare.

This species has often been confused with *P. berchtoldii*, which differs in having very conspicuous nodal glands.

W. Cromford Canal nr. Matlock 25.

S. Bradley Dam 24.

Linton's records for '*P. pusillus*' are probably mostly of *P. berchtoldii* and are therefore omitted.

P. obtusifolius Mert. & Koch   *Grassy Pondweed*

Native. Canals and old ponds. Very rare.

S. Canal nr. Chellaston 3729.

Linton: Hardwick 46.

Also old specimens (1842 & 1843) from Netherseal 21.

P. berchtoldii Fieb. ('*P. pusillus* L.')   *Small Pondweed*

Native. Ponds, canals, streams, ditches; tolerant of both calcareous and very acid water. Occasional.

W. Umberley Brook 27, River Lathkill nr. Youlgreave in 26.

S. Bradley Dam in 24.

Also an old specimen (1861) from Repton rocks 32.

There is little doubt that many of the records for '*P. pusillus*' in Linton's Flora are for *P. berchtoldii*, but both species have been shown by J. E. Dandy and G. Taylor to occur in the county (see *Atlas of the British Flora*).

P. compressus L. (*P. zosterifolius* Schumach.)   *Grass-wrack Pondweed*

Native. Canals and slow streams. Occasional.

W. Reservoirs, Cromford 2957.

E. Canal, Cromford 3056, canal at Whatstandwell 3354, Thrumpton 5051.

S. Ashbourne Pond 1846, canal, Derby 3730, Trent Lock 4831, Sawley 4731, Trent and Mersey Canal 3728.

Linton: Willington 22, Drakelow 21, nr. Burton 22.

Also an old record (1862) from Melbourne Pool 32.

P. crispus L.   *Curled Pondweed*

Native. Streams, canals, ditches and ponds. Common.

W. Recorded for 07, Newton 9984, Miller's Dale 1573, millstream, Calver 2475, recorded for 16, Lathkill Dale 2165, Alport 2264, Bradford Dale 2063, Tissington 1752.

E. Ford Pond 4080, Renishaw 4378, Hardwick Ponds 4764, Pleasley Vale 5265, recorded for 35, Dale Abbey 3943.

S.   Recorded for all squares except 23, 43.

Linton: 'Common. Generally distributed'.

Also old specimens from 'reservoir in stream, Cromford' (1884) and Borrowash canal (1893).

'var. *serratus* (Hudson)' is merely a young growth-form of *P. crispus.*

P. × **cooperi** (Fryer) Fryer (*P. crispus* × *P. perfoliatus*)

Native. In a canal with the parents. Very rare.

S.   Willington 2928.

Linton: Canal nr. Chaddesden 33.

Also an old specimen (1884) from the 'canal between Borrowash and Chaddesden' 43.

P. × **lintonii** Fryer (*P. crispus* × *P. friesii*)

Native. In canals with the parents. Very rare.

E.   Canal, Renishaw 47.

S.   Trent Lock 43, Weston-on-Trent 4127.

Linton: Canal, Renishaw 47.

Also old specimens from the Killamarsh canal 48 (1897; recorded as *P. obtusifolius* in Linton's Flora), from R. Trent near Drakelow 21 (1895).

P. pectinatus L. (incl. *P. interruptus* Kit.)   *Fennel-leaved Pondweed*

Native. Rivers, canals and ponds, chiefly in lowland base-rich waters. Not infrequent.

P. interruptus Kit. was a name given to robust plants much branched above, the branches spreading like a fan. This seems only a growth-form of *P. pectinatus.*

W.   Bonsall Moor 2559.

E.   Hardwick Ponds 4563.

S.   Henmore Brook, Ashbourne to Clifton 1846 etc., Yeldersley Pond 2144, Trent Lock 4831, Sawley 4731, Thrumpton 5031, Scropton 1829, Egginton 2727, canal, Swarkestone 3327, Weston-on-Trent 4127.

Linton: Beighton 48, nr. Chesterfield 37, nr. Renishaw 47, Whatstandwell 35, above Allestree 34, Shipley Gate and Kirk Hallam 44, Breadsall 33, New Stanton 43.

GROENLANDIA   Gay

G. densa (L.) Fourr. (*Potamogeton densus* L.)   *Opposite-leaved Pondweed*

Native. Streams, canals, ditches and shallow ponds; most commonly in clear swiftly running water.   Rare.

E.   Chesterfield Canal 4475, Markland Grips 5074.

S.   New Stanton 4639.

Linton: Cromford 25, Scarcliffe Park Wood and Nether Langwith 57, canal, Whatstandwell 35, Osmaston Lake 24, nr. Calke Abbey and Old Trent, Repton 32.

ZANNICHELLIACEAE

ZANNICHELLIA L.

Z. palustris L.   *Horned Pondweed*

Native. Streams, ditches, pools. Occasional.

W.   Pool, Townend, Chapel-en-le-Frith 0681, Pavilion Gardens, Buxton 0573, Lathkill Dale 16.

S.   River Dove, Hanging Bridge 1546, Yeldersley Brook 2144, canal, nr. Weston-on-Trent 4027.

Linton: Miller's Dale 17, large pond at Chatsworth 27, River Wye at Ashford 16, Creswell Crags and Markland Grips 57, Langwith Wood 56, Foston 13, Gill's Park, Ticknall 32.

LILIACEAE

NARTHECIUM Hudson

N. ossifragum (L.) Hudson   *Bog Asphodel*

Native. Bogs, wet heaths and moors and wet acid places on mountains. Locally frequent.

W.   Recorded for 08, Jaggers Clough bog 1687, Blacka Moor 2980, Abney Moor 2080, by water-tower, Burbage 2680, recorded for 07, Ramsley Wood 2976, Owler Bar 2877, Smeekley Wood 2976, Hay Wood 2477, Bar Brook 2775, Axe Edge 0369, Washgate Valley 0467, Umberley Brook 2869.

E.   Hipper Sick 3068.

S.   Hulland Moss 2446.

Linton: East Moor 37, Dethick 35.

CONVALLARIA L.

C. majalis L.   *Lily-of-the-Valley*

Native. Woods on limestone scree in the dales and often persisting after the woods have been opened up or destroyed; commonly cultivated and sometimes escaping. Locally abundant.

Often associated with *Melica nutans, Rubus saxatilis, Geranium sanguineum* and a species-rich shrub layer and then probably marking sites where woodland has persisted for very long periods.

W.   Taddington Dale 1670/1, Chee Tor 1273, Cressbrook Dale 1773, Monsal Dale 1771, 1871, Coombs Dale 2274, Lathkill Dale 1865, Dovedale 1453, Via Gellia 2556, 2656.

E.   Nethermoor 4464, Pleasley Vale 5265.

Linton: Great Shacklow Wood 16, High Tor 25, Brassington Rocks 25, Cresswell 57, Dale 43, Anchor Church 32.

POLYGONATUM Miller

P. odoratum (Miller) Druce   *Angular Solomon's Seal*
Native. Ledges of limestone cliffs; tolerant of shade. Very rare.

W. Deep Dale 0971, Water-cum-Jolly 1672, The Nabbs, Dovedale 1453, Brassington Rocks 2154.

E.   Shining Cliff Woods 3352.

P. multiflorum (L.) All.   *Solomon's Seal*
Native. Woods, especially on limestone. Rare.

W. Recorded for 07, The Gorge, Cressbrook Dale 1773, nr. Wormhill 1274, below Curbar Edge 2475, Youlgreave 1964, Via Gellia 2556.

E.   Pleasley Park Wood and Pleasley Vale 5164/5.

S.   Ladyhole Lane 2044.

Linton: Duffield 34, Foremark 32.

RUSCUS L.
R. aculeatus L.   *Butcher's Broom*
Introduced. Grown in gardens and sometimes escaping. Rare.

S.   A little in a plantation, Blackwall nr. Biggin 2549, established in a garden at Littleover 3334, and by the Council House, Derby 3536, Winshill 2723.

LILIUM L.
L. martagon L.   *Martagon Lily*
Probably introduced. Commonly grown in gardens and naturalized in a few localities. Rare.

W. Naturalized in the wood below Hopton Hall 2552.

Linton: Not recorded.

TULIPA L.
T. sylvestris L.   *Wild Tulip*
Introduced. Naturalized in old pastures, near streams. Very rare.
No recent records.

Linton: Kedleston 34, in the Holmes, Derby 33, Sudbury Hall 13.

GAGEA Salisb.

G. lutea (L.) Ker-Gawler (*G. fascicularis* Salisb.)   *Yellow Star-of-Bethlehem*
Native. Woods, copses and pastures, especially on limestone. Local and infrequent.

W. Cressbrook Dale 1774, Ravensdale 1773, Lathkill Dale 2066, nr. Ible 2456, Brassington 2154.

E. Creswell Crags 5373.

Linton: Meadow Place Wood, Youlgreave 16, Bradford Valley 16, Beighton 56, by Erewash nr. Langley Mill 45, Derby 33.

ORNITHOGALUM L.

O. umbellatum L.   *Star-of-Bethlehem*
Probably introduced. Meadows and pastures. Rare.

W. Water-cum-Jolly 1572, Havenhill Dale 2152.

E. Langer Lane, Wingerworth 3767.

S. Hulland Moss 2546.

Linton: Nr. Haddon Hall 26, The Holmes, Derby 33.

O. nutans L.   *Drooping Star-of-Bethlehem*
Introduced. Meadows. Very rare; no recent records.

Linton: Meadows near Derby.

ENDYMION Dumort.

E. non-scriptus (L.) Garcke (*Scilla festalis* Salisb.)   *Bluebell*
Native. Woods and hedgebanks, and pastures on sites where there has been woodland in the fairly recent past; thriving on a wide range of soil-types but absent from very acid and markedly podsolized soils. Common and widespread. White-flowered plants occur here and there.

Recorded for all squares.

Linton: 'Common. Generally distributed'.

ALLIUM L.

A. scorodoprasum L.   *Sand Leek*
Native. Grassland and scrub on dry soils. Very rare.

W. By roadside, Alport 2365.

S. Canal side, nr. Swarkestone 3829.

Linton: Not recorded.

A. vineale L.   *Crow Garlic*
Native. Cliff-ledges, walls, banks, meadows, field-borders and roadsides. Local but not infrequent.

Two varieties occur in the county: var. *vineale*, with rather lax inflorescences of mixed flowers and bulbils; and var. *compactum* (Thuill.) Boreau, with compact inflorescences of bulbils only.

Var. **vineale**

W. Recorded for 07, Miller's Dale 1473, Upper Dale 1872, Monk's Dale 1274, Peter Dale 1374, Tideswell Dale 1573, Coombs Dale 2274, rocks, Hartington 1360, Mill Dale 1455, High Tor, Matlock 2958.

E. Eckington 4379, Pleasley Vale 5165.

S. Roadside, nr. Cat and Fiddle Mill, Dale Abbey 4339.

Linton: Barlow Lees 37, Streetley 57.

Var. **compactum** (Thuill.) Boreau.

No recent records

Linton: Castleton 18, High Tor, Matlock Bath 25, Bakewell 26, Dovedale 15, Calver 27, Ashover 36, Morley 34, Breadsall 33, meadows by river at Sawley 43, between Stapenhill and Cauldwell 21.

A. oleraceum L.   *Field Garlic*

Native. Ledges on limestone cliffs, dry rough pastures and field-borders. Rare.

W. The Winnats 1382, Cave Dale 1582, Conies Dale 1280, Deep Dale, Buxton 0971, Miller's Dale 1573, Topley Pike 1072, Monk's Dale 1375, Long Dale 1361, Hartington 1360, The Nabbs, Dovedale 1453, Matlock 2959.

E. Barlborough 4777, nr. Clowne 4975, Steetley Quarry 5478, Glapwell 4966, Langwith Wood 5068.

Linton: Canal path between Sandiacre and Long Eaton 43, nr. Repton, nr. Ingleby and Knowle Hills 32.

A. ursinum L.   *Ramsons*

Native. Damp woods and shady places; often forming dense stands in valley-bottom woods on base-rich soils where a seasonally high water-table leaves them moist but well-aerated during the growing season. Locally abundant.

Recorded for all squares except:

> **W.** 09, 19.
> **E.** 45.
> **S.** 33.

Linton: Newton Wood 45.

COLCHICUM L.

C. autumnale L.   *Autumn Crocus, Meadow Saffron*

Native. Damp meadows and woods on calcareous and neutral soils. Local; infrequent and diminishing.

E. Allestree/Duffield 3541.

S. Nr. Cubley 1538.

Linton: Nr. the Derwent above Derby 34, Breadsall 33, Swarkestone and Ingleby 32, Anchor Church 32.

TRILLIACEAE

PARIS L.

P. quadrifolia L.   *Herb Paris*

Native. Damp woods, especially on calcareous soils. Local and infrequent.

W. Water-cum-Jolly Dale 1672, Monk's Dale 1374, 1375, 1275, The Nabbs
and the Doveholes, Dovedale 1453, Via Gellia 2656, 2556, nr. the Black
Rocks 2955, Brassington Rocks 2154.

E. Whitwell Wood 5279.

Linton: Nr. Mellor 98, Buxton 07, Whitwell 57, Pinxton and Newton Wood
45, Shirley Wood 24, east of Alkmonton 13, Long Lane and Dalbury Lees 23,
Darley and Breadsall 33, Spondon and nr. Dale 43, Burton and Bretby 22,
Brian's Copse nr. Ticknall, Calke Park and Repton Rocks 32.

JUNCACEAE

JUNCUS L.

J. squarrosus L.   *Heath Rush*

Native. Moist heaths and moorland; especially characteristic of closely grazed
or trampled upland pastures on peat-podsols. Locally abundant.

W. Recorded for all squares.

E. Slagmill Plantation 3068, Holloway etc. 35, Breadsall 3841.

S. Ashbourne 1946, Hulland Moss 2546, Repton Rocks 3221, Ticknall
3523.

Linton: Gresley 21.

J. tenuis Willd.

Introduced. Sides of paths and waste places on acid soils. Occasional and
spreading.

W. Fernilee 0178, Via Gellia 2655.

E. Shining Cliff Wood 3352.

S. Canal side, Willington 3029.

Linton: Not recorded.

J. compressus Jacq.   *Round-fruited Rush*

Native. Wet meadows, marshes, margins of pools. Very rare.

E. Whitwell Wood 5278, Hardwick Ponds 4563/4, Griff Wood, Ault
Hucknall 4665.

S. Trent 5031.

Linton: Cromford Moor 25, Little Chester 33, Ockbrook 43.

J. bufonius L.   *Toad Rush*

Native. Moist bare places on arable land, in woods, by roads and paths and on moorland. Common and widespread.

W. Alport Dale 1390/1, Hayfield 08, Cat and Fiddle 0071, Monk's Dale 1374, Big Moor 2877, Brand Top 0468, Hartington 1261, Fenny Bentley 1850, Kniveton 2050.

E. Beauchief Abbey 3381, Brierley Wood 3775, recorded for 47, Whitwell Wood 5278, Ogston 3760, Bramley Vale 4666, Pleasley Vale 5265, Holloway etc. 35, Alfreton Brook 4056, Breadsall 3841.

S. Recorded for all squares except 32.

Linton: 'Generally distributed and plentiful'.

J. inflexus L. (*J. glaucus* Sibth.)   *Hard Rush*

Native. Marshes, wet meadows and damp waysides, especially on heavy base-rich substrata. Locally common.

Recorded for all squares except 09, 19, 34.

Linton: 'Common'.

J. effusus L.   *Soft Rush*

Native. Marshes, wet meadows and pastures, streamsides, flushes, ditches, damp waysides etc.; especially on acid soils. Common and widespread.
Recorded for all squares.

Linton: 'Common'.

J. subuliflorus Drej. ('*J. conglomeratus* L.')   *Common Rush*

Native. In similar situations to *J. effusus* but less tolerant of high acidity and more tolerant of dry conditions; often on higher ground marginal to flushes dominated by *J. effusus*. Common and widespread but less gregarious than *J. effusus*.

Sometimes confused with forms of *J. effusus* having compact inflorescences but readily distinguishable by the dull (not glossy) stem with prominent ridges just below the inflorescence and by the small elevation in the hollowed top of the capsule. Intermediates between the two species are seen occasionally. They are commonly sterile and are probably hybrids.

Recorded for all squares except:

> **W.** 09, 17.
> **E.** 48, 56, 45.

Linton: 'Generally distributed'.

J. subnodulosus Schrank (*J. obtusiflorus* Ehrh.)   *Blunt-flowered Rush*

Native. Fens and marshes, especially on calcareous substrata. Very rare.

E. Nr. the Walls, Whitwell 5078.

Linton: Scarcliffe Park Wood 57, Shirley 24, Scropton 13, Burton 22.

J. acutiflorus Ehrh. ex Hoffm.   *Sharp-flowered Rush*

Native. Wet meadows, swampy woodlands and moorland flushes; chiefly on acid substrata. Widespread and locally abundant.

W. Warhurst Fold 9995, Win Hill 1984, Ringinglow 2883, recorded for 07, Froggatt Edge 2476, White Edge 2677, Hartington 1261, Elton 2161, Fenny Bentley 1750, Bonsall Moor 2559.

E. Killamarsh 4580, Holmesfield 3277, recorded for 47, Markland Grips 5074, Wingerworth Great Pond 36, Pleasley Vale 5265, Dale 4438, Belper 34.

S. Yeaveley 1839, Ashbourne Green 1847, Alkmonton Bottoms 2073, Osmaston 2143, Bradley Brook 2244, Hulland Moss 2546, Melbourne 32, recorded for 21.

Linton: 'Generally distributed and often very abundant'.

J. articulatus L. (*J. lamprocarpus* Ehrh.)   *Jointed Rush*

Native. Meadows, marshes, ditches, wet paths and tracks and margins of ponds. Common and widespread.

Very variable in habit and in the size and form of the inflorescence.

W. Recorded for all squares except 15.

E. Recorded for all squares except 48, 44.

S. Cubley Common 1639, Ashbourne Green 1847, Hulland Moss 2546, Mercaston 2642, Swarkestone 3628, Drakelow 2318.

Linton: 'Common'.

J. bulbosus L. (*J. supinus* Moench)   *Bulbous Rush*

Native. Moist places on heaths and moors, bogs, cart-tracks and woodland rides; always on acid substrata. Common and widespread.

More robust forms with six stamens and sharply angled capsules, sometimes separated as *J. kochii* F. W. Schultz, have been recorded from a few localities.

W. Alport Dale 1390, Roych 0783, Derwent 1889, Hathersage 2582, recorded for 07, Grindleford 2578.

E. Hardwick Hall 4563, nr. Stanley 4040.

S. Spinnyford Brook 2445.

Linton: Goyts Clough and Axe Edge 06/07.

LUZULA DC.

L. pilosa (L.) Willd. (*L. vernalis* (Reichard) DC.)   *Hairy Woodrush*

Native. Woods, thickets and hedgebanks. Frequent.

W. Recorded for all squares except 15 (although abundant in Hall Dale 1353, nr. Dovedale on the Staffordshire side).

E. Beauchief 3381, Ford nr. Eckington 4081, Frith Wood, Dronfield 3679, also Kitchinflat Wood, Cutthorpe, Linacre and Troway in 37, below Birkinshaw Wood 3369, recorded for 46, Pleasley Vale 5265, Ambergate 3451.

S.   Woodcock Dumble 1848, Snelston 1644, Shirley Wood 2042, Hulland Moss 2546, recorded for 23.

Linton: Derwent Moor 28, Dovedale 15 (see above), Whitwell Wood 57, Newton Wood 45, Abbots Clownholme 13, Bretby Wood 22, Calke and Repton Shrubs 32.

L. sylvatica (Hudson) Gaudin (*L. maxima* (Reichard) DC.)

Greater Woodrush

Native. Woods and rough pastures on flushed acid mulls and especially on rocky ground near streams. Local but often prominent in mountain cloughs.

W.   Alport Castles 1490/1, nr. Malcoff 0783, Goyt Valley 0774, Longshaw 2578, Washgate Valley 0567, between Glutton Bridge and Hollinsclough 0766, Chatsworth 26/27, Via Gellia 2656.

E.   Beauchief 3381, Ford nr. Eckington 4081, Linacre 3170, Cordwell 3276, Chanderhill 3369, recorded for 35, West Hallam 4341.

S.   Edlaston Coppice 1743, by River Dove, Norbury 1342, Biggin-by-Hulland 2547, Henmore Brook above Sturston Mill 2046, recorded for 33, Dale Abbey 4338, Anchor Church 3327, Robin Wood 3525.

Linton: Charlesworth 09, Mellor 98, Edale and Black Tor 18, Hood Brook 28, Dovedale 15, Hardwick 46, Handley Wood, Shottle 34, Drakelow 22.

L. campestris (L.) DC.   *Field Woodrush*

Native. Grassland on acid or surface-leached soil, heaths, waysides. Common and widespread.

Recorded for all squares.

Linton: 'Common. Widely prevalent and abundant'.

L. multiflora (Retz.) Lej. (*L. erecta* Desv.)   *Many-headed Woodrush*

Native. Heaths, moorlands and woods on damp acid peaty soils. Frequent.

Plants with very compact inflorescences and larger seeds, variously called var. *congesta* (Thuill.) Buchenau or *L. congesta* (Thuill.) Lej., have been recorded from localities in all parts of the county. More observations are needed on the distinctness, both morphologically and ecologically, of these forms, which may well merit subspecific rank.

W.   Recorded for 09, Roych Clough 0883, Fairholmes, Derwent 1889, Dore Moor 2982, Withinleach Moor 9976, Burbage Edge 0372, Lover's Leap 0772, recorded for 17, Chatsworth Park 2870, also under White Edge and Leash Fen in 27, Dove Head 0368, Birchover (*v. pallescens*) 2462, White Springs, Rowsley 2865.

S.   Fenny Bentley Brook 1748, Bradley Wood 2046, Bradley Brook 2244, Grange Wood, Lullington 2714.

Linton: Ashover Hay and Tansley Moor 36, Horsley Carr 34, Burton and Willington 22, Calke and Repton 32.

AMARYLLIDACEAE

GALANTHUS L.

G. nivalis L.   *Snowdrop*

Probably introduced. Much planted and readily becoming naturalized in woods, plantations, damp meadows and hedgebanks. Frequent.

W. Calver to Baslow 2473, Rowsley 2566, Thorpe 1550, Beresford Dale 15.

E. Unthank 3076, Eckington, Valleys of Rivers Moss and Rother 47, Wessington 35, Lindway Springs 3557.

S. Edlaston 14, Nether Biggin 2548, nr. Elvaston Castle 4032, Borrowash 4134, Repton 32.

Linton: Cordwell Valley 37, Matlock, Cromford 25, Kirk Ireton 24, Shirley 24, Duffield 34, Breadsall 34, Morley 34, Ockbrook 43, by the Old Trent, Repton 32.

NARCISSUS L.

N. pseudonarcissus L.   *Wild Daffodil, Lent Lily*

Almost certainly native. Damp woods, copses and grassland. Once locally abundant but now lost from most of its former localities.

W. Taddington Wood by Taddington Dale 1670, Stand Wood 2670, Rowsley 2566, Lindup Wood 2567.

E. Pratt Hall 3273, Unthank 3076, Brierley Bottoms 3172, Linacre reservoirs 3372, Scarcliffe Park Wood 5270, Bradley Hollow 3362, Stony Houghton 4966, Langwith Wood 5068, Horsley Castle 3743, Shipley Hall 4444.

S. Naturalized at Snelston 1543, and Shirley Wood 2042, fields nr. Mercaston Old Hall 2472.

Linton: Chapel 08, Goyt Valley 07, Cromford 25, Matlock Bridge, Whatstandwell, Lea Wood and Alderswasley 35, Borrowash–Draycott and Spondon 43, Calke, Repton and Swarkestone 32, Drakelow 21.

N. majalis Curtis ('*N. poeticus* L.')   *Pheasant's Eye*

Introduced. Escaping from gardens and becoming established. Very rare.

S. One clump in a field nr. Rodsley 1941.

Linton: Not recorded.

N. × biflorus Curtis (? *N. majalis* Curtis × *N. tazetta* L.)   *Primrose Peerless*

Introduced. Naturalized in meadows. Very rare.

No recent records.

Linton: Starkholmes, Matlock Bath 25, Matlock 35, Heanor 44.

Y*

IRIDACEAE

SISYRINCHIUM L.

S. bermudiana L.   *Blue-eyed Grass*

Introduced. Naturalized in marshy meadows. Very rare.

W. Peak Forest canal nr. Bugsworth 0182, Axe Edge by the roadside 0270, 0269.

Linton: Not recorded.

IRIS L.

I. foetidissima L.   *Gladdon, Stinking Iris*

Probably native. Open woods and hedgebanks, usually on calcareous soils. Very rare; no recent records.

Linton: Nr. Willersley Park 25.

I. pseudacorus L.   *Yellow Flag*

Native. Marshes, wet meadows, margins of pools, canals and streams. Frequent but chiefly in the south of the county.

W. Hayfield 08, Calver New Bridge 2475, Alport 2164, Tissington Dam 1751.

E. Recorded for all squares except 34, 44.

S. Recorded for all squares.

Linton: Breadsall 34.

CROCUS L.

C. nudiflorus Sm.   *Autumn-flowering Crocus*

Introduced. Naturalized in damp meadows and pastures. Rare.

W. Chinley 0483.

E. Dronfield 3478, 3578, Brierley Wood 3775.

S. Rodsley 2040.

Linton: Duffield 34, Derwent at Derby 33, Holmes and Siddals, Derby 33, Repton 32.

DIOSCOREACEAE

TAMUS L.

T. communis L.   *Black Bryony*

Native. Hedgerows, wood-margins, copses and scrub, especially on moist but well-drained fertile soils; absent from the north-west of the county. Frequent.

W. Cressbrook Dale 1773, Thorpe 1650, Kniveton 2150, Ireton 2650.

E. Recorded for all squares.

S. Recorded for all squares.

Linton: 'Common. Abundant'.

ORCHIDACEAE

CYPRIPEDIUM L.

C. calceolus L. *Lady's Slipper*
Formerly native. Wooded limestone cliffs. Very rare; no recent records.
Linton: 'Formerly found on the Heights of Abraham' 25.

CEPHALANTHERA L. C. M. Richard

C. damasonium (Miller) Druce (*C. pallens* L. C. M. Richard)
White Helleborine
Native. Limestone woods. Very rare; no recent records.
Linton: Newton Wood 22.

C. longifolia (L.) Fritsch (*C. ensifolia* (Schmidt) L. C. M. Richard)
Long-leaved Helleborine
Native. Limestone woods. Very rare; no recent records.
Linton: Markeaton 33.

EPIPACTIS Zinn

E. palustris (L.) Crantz *Marsh Helleborine*
Native. Fens and fen-woods. Very rare; lost from several of its former localities.
E. Marsh at Whitwell Wood 5278.
Linton: Matlock 25, Scarcliffe Park Wood 57, Woodeaves nr. Ashbourne 15.

E. helleborine (L.) Crantz (*E. latifolia* (L.) All.) *Broad Helleborine*
Native. Woods, wood-margins, hedgebanks etc., on a wide range of soil-types but avoiding strongly acid and podsolized soils. Locally frequent.
Very variable in stature, width of leaves, flower-colour, shape of labellum and number, size and roughness of the labellar bosses. *E. atroviridis* Linton, with the broad rounded leaves of *E. helleborine* 'but rather more numerous lanceolate leaves between the lower leaves and the flowering spike' and with flowers not or only slightly rose-coloured, does not seem to merit specific rank.
W. Hathersage 2381, Deep Dale 0971, Wye Valley 1072, Priestcliffe Lees 1572, Cressbrook Dale 1773, Hay Dale 1772, Eyam 2176, Baslow Hill 2673, Congreave, Rowsley 2465, Biggin Dale 1458, nr. Horsley, Fenny Bentley 1950, Brassington Rocks 2154, Black Rocks 2955, Via Gellia 2656 (including 'E. atroviridis Linton').
E. Twentywell Lane 3280, Beauchief 3381, Ford nr. Eckington 4081, Holmesfield 3075, Pebley Pond 4878, Whitwell Wood 5278, Stubbin Court 3567, Wingerworth Great Pond 3667, Pleasley Vale 5265, Ogston Carr 3659 and Brackley Gate 3942, both 'E. atroviridis Linton'.

S.   Yeldersley 2144, Biggin-by-Hulland 2548.

Linton: Bradwell 18, Osmaston 14, Spondon 43, Bretby and Repton Shrubs 32; also, as '*E. atroviridis* Linton', Bar Brook, Baslow 27, Chatsworth Park 27, Monk Wood, Sheepbridge 37, Newbold 37, Linacre Wood 37, Pebley Pond 47, Ashover Hay 36.

E. phyllanthes G. E. Sm.   *Green Helleborine*

Native. Damp places in woods on limestone. Very rare.

E.   Whitwell Wood 5278.

Linton: Not recorded.

E. atrorubens (Hoffm.) Schultes   *Dark-red Helleborine*

Native. Limestone rocks and screes, in woods or in the open. Rare, but with many plants in some of its localities.

W.   Ravensdale, Cressbrook Dale 1773, Coombs Dale 2274, Longstone Edge 1972, 2072, Biggin Dale 1458, Via Gellia 2656.

SPIRANTHES L. C. M. Richard

S. spiralis (L.) Chevall. (*S. autumnalis* L. C. M. Richard)
                                                      Autumn Lady's Tresses
Native. Moist meadows and pastures, usually on limestone. Very rare.
W.   Bonsall 25.

Linton: Matlock 35, Dovedale 15, New Lake, Osmaston Manor 24, Repton 32.

LISTERA R. Br.

L. ovata (L.) R.Br.   *Twayblade*

Native. Moist woods, pastures, hedgebanks and waysides on base-rich soils. Common and widespread.

W.   Recorded for all squares except 09, 19, 18.

E.   Recorded for all squares except 46, 45, 34, 44.

S.   Railway banks, nr. Ashbourne 1747, recorded for 23, Ticknall 3523, Bretby 2922.

Linton: 'Common. Generally distributed'.

L. cordata (L.) R. Br.   *Lesser Twayblade*

Native. Under heather on moist moorland and on wet rock-ledges of shale or gritstone. Very rare.

W.   Between Grindleford and Hathersage 2680, Cupola nr. Hathersage 2582.

Linton: Kinder Scout 08, Beeton Rod, Hassop 27.

NEOTTIA Guett.

N. nidus-avis (L.) L. C. M. Richard   *Bird's-nest Orchid*
Native. Shady woods especially on limestone but occasionally in rocky woods elsewhere. Very rare.
W. Lathkill Dale 2066.
E. Whitwell Wood 5279.
Linton: Nr. Mottram 09, Meersbrook 38, Dovedale 15, nr. Black Rocks and Matlock Bath 25, Wyaston 14, Repton Shrubs 32.

COELOGLOSSUM Hartman

C. viride (L.) Hartman (*Habenaria viridis* (L.) R.Br.)   *Frog Orchid*
Native. Pastures, especially on limestone. Local.
W. Houndkirk Hill 2881, Brook Bottom, Buxton 0671, Grin Plantation 0472, Ashwood Dale 0972, Diamond Hill 0570, Harpur Road, Ladmanlow 0471, Ravensdale 1774, Longstone Edge 1972, High Rake 2172, Brierlow Bar/ Earl Sterndale Moor 0969, nr. Newhaven 1561, Hartington 1259, above Matlock Bath 2858, pastures, Hipley Toll Bar 2154, Via Gellia 2557, Kniveton 2150, Grange Mill 2357, Harborough Brick Works 2955.
E. Crich 3455.
Linton: Charlesworth Coombs 09, Chapel 08, Bradwell 18, Ashopton 28, Eyam Moor 27, Whitwell Wood 57, Stanton-by-Dale 43, Yeldersley 24, Calke 32, Foremark 32.

GYMNADENIA R. Br.

G. conopsea (L.) R.Br. (*Habenaria conopsea* (L.) Bentham)   *Fragrant Orchid*
Native. Grassland, especially on limestone; fens and marshes. Locally frequent.
Var. *densiflora* (Wahlenb.) G. Camus, Bergon and A. Camus, with broad basal leaves and flowering spike usually exceeding 10 cm. in length, the flowers rose-red and clove-scented, occurs rarely in fens.
W. Ashwood Dale 0972, Countess Cliff 0570, Abney Moor 1879, above Chee Dale 1272, Miller's Dale 1373, High Rake, Hucklow 1677, Ravensdale 1774, Monk's Dale 17, Wormhill 17, Litton 17, Longstone Edge 2273, Washgate Valley 0567, railway bank nr. Hartington Station 1461, nr. Brightgate 2659, Via Gellia 2656, Grange Mill 2357, pastures nr. Hipley Toll Bar 2154.
E. Whitwell Wood 5279.
Linton: Westend Farm 19, nr. Hayfield 08, Bradwell 18, Lathkill Dale 26, Markland Grips 57, Scarcliffe Park Wood 57, Pleasley 56, nr. Dale 43, nr. Osmaston Church 24.

LEUCORCHIS E. Mey.

L. albida (L.) E. Mey. ex Schur (*Habenaria albida* (L.) R.Br.)
                                        Small White Orchid
Native. Hilly pastures. Very rare.
W. Goyt Valley 0176.
Linton: Nr. Glossop 09.

PLATANTHERA L. C. M. Richard

P. chlorantha (Cust.) Reichenb. (*Habenaria chloroleuca* Ridley)
Greater Butterfly Orchid
Native. Woods and grassy slopes on base-rich or calcareous soils. Occasional.
W. King Sterndale (one plant, 1934) 0871, nr. Calver New Bridge 2475, meadow by road nr. Calver Sough 2374, Froggatt 2476, Highlow Brook, Hathersage 2179, upper reaches River Dove 0467.
E. Cordwell Valley 3176, Whitwell Wood 5278, between Whatstandwell and Wigwell Grange 3154.
Linton: Stirrup 99, Charlesworth Coombs 09, Westend Farm 19, Bradwell 18, Lathkill Dale 16/26, Dovedale and Thorpe 15, Matlock Bath 25, Snelston Park 14, between Edlaston and Cubley 14/13, Osmaston 24, Calke 32.

P. bifolia (L.) L. C. M. Richard (*Habenaria bifolia* (L.) R.Br.)
Lesser Butterfly Orchid
Native. Grassy or heathy hillsides and open woods on at least moderately base-rich soils. Very rare; many early records were probably *P. chlorantha*.
W. Goyt Valley 0178.
Linton: Newfields nr. Chapel 08, Buxton 07, Ferriby Brook, Morley 34, nr. Dale 43, between Ashbourne and Woodeaves 14/15, Milton 32.
(Note by Linton: 'None of above are certainly *H. eu-bifolia*; the entry in Top. Bot. however, evidences its occurrence in the County'.)

OPHRYS L.

O. apifera Hudson   *Bee Orchid*
Native. Pastures, field-borders, banks, quarry-spoil etc., especially on recently disturbed limestone soil, scree or quarry-waste; tolerant of some shade in scrub or open woodland. Local.
W. Longstone Edge 2073, Hipley Toll Bar 2154, old quarry, Middleton-by-Wirksworth 2755, sand-pit nr. Brassington 2355.
E. Pebley Pond 4880, 4878, Markland Grips 5074, Bolsover 5071, Fallgate 3562, Quarry, Crich 3455.
S. Ticknall 3523.
Linton: Bradwell Dale 18, Monsal Dale 17, Masson Hill 25.
Var. *trollii* (Heg.) Druce: Ticknall 3523.

O. insectifera L. (*O. muscifera* Hudson)   *Fly Orchid*
Native. Open limestone woods or scrub and quarry spoil-heaps. Local.
W. Via Gellia 2656.
E. Langwith Wood 4967, Whitwell Wood 5278, Fallgate, old quarry 3562.
Linton: Monsal Dale 17, nr. Hartington 16, Longstone Edge 27, Dovedale 15.

ORCHIS L.

O. ustulata L. *Burnt Orchid, Dark-winged Orchid*

Native. Limestone pastures, dry banks. Very rare but plentiful in a few localities.

W. Hipley Toll Bar 2154, Longcliffe 2255, Via Gellia 2655, 2656.

Linton: Ashwood Dale 07, Alfreton 45, South Normanton 45.

O. morio L. *Green-winged Orchid*

Native. Meadows and pastures, especially on base-rich soils. Occasional.

W. Bonsall 2758, Matlock Bath 2958.

E. Troway 3980, Ford nr. Eckington 4081, Barlborough 4777.

S. Nr. Osmaston 1844.

Linton: Ashwood Dale 07, Miller's Dale 17, Dovedale 15, Renishaw 47, Crich 35, Morley 34, Yeldersley, Shirley and Hollington 24, Ockbrook 43, Ticknall 32, Repton 32.

O. mascula (L.) L. *Early Purple Orchid*

Native. Woods, copses and open pastures, especially on base-rich soils and very characteristic of the limestone dales. Locally frequent.

W. Recorded for 08, Deep Dale 0970, Cunning Dale 0872, Woo Dale 0972, Miller's Dale1 473, Monk's Dale 1373, Cressbrook Dale 1574, Coombs Dale 2173, Longstone Edge 2073, Hartington 1360, Lathkill Dale 2066, Biggin Dale 1457, Hipley Toll Bar 2154.

E. Ford nr. Eckington 4081, Linacre 3472, recorded for 47, Markland Grips 5074, Whitwell Wood 5278, Fallgate 3561, Pleasley 5265, Windley 3045, West Hallam 4341.

S. Breadsall 3739, recorded for 32, 21 and 31.

Linton: Bradwell 18, Crich 35, Wyaston 24, Edlaston 24, Alkmonton 13, Radbourne 23, Ockbrook 43.

ACERAS R. Br.

A. anthropophorum (L.) Aiton f. *Man Orchid*

Native. Open woods and scrub on limestone. Very rare.

E. Ashover 36.

Linton: Not recorded.

DACTYLORHIZA Nevski

D. incarnata (L.) Soó ('*Orchis latifolia*', in part) *Meadow Orchid*

Native. Wet meadows and marshes. Local and rather rare.

The Derbyshire plant is ssp. *incarnata.*

No recent records.

D. fuchsii (Druce) Soó ('*Orchis maculata* L.', in part)   *Common Spotted Orchid*
Native. Woods, grassy slopes and meadows on damp base-rich soils and in marshes and fens. Common and widespread.
W. Recorded for all squares except 09, 19, 28.
E. Troway 3980, Frith Hall 3270, Pebley Pond 4878, Whitwell Wood 5278, Wingerworth Great Pond 3662, Pleasley Vale 5265, by Ambergate Canal 3452, West Hallam 4341.
S. Recorded for all squares except 43, 21.
Linton: 'Common and abundant everywhere'.

D. fuchsii × D. praetermissa (?'*Orchis latifolia* L.', in part)
Native. A very handsome tall plant with ring-spotted leaves, often seen where both parents occur. Local.
W. Between Froggatt and Calver 2375, 2476, opposite Dalebrook House, East Moor 2971, Owler Bar 2977.

D. maculata (L.) Soó ssp. **ericetorum** (Linton) P. F. Hunt and Summerhayes ('*Orchis maculata* L.', in part)   *Heath Spotted Orchid*
Native. Moist acid peaty substrata, in open woods or on heaths and moors; often by rills and streamlets on the gritstones and shales of the northern moorlands. Locally frequent.
W. Recorded for 08, Win Hill 1984, nr. Clarion Clubhouse 2881, Grin Plantation 0572, Leash Fen 2874, under White Edge 2677, Salter Sitch 2877, recorded for 06, Parsley Hay 1562, Birchover 2361.
S. Bradley Brook 2244.
Linton: Grouse Inn, between Hayfield and Glossop.

D. praetermissa (Druce) Soó ('*Orchis latifolia* L.', in part)   *Fen Orchid*
Native. On wet base-rich peat in fens. Local and rather rare.
W. Hathersage 2383, Monsal Dale 1770, between Froggatt and Calver 2375, 2476, Owler Bar 2977, Ashford 1769, Brick Works, Harborough Rocks 2455.

D. purpurella (T. & T. A. Steph.) Soó   *Northern Fen Orchid*
Native. Fen peat. Very rare and only in the north-west of the county.
D. praetermissa and *D. purpurella* rarely occur together, the former being found to the south and the latter to the north of a line passing through North Derbyshire. The two species thus behave like mutually exclusive counterparts.
W. Grin Plantation 0472.

ANACAMPTIS L. C. M. Richard

A. pyramidalis (L.) L. C. M. Richard (*Orchis pyramidalis* L.) *Pyramidal Orchid*
Native. Grassy limestone slopes and banks. Local and rather rare.

W. Cressbrook Dale 1774, Longstone Edge 2073, Hassop Mines 2272, Pastures nr. Hipley Toll Bar 2154, Via Gellia, west of Slaley 2656.

E. Ashover 3562.

Linton: Monsal Dale 17, below Alport 26, Dovedale 15, Creswell and Markland Grips 57, Newton Wood 35, Ockbrook 43.

ARACEAE

ACORUS L.

A. calamus L.   *Sweet Flag*

Introduced, but completely naturalized. In shallow water at the margins of ponds, canals and rivers. Local, and chiefly at low altitudes.

E. Upper Newbold 3573, Stubbing Pond 3567, nr. Hardwick 4563, canal, Ambergate 3452, Breadsall Priory 3841.

S. Osmaston Ponds 2042, Kirk Langley 2839, Shelton Lock 3731, nr. Stenson 3230, Trent Lock 4931, canal nr. Wilne 4530, nr. Egginton 2727, Anchor Church 3327, Trent-side nr. Walton Wood 2017.

Linton: Norton 38.

ARUM L.

A. maculatum L.   *Lords-and-Ladies, Cuckoo-pint*

Native. Woods and hedgebanks on good fertile soils; very shade-tolerant. Common and widespread.

Recorded for all squares except:

W. 09, 19, 28.

Linton: 'Common. General and abundant'.

A. italicum Miller

Introduced. A rare garden escape.

E. Naturalized in Whirlow Park, Sheffield 3283.

Linton: Not recorded.

LEMNACEAE

LEMNA L.

L. polyrhiza L.   *Greater Duckweed*

Native. In still water of ponds, canals and ditches. Local and infrequent.

E. Canal, Renishaw 4378.

S. Radbourne 3635, canal, Allenton 3731, Swarkestone 3728.

Linton: Thrumpton Ferry 53, west of Willington 22.

L. trisulca L.   *Ivy Duckweed*

Native. In still water of ponds, canals and ditches. Frequent.

W. Ashford 1769, reservoir nr. Wash Brook, Tissington 1651.

E. Canal, Killamarsh 4681, Pebley Pond 4878, nr. Rowthorne 4864, canal nr. Cromford 3056.

S. Recorded for all squares except 23, 22, 21.

Linton: Radbourne and Etwall 23, Marston-on-Dove and Willington 22.

L. minor L.   *Duckweed*

Native. Ponds, canals, ditches. Common and widespread.
Recorded for all squares except:

W. 09, 28.

E. 34.

Linton: 'Common. General and abundant'.

L. gibba L.   *Gibbous Duckweed*

Native. In still water of ponds and ditches. Rare, and chiefly in the south of the county.

E. Canal, Renishaw 4378.

S. Canal, Chellaston 3731, Sawley 4731, Repton 3125.

Linton: Loscoe 44, Sudbury 13, Drakelow and Egginton 22.

SPARGANIACEAE

SPARGANIUM L.

S. erectum L. (*S. ramosum* Hudson)   *Bur-reed*

Native. On mud or in shallow water of ponds, ditches, canals and streams and in ungrazed marshes. Common and widespread.

There are three subspecies recorded for Derbyshire:

Ssp. **erectum**
Recorded for all squares except:

W. 09, 19, 17, 27.

Ssp. **neglectum** (Beeby) Schinz & Thell. (*S. neglectum* Beeby)

E. Ambergate Canal 3452.

S. Spinnyford Brook 2443, Mercaston 2643, canal nr. Spondon 3835, canal nr. Weston-on-Trent 4027, Cauldwell 2517.

Ssp. **microcarpum** (Neuman) Hylander ('*S. ramosum* var. *microcarpum* Neuman')

No recent records.

Linton: Dovedale 15, Brough 18, Ley's Fen, Old Brampton 37, Tibshelf 46, pond between Yeaveley and Cubley 13.

S. emersum Rehm. (*S. simplex* Hudson, in part)   *Unbranched Bur-reed*
Native. Shallow water of streams and ponds. Local.

W. Swann's Canal, Turncliff 0470, Longstone Moor 1974, Emperor Stream, Chatsworth 2771, Beresford Dale 1259, Dovedale 1453.

E. Brierley Wood 3775, canal at Renishaw 4477, Wingerworth Pond 3766, canal, Ambergate 3452, nr. Langley Mill 4647.

S. Clifton 1746, Bradley Brook 2443, Brailsford Brook 2441, railway cutting, Breadsall 3839, Egginton 2628, Swarkestone 3728, Weston-on-Trent 4027.

Linton: Ashford 16, Meynell Langley 23, Sandiacre 43, Long Eaton 43.

S. angustifolium Michx.   *Floating Bur-reed*
Probably a recent arrival. Peaty ponds in upland areas. Very rare.

W. Longshaw Pond 2679.

Linton: Not recorded.

TYPHACEAE

TYPHA L.

T. latifolia L.   *Bulrush, Reedmace or Cat's-tail*
Native. Ponds, canals, slow-flowing streams, reed-swamp. Widespread at low altitudes.

Recorded for all squares except:

        W. 09, 19, 17, 26, 28.
        E. 34.

Linton: River Etherow 09.

T. angustifolia L.   *Lesser Reedmace*
Native. Ponds, canals, slow-flowing streams; chiefly on organic substrata. Uncommon.

E. Holbrook nr. Killamarsh 4481, Pebley Pond 4879, Bramley Vale 4667, Stanley Lane, Morley 3940.

S. Sudbury Lake (planted?) 1532, Egginton 2727, Repton Shrubs 3123, Cauldwell 2517.

Linton: 'Rare'.

CYPERACEAE

ERIOPHORUM L.

E. angustifolium Honck.   *Common Cotton-grass*
Native. Wet parts of bogs and acid fens; a colonist of bare wet peat. Locally abundant.

W. Nr. Glossop 09, Alport Dale 1390, Napkin Piece 9987, recorded for 08, Derwent 1788, Withinleach Moor 9976, Longstone Moor 1974, Leash Fen 2874, Washgate Valley 0467, Friden 1660.

E. Foxhouse 2680, Totley Bents 3080, Killamarsh 4481, Pebley Pond 4978, Scarcliffe Park Woods 5270, Hipper Sick 3068, recorded for 35.

S. Bradley Brook 2244, Hulland Moss 2546.

Linton: West of Dore 28, Pleasley 56, Ockbrook 43.

E. latifolium Hoppe   *Broad-leaved Cotton-grass*

Native. On calcareous mud or peat, in woods or in the open. Very rare.

E. Whitwell Wood 5279.

Linton: Scarcliffe Park Woods 57, Morley Moor 34.

E. vaginatum L.   *Cotton Grass, Hare's Tail*

Native. Damp acid peaty places; dominant over great areas of high-level bog in northern Derbyshire.

W. Nr. Glossop 09, nr. Bleaklow 1296, Roych Clough 0783, Derwent Edge 1988/9, Ringinglow 2883, Umberley Sick 2969, Withinleach Moor 9976, Upper Goyt Valley 0173, Longstone Moor 1974, Eyam Moor 2278, Curbar Edge 2575, Cisterns Clough 0369, Washgate Valley 0467, Farley Moor 2962.

E. East Moor 3071, Stannage Golf Course 3369, Matlock Moor nr. Flash Dam 3064.

Linton: Charlesworth 98, Pleasley 56, Dethick Moor 35, Ockbrook 43.

TRICHOPHORUM Pers.

T. cespitosum (L.) Hartman (*Scirpus caespitosus* L.)   *Deer Sedge*

Native. Heaths and moorland. Locally common but only in the moorland areas.

W. Alport Dale 1390/1, Kinder Scout 0987, Derwent 1788, Owler Tor 2580, Goyts Moss 0172, Salter Sitch 2877, Axe Edge 0369.

Linton: 'Locally common. Common on all the moorland areas, rare in the rest of the county'. Allestree Park 34.

ELEOCHARIS R. Br.

E. acicularis (L.) Roemer & Schultes   *Slender Spike-rush*

Native. Marshy meadows and margins of ponds and canals. Very rare.

E. Pebley Pond 4879, nr. Hardwick 4564.

Linton: Ockbrook 43, Old Trent, Repton 32.

E. quinqueflora (F. X. Hartmann) O. Schwarz (*Scirpus pauciflorus* Lightf.)
                                                    Few-flowered Spike-rush

Native. Damp peaty places which are moderately base-rich. Very rare.

E.   Whitwell Wood 5278.

Linton: Umberley Brook nr. Chatsworth 27, Repton Rocks 32.

E. palustris (L.) Roemer & Schultes   *Common Spike-rush*
Native. Marshes, ditches and margins of ponds. Locally abundant and widespread.

W.   Warhurst Fold 9993, Hope 1683, Stanley Moor 0471, Monsal Dale 1771, Calver 2475, Bakewell 26, Dovedale 1551.

E.   Recorded for all squares except 38, 44.

S.   Recorded for all squares except 23.

Linton: Reservoir, Chapel 08, above Rowarth 08, nr. Hope 18, Lathkill Dale 16.

SCIRPUS L.

S. sylvaticus L.   *Wood Club-rush*
Native. Sides of ponds and streams, wet places in woods; usually on peaty soils subject to periodic flooding. Local.

W.   Watford Bridge 0086, Calver New Bridge 2475, Moss Carr 0665, Clough Wood 2561, Dovedale 1551, Fenny Bentley to Bradbourne 1950, 1850, Black Rocks 2955.

E.   Ford Valley 4280, Barlborough Park 4778, Alfreton Brook 4056.

S.   Nr. Sturston 1846, nr. Clifton 1544, nr. Norbury 1242, Henmore Brook 2046.

Linton: Monsal Dale 17, by River Dove above Hartington 16, Bubnell Cliff 27, Baslow 27, Marston-on-Dove 22, Calke Park 32, Repton 32, Drakelow 21.

BLYSMUS Panzer

B. compressus (L.) Panzer ex Link (*Scirpus caricis* Retz.)   *Broad Blysmus*
Native. Marshy spots in limestone dales, usually in small quantity only. Local.

W.   Taddington Dale 1770, Chee Dale 1273, Monk's Dale, Peter Dale, Miller's Dale, Tideswell, Monsal Dale and other dales in square 17.

Linton: 'Just below the bridge at Bakewell' 26.

SCHOENOPLECTUS (Reichenb.) Palla

S. lacustris (L.) Palla (*Scirpus lacustris* L.)   *Bulrush*
Native. Rivers, canals and ponds. Local and rather uncommon.

E.   Barlborough Park 4778.

S.   Osmaston 2041, Trent 4931, Sawley 4731, Stanton-by-Dale 4638, Willington 2927.

Linton: Whaley Mill 47, Pebley Pond 56, Derby 33, Melbourne 32, Calke 32, Drakelow 21.

S. tabernaemontani (C. C. Gmelin) Palla (*Scirpus tabernaemontani* C. C. Gmelin)   *Glaucous Bulrush*
Native. Pools. Very rare.
S.   Pool at Meynell Langley Hall 3039, railway pools, Trent 4931.
Linton: Pools, Trent Station 43.

ISOLEPIS R Br.

I. setacea (L.) R.Br. (*Scirpus setaceus* L.)   *Bristle Scirpus*
Native. Marshes and beside rivers and lakes. Local.
W. Kinder Scout 0889, Win Hill 1984, Foxhouse 2680, recorded for 07, under Baslow Edge, 2673, recorded for 06.
E.   Wingerworth 36, Whatstandwell 3354.
S.   Sandybrook 1848, Osmaston 1943, Shirley Wood 2141, Bradley Brook 2244, Trent 4931, Sawley 4731.
Linton: Derwent Dale 19, Wye Valley 17, Dovedale 15, Cromford 25, Barlow 37, Harlesthorpe 47, Cubley 13, Findern 33, Bretby 32, Repton 32, Drakelow 21.

CAREX L.

C. laevigata Sm.   *Smooth Sedge*
Native. Marshes and damp woods, usually on acid but nutrient-rich soils. Local.
W. Ladybower 1693, Derwent 1889, nr. Hathersage 2381, Blacka Moor 2880, Axe Edge 0270, Bretton Clough 2179, Nether Padley 2578, Sherriff Wood 2778, Axe Edge 0369, Bakewell 2268.
E.   Ecclesall Wood 3282, Monk Wood 3475, Tansley Moor 3261.
S.   Ashbourne 1745, Hulland Carr 2645, Recorded for 43.
Linton: Charlesworth 98.

C. hostiana DC. (*C. speirostachya* Sm.)   *Tawny Sedge*
Native. Fens and calcareous spring-lines and flushes. Local.
W. Hood Brook, Hathersage 2382, Brampton East Moor 2871.
E.   Whitwell Wood 5278.
S.   Bradley Brook 2244.
Linton: Not recorded.

C. hostiana × C. demissa
E.   Whitwell Wood 5278.
S.   By Bradley Brook 2244.
Lintin: Not recorded.

C. binervis Sm. *Ribbed Sedge*

Native. Open woodland, heaths, damp waysides, rough pastures and moorland on acid soils. Locally frequent, especially on the Millstone Grit and Edale Shales.

W. Alport Dale 1390, Crowden Brook 0986, recorded for 18, recorded for 07, Wye Valley 17, recorded for 06, Harthill Moor 2263, Farley Moor 2962.

E. Shining Cliff Woods 3352.

Linton: Stannage Edge 28, Harland Edge, Beeley 26, Tansley Moor 36, Shirley Wood 24, Repton Rocks 32, Burton-on-Trent 22.

C. flava agg. *Yellow Sedge*

The *C. flava* aggregate includes *C. lepidocarpa* Tausch, *C. demissa* Hornem. and *C. serotina* Mérat. The records listed below are for the aggregate, the species not having been specified.

W. Alport Dale 1390, recorded for 08, Derwent 1889, Shatton 1982, Burbage Water Tower 2680, Goyt Valley 07, Monk's Dale 1374, recorded for 06, nr. Brassington 2355.

E. Pebley Pond 4878, Whitwell Woods 5278, Duffield 34.

S. Tinker's Inn 1744, Bradley Brook 2244, Hulland Moss 2546, Shirley Wood 2141, recorded for 33, 43.

Linton: Charlesworth Coombs 98, Baslow 27, Foremark Park 32, Repton Rocks 32.

C. lepidocarpa Tausch ('*C. flava* var. *elatior* Schlecht.')
Long-stalked Yellow Sedge

Native. Calcareous fens, spring-lines and flushes. Occasional.

W. Monk's Dale 1374, Chee Dale 1273.

E. Pebley Pond 4878, Whitwell Wood 5278.

Linton: Records sub. nom. *flava.*

C. demissa Hornem. ('*C. flava* L.') *Common Yellow Sedge*

Native. Spring-lines, flushes, streamsides, marshes, fens etc., usually where there is some movement of moderately mineral-rich but non-calcareous water; occasionally with *C. lepidocarpa* in calcareous habitats. Frequent.

W. Crooked Clough 0994, Alport Dale 1390, recorded for 08, Grainfort Clough 1988, Burbage Water Tower 2680, recorded for 07, Monk's Dale 1374, Brampton East Moor 2871, recorded for 06, Nr. Brassington 2355.

E. Pebley Pond 4978, Whitwell Wood 5278.

S. Tinker's Inn 1744, Bentley Brook 2244, Shirley Wood 2141.

Linton: sub. nom. *flava.*

C. serotina Mérat ('*C. flava* var. *oederi* Retz.') *Small-fruited Yellow Sedge*
Native. Damp places on base-rich substrata. Very rare.

E. Whitwell Wood 5278.

Linton: Sub. nom. *flava.*

C. sylvatica Hudson   *Wood Sedge*

Native. Woods, moist hedge banks, scrub and grassland on the site of former woodland etc., especially on heavy moist soils. Frequent.

W. Recorded for o8 and o7, Cressbrook Dale 1773, Stand Wood 2670, recorded for o6, Lathkill Dale 1865, Matlock Bath 2957, nr. Bradbourne 2052.

E. Pebley Pond 4878, Markland Grips 5075, Whitwell Wood 5278, Stainsby Pond 4564, Pleasley Vale 5265, nr. Brackenfield 3758.

S. Fenny Bentley 1849, Offcote 1848, Agnes Meadow, Kniveton 2147, between Bradley and Atlow 2247, Radbourne 2836, Smisby 3419.

Linton: Charlesworth 99, Derwent Dale 19, New Mills 98, Ollerbrook, Edale 18, Great Shacklow Wood 26, Dovedale 15, Via Gellia 25, Scarcliffe Park Wood 57, Sheepbridge 37, Ault Hucknall 46, Morley 34, Cubley Common 13, Calke 32, Repton Shrubs 32.

C. pseudocyperus L.   *Cyperus Sedge*

Native. Sides of ponds, ditches and slow streams. Local and rare.

S. The Dumble, Hulland 2446, ponds between Yeaveley and Cubley 1839, Meynell Langley Park 3039, Willington Junction 3029, Repton Shrubs 3123.

C. rostrata Stokes   *Beaked Sedge, Bottle Sedge*

Native. Margins of pools and canals and wet peaty places with a permanently high water-level, especially in moderately base-poor and acid water but occasionally on calcareous substrata. Frequent.

W. Derwent 1789. Combs Reservoir 0379, Baslow Edge 2673, Lathkill Dale 1865, River Lathkill above Alport 2164.

S. Below Bradley Wood 1946.

Linton: Miller's Dale 17, Chee Tor 17, Via Gellia 25, Old Brampton 37, Mickleover 33, Swarkestone 32, Repton 32.

C. vesicaria L.   *Bladder Sedge*

Native. Margins of streams and ponds, marshes, wet copses etc.; commonly on non-peaty and at least moderately nutrient-rich soils, but sometimes on peat. Occasional.

W. Derwent 1789, Combs Reservoir 0379, Longshaw 2578.

E. Pebley Pond 4878, Park Hall Woods, Barlborough 57, Stubbing Court Pond 3667, Great Pond, Hardwick 4563.

S. Swarkestone 3728.

Linton: Stirrup Wood 98, Miller's Dale 17, Elvaston 43.

C. riparia Curtis   *Great Pond-sedge*

Native. By pools, canals and slow-flowing streams, in ditches and occasionally in wet woods. Frequent.

W. Bakewell 2269, Bradbourne Mill 2052.

E.   Glapwell Wood 4766, Cromford Canal 3255.

S.   Spinnyford Brook 2444, Clifton 1544, Radbourne 2835, Stanton-by-Bridge 3727.

Linton: Brough 18, Harlesthorpe nr. Clowne 57, Scarcliffe Park Wood 57, canal, Kirk Hallam 44, Breadsall 33, Egginton 22, Drakelow 21.

C. acutiformis Ehrh.   *Lesser Pond-sedge*

Native. Pools, river-banks, ditches, fens, fen-woods etc., on fen-peat or fairly nutrient-rich inorganic soil.   Frequent.

W.   Recorded for all squares except 09, 19, 28, 25.

E.   Killamarsh 4481, Pebley Pond 4878, Scarcliffe Park Wood 5170, Wingerworth 3667, Hardwick Hall 4563, Pleasley Vale 5265.

S.   Clifton 1544, Bradley 2244, Longford Park 2138, Darley Abbey 3538, Grove Farm 2318.

Linton: Sawley 43, Marston-on-Dove 22.

C. pendula Hudson   *Pendulous Sedge*

Native. Damp woods and shady streamsides and hedgebanks, especially on heavy moist soils. Rare.

W.   Walks, Matlock Bath 2957.

S.   Cauldwell 2517.

Linton: Mellor 98, Miller's Dale 17, Ambergate 35, Repton Shrubs 32.

C. strigosa Hudson   *Slender-spiked Wood Sedge*

Native. Damp woods, especially by streams, and shady banks on base-rich soils. Very rare.

No recent records.

Linton: Lover's Walk, River Derwent, Newton Wood, Ockbrook.

C. pallescens L.   *Pale Sedge*

Native. Damp open woodland and damp pastures. Local and infrequent.

W.   Hathersage 28, east of Suspension Bridge over Goyt Reservoir 0775, Calver Bridge 2475.

E.   Whitwell Wood 5278.

S.   Osmaston 1844, Shirley Brook 2141, Bradley Brook 2244.

Linton: Derwent Dale 19, Westend Farm 19, River Sett nr. Hayfield 08, Edale 15, Heeley–Norton Lees 38, Langwith Wood 56, Tibshelf 46, Mickleover 33, Calke 32, Repton Rocks 32.

C. panicea L.   *Carnation-grass*

Native. Fens, mineral-rich flushes and moist limestone grassland, as on north- or west-facing slopes in the dales. Common and widespread.

W.   Recorded for all squares.

E.   Nr. Eckington 4081, Brierley Wood 3775, Pebley Pond 4878, Markland Grips 5074, Wingerworth 3667, Pleasley Vale 5265, Crich Quarry 3555.

S.   Ashbourne 1746, Hulland Moss 2546, Hungry Bentley 1838.

Linton: 'General and abundant'.

C. flacca Schreber   *Glaucous Sedge*

Native. Dry calcareous grassland, rocky limestone slopes, meadows, waysides. Common and widespread.

W.   Recorded for all squares except 09, 19.

E.   Povey 3580, recorded for square 48, Brierley Wood 3775, Pebley Pond 4878, Scarcliffe Park Pond 5270, Ogston 3760, Pleasley Vale 5265, nr. Quarndon 3240.

S.   Ashbourne 1850, Bradley 2046.

Linton: 'The commonest of our sedges, everywhere abundant'.

C. hirta L.   *Hairy Sedge*

Native. Damp meadows and pastures, marshes, sides of ditches and moist open woods. Common and widespread.

W.   Recorded for all squares except 09, 19, 08.

E.   Killamarsh 4481, Brierley Wood 3775, Scarcliffe Park Wood 5270, Stainsby Pond 4564, Pleasley Vale 5265, Alfreton Brook 4056.

S.   Ashbourne 1847, Bradley 2245, Sudbury 1534, Dale Abbey 4338, Scropton 1829, Weston-on-Trent 4128, Grove Farm 2318.

Linton: 'General in all the districts'.

C. pilulifera L.   *Pill Sedge*

Native. Rough grassland, heaths and moorland on acid, sandy or peaty soils; rarely in open woods. Occasional.

W.   Win Hill 1885, Goyt Valley 07, Wye Valley 17, Coombs Dale 2274, Harthill Moor 2263.

E.   Pleasley Vale 5265.

S.   Bradley Wood 2046, Shirley Wood 2042, Sandhills, Cubley 1539.

Linton: Charlesworth Coombs 98, River Sett, Hayfield 08, Jaggers Clough 18, Middleton Dale 27, Darley Dale 26.

C. ericetorum Poll.

Native. In short open turf of dry limestone grassland on shallow soil and on ledges of limestone outcrops. Very rare, and only on the Magnesian Limestone.

E.   Markland Grips 5074.

Linton: Not recorded.

C. caryophyllea Latour.   *Spring Sedge*

Native. In short turf of dry grassland, especially over limestone; heaths, banks, waysides. Common and widespread.

W. Recorded for all squares except 09, 19.

E. Ford nr. Eckington 4081, Bradway 3379, recorded for 47, Markland Grips 5074, Scarcliffe Park Wood 5270, Ashover 3463, Pleasley Vale 5265, Alfreton Brook 4056.

S. Ashbourne 1846, Offcote 2048, Hungry Bentley 1838, Kirk Langley 2939.

Linton: 'Generally distributed'.

C. montana L.   *Mountain Sedge*

Native. Rough grassy places and open woods on the Magnesian Limestone. Very rare.

E. Markland Grips 5074.

C. digitata L.   *Fingered Sedge*

Native. In short open limestone turf on slopes and rocky terraces and on ledges of limestone cliffs and outcrops; shade-tolerant and sometimes in open woods. Local and infrequent.

W. Taddington Dale 1671, Monsal Head 1871, Coombs Dale 2274.

C. ornithopoda Willd.   *Bird's-foot Sedge*

Native. In short turf of limestone slopes on shallow soil; only on Carboniferous Limestone in the dales. Very local, but in some quantity where present.

W. Cressbrook Dale 1774, above Litton railway tunnel 1672.

C. elata All.   *Tufted Sedge*

Native. By fen streams and ditches. Rare.

S. New Stanton 4639,

C. acuta L.   *Slender Tufted Sedge*

Native. By ponds, streams and canals and in marshy places with a permanently high water-level. Local and infrequent.

W. Recorded for 07, Wye Valley 17, Ashford 1969, Bakewell 2269, nr. Woodeaves 1950, pond nr. Bradbourne Church 2053.

E. Lakeside, Renishaw Park 4478.

S. River Dove, Clifton 1544, Sutton Brook 2234, River Trent, Sawley 4731, Weston-on-Trent 4127.

Linton: Ashford 16, between Shirley Wood and Rodsley 24, Swarkestone 32, Burton 22.

C. nigra (L.) Reichard (*C. goodenovii* Gay)   *Common Sedge*

Native. Marshes, wet pastures and moorlands, moist waysides; chiefly but not exclusively on acid substrata, but on high moorland and bog more or less confined to streamsides, flushes and spring-lines, where it may be very abundant. Common and widespread.

A markedly tussocky form occurs locally and has been mistakenly named var. *juncella* Fries. It appears to be *C. subcaespitosa* (Kükenth.) Wiinstedt, but its taxonomic rank requires further investigation: it may be merely a growth-form.

Recorded for all squares except:

> **E.** 46.
> **S.** 23, 43, 22.

Linton: 'Frequent. Generally distributed'.

C. paniculata L.   *Greater Tussock Sedge*

Native. Fens and fen-woods on peaty base-rich substrata with at least seasonally high water-level. Locally frequent.

W. Fox House 2680, Leash Fen 2874, Bleakley 2163, below Robin Hood's Stride 2362, Hipley Toll Bar 2154.

E. Brierley Wood 3775, Alfreton Brook 4056, Dale 4438.

S. Ashbourne 1846, Sturston 1946, Tinker's Inn 1744, Wyaston Brook 1942, nr. Norbury 1141, Shirley Brook 2244, Brailsford Brook 2441.

Linton: Charlesworth Coombs 98, Snake Inn 19, Renishaw Canal 47, Mastin Moor 47, Ashover 36, Kedleston 34, Marston Montgomery 13, Foremark Bottoms 32, Repton Rocks 32, Drakelow 21.

C. diandra Schrank (*C. teretiuscula* Good.)   *Lesser Tussock Sedge*

Native. Wet peaty meadows and by peaty pools. Very rare.

No recent records.

Linton: Between Shirley Wood and Rodsley 24.

C. otrubae Podp. ('*C. vulpina* L.')   *False Fox-sedge*

Native. Sides of streams, ditches and canals, damp waysides and hedgebanks, usually on heavy non-peaty soils. Common.

W. Bentley Brook 1850.

E. Canal, Eckington 4480, canal, Whittington 3974, Pebley Pond 4878, Ault Hucknall 4665, Cromford Canal 3353, Alfreton Brook 4056, canal nr. Langley Mill 4647.

S. Nr. Yeaveley 1841, Roston 1241, Mercaston Brook 2742, Cubley Common 1639, canal, Breadsall 3639, Marston-on-Dove 2329, nr. Egginton 2727, Drakelow 2318.

Linton: 'Generally distributed at low levels'.

C. disticha Hudson
Brown Sedge

Native. Marshes and wet meadows, streamsides etc. Local.

W. Recorded for 16, Lathkill Dale above Conksbury Bridge 2165.

E. Whatstandwell 3353, Alfreton Brook 4056.

S. Sudbury 1534.

Linton: Osmaston 24, Morley 34, Mickleover 33, Ockbrook 43, Upper Pond Calke, and Repton 32, Drakelow 21.

C. divulsa Stokes
Grey Sedge

Native. Open woods, banks, waysides, waste ground. Very rare.

E. Whitwell Wood 5278.

S. Roadside between Rodsley and Shirley Mill 2040.

Linton: Dovedale 15.

C. spicata Hudson (*Carex contigua* Hoppe; '*C. muricata* L.')

Native. By ditches, canals, streams and ponds and also in rough pasture and on hedgebanks. Frequent.

W. Monk's Dale 1374, nr. Froggatt Bridge 2475, Alport 2164, Parwich Hill 1954, Fenny Bentley 1850, Kniveton 2150.

E. Pebley Pond 4878, Nethermoor 4260.

S. Clifton 1644, Ashbourne Green 1948, Brailsford 2540.

Linton: Probably under *C. muricata*.

C. muricata L.
Prickly Sedge

Native. Dry grassy and heathy places and amongst scrub. Local.

There are two subspecies in the county:

Ssp. **muricata** (*C. pairaei* F. W. Schultz), with the spikelets more or less contiguous and fruit 3 to 4 mm. long, grows on acid sandy or gravelly as well as calcareous soils. Local: recorded only for the north-west and south-east of the county.

W. Recorded for 08, 07, Monk's Dale 1375, Longstone Edge 2373.

E. Whatstandwell 3353, Crich 3554.

S. Clifton 1645, Yeldersley Rough 2243, nr. Hulland Moss 2546, Swarkestone Bridge 3628.

Linton: (Probably including records for *C. spicata* Huds.). Alport Dale 19, Monsal Dale 17, Bolsover 47, Lathkill Dale and Haddon 26, Tibshelf, Sutton Scarsdale and Ault Hucknall 46, Hartington to Beresford Dale 15, Wirksworth, Via Gellia, Bonsall to Matlock Bridge 25, Denby Common 44, Cubley, Marston Montgomery and Boylestone 13, Longford 23, Mickleover 33, Ockbrook, Kirk Hallam and Trent Lock 43, Willington and Egginton Common 22.

Ssp. **leersii** Ascherson & Graebner (*C. polyphylla* Kar. & Kir.), with usually distant spikelets and fruit 4 to 5 mm. long, is more closely restricted to dry calcareous soils, and is found on both the Carboniferous and Magnesian Limestones. Local and uncommon.

W. Coombs Dale 2274, roadside nr. Alport 2264, Lathkill Dale 2066, hedgebank, Cromford 2956.

Linton: Not recorded.

C. echinata Murray   *Star Sedge*

Native. Marshy places, meadows and damp woods on acid soil; flushes, spring-lines and streamsides on upland moors and bogs. Common and widespread.

W. Alport Dale 1390, recorded for 08, Win Hill 1984, Burbage Water Tower 2680, recorded for 07, White Edge 2677, recorded for 06, Grange Mill 2357.

E. Recorded for 48, 57, Spitewinter 3366.

S. Bradley Brook 2244, nr. Shirley Wood 2141.

Linton: Charlesworth Coombs 98, Kinder Scout 08, Bamford 28, Froggatt 27, Brackenfield 35, Morley 34, Calke 32, Repton Rocks 32.

C. remota L.   *Remote Sedge*

Native. Wet shady places on acid soil or peat; banks of ditches and canals. Locally common.

W. Chew Wood 9992, Roych Clough 0783, below Back Tor 1385, below Burbage Edge 0372, Calver Bridge 2475, Washgate Valley 0467, Fenny Bentley 1850.

Recorded for all squares except:

E.   56, 44.
S.   43, 22, 32.

Linton: Charlesworth 98, Westend Farm 19, Hood Brook, Hathersage 28, Ockbrook 43, Egginton 22, Repton 32.

C. curta Good.   *White Sedge*

Native. Bogs, moors, moorland flushes; restricted to acid substrata. Locally abundant.

W. Recorded for 08, Cheeks Hill 0270, Goyts Moss 0172, Totley Moss 2778, Leash Fen 2874.

S. Snelston 1642, Weston-on-Trent 4028.

Linton: Snake Inn 19, Bamford 28, Redcar Brook 38, Chee Tor 17, Shirley 24, Repton 32.

C. ovalis Good.   *Oval Sedge*

Native. Rough grassy places, tracks and woodland rides on moist acid soils. Locally common.

W. Recorded for all squares except 16.

E. Beauchief 3381, Cowley Bar 3278, Ogston 3760.

S. Ashbourne Green 1847, Bradley Brook 2244, Sudbury 1534, Kirk Langley 2839, Drakelow 2318.

Linton: 'Generally distributed'.

C. pulicaris L. *Flea Sedge*

Native. Damp calcareous grassland, fens, base-rich flushes and spring-lines on heaths and moors as well as in limestone country. Locally frequent, and a feature of grassland on north-facing slopes of the limestone dales.

W. Recorded for all squares except 09, 08, 26.

E. Whitwell Wood 5278.

S. Hulland Moss 2546.

Linton: Foremark Park 32.

C. dioica L. *Dioecious Sedge*

Native. Base-rich flushes, fens etc. Rare.

W. Brampton East Moor 2871.

E. Whitwell Wood 5278.

S. Hulland Moss 2546.

Linton: Above Grindsbrook 18, Umberley Brook 27.

GRAMINEAE

PHRAGMITES Adanson

P. australis (Cav.) Steudel (*P. communis* Trin.) *Reed*

Native. Rivers, pools, swamps, damp copses. Infrequent, except locally in the south of the county.

W. Watford Bridge 0086, Bradwell 1781, Monsal Dale 1771.

E. Colliery, Killamarsh 4481, Chesterfield Canal 3974, Scarcliffe Park Wood 5270, recorded for 36, 35, Belper 3447, nr. Langley Mill 4646.

S. Recorded for all squares except 22, 21.

Linton: Stirrup Wood 99, Charlesworth 09 (possibly both refer to the same locality).

MOLINIA Schrank

M. caerulea (L.) Moench (*M. varia* Schrank) *Purple Moor-grass*

Native. Characteristic of gently sloping moorland through which water moves; often dominating extensive areas receiving run-off or seepage water from steeper slopes and also on well-watered flats beside streams; makes higher demands on mineral nutrients than either *Nardus* or *Eriophorum vaginatum,* and is commonly rare or absent in vegetation dominated by them; not infrequently occupying an intermediate zone between blanket bog and hill grassland. Locally abundant.

W. Recorded for all squares except 16, 25.

E. Whitwell Wood 5278.

S. Hulland Moss 2546, Bradley Brook 2244.

Linton: Tansley Moor 35.

SIEGLINGIA Bernh.

S. decumbens (L.) Bernh.   *Heath-grass*

Native. In acid grassland and locally on moist calcareous substrata. Frequent.

W. Recorded for all squares except 09, 28.

E. Beauchief Abbey 3381, Unthank 3075, Whitwell Wood 5278, Hardwick Hall 4563, Fritchley 3653.

S. The Holts, Clifton 1743, Brailsford Brook 2441, Kirk Langley 2939, Dale Abbey 4338.

Linton: Troway 47, Elmton 57, Tansley Moor 35, Shirley 24, Cubley 13, Findern by the railway 32, Calke Park 32.

GLYCERIA R. Br.

G. fluitans (L.) R. Br.   *Floating Sweet-grass, Flote-grass*

Native. Brooks, canals, ponds. Common.

W. Derwent Reservoir 1691, recorded for 08, Win Hill 1984, recorded for 07, Longstone Moor 1974, Rowsley Moor 2874, recorded for 06, Lathkill Dale 2065, Fenny Bentley 1850, Carsington 2453, Kirk Ireton 2650.

E. Beauchief Abbey 3381, Beighton 4582, Pebley Pond 4878, Linacre 3272, Wingerworth Great Pond 3667, Nethermoor 4464, Pleasley Vale 5265, Whatstandwell 3354, Alfreton Brook 4056, Little Eaton 3641.

S. Ashbourne 1846, Spinnyford Brook 2443, Yeaveley 1839, Derby 3637, recorded for 43, Scropton 1829, Marston-on-Dove 2329, Knowle Hills 3725, Swarkestone 3728, Weston-on-Trent 4027.

Linton: 'Generally distributed'.

G. plicata Fries   *Plicate Sweet-grass*

Native. Brooks, canals, ditches, ponds. Frequent.

W. Chapel-en-le-Frith 0681, Swann's Canal 0470, Monk's Dale 1374, Calver New Bridge 2574, Hartington 1261, Lathkill Dale 2164, Dovedale 1451, Kniveton 2050.

E. Totley Bents 3080, recorded for 48, 47, 57, Ogston 3760, Pleasley Vale 5265.

S. Ashbourne 1847, Sturston 24, Yeaveley 1839, Sutton Brook 23, Dale Abbey 4338, Drakelow 21.

Linton: Mickleover 33, Markeaton 33.

G. × pedicellata Townsend (*G. fluitans* × *G. plicata*)   *Hybrid Sweet-grass*

Native. Brooks, canals, ditches, ponds; often, but by no means invariably, with one or both parents. Frequent.

W. Recorded for squares 06, 07.

E. Brierley Wood 3775, Markland Grips 5075, Whitwell Wood 57.

S. Marston-on-Dove 22.

Linton: Alport Dale 19, Hope 18, Miller's Dale–Monsal Dale 17, Calver 27, Cromford 35, Mugginton 24, Cubley 13, Derby, Chaddesden 33.

G. declinata Bréb.   *Glaucous Sweet-grass*

Native. Shallow water of streams, ditches and ponds and on their muddy margins; tolerating drier conditions than the preceding species. Not infrequent.

W. Derwent Reservoir 1391, nr. Hayfield 08, Longstone Moor 1974, Hartington 1260, Dovedale 1454, Carsington 1453.

E. Recorded for 48, 57, Alfreton Brook 4056, Crich 3654.

S. Ashbourne Green 1947, Edlaston 1842, Shirley 2141, ponds nr. Yeaveley 1839, Hollington 2239.

Linton: Not recorded.

G. maxima (Hartman) Holmberg (*G. aquatica* (L.) Wahlenb.) *Reed Sweet-grass*

Native. Margins of slow-flowing rivers, canals, ponds etc., and in places subject to periodic flooding from streams; chiefly lowland. Locally frequent.

W. Calver 27, Bakewell, Chatsworth etc.. 26.

E. Eckington 4081 Ford 4080, Whittington 37, Killamarsh 47, recorded for 35.

S. Yeldersley 2144, Brailsford 2540, Culland 2440, Markeaton 3327, canals at Derby 33, Borrowash 4034, Scropton 1829, Swarkestone 3728, Twyford 3523.

Linton: Cromford 25, Fritchley 35, Sudbury 13, Kirk Hallam 43, Thrumpton Ferry 53, Marston-on-Dove 22, Egginton 22.

CATABROSA Beauv.

C. aquatica (L.) Beauv.   *Water Whorl-grass*

Native. Shallow canals, ditches and drains, and muddy margins of ponds. Rare.

S. Shirley Wood 2141, Hollington/Shirley 2240, Peathayes 2139, Dale Moor 4538.

Linton: Ebbing and Flowing Well, Buxton 07, Wardlow Mires 17, Whitwell 57, Swarkestone 32.

FESTUCA L.

F. pratensis Hudson   *Meadow Fescue*

Native. Moist meadows and pastures, grassy waysides. Frequent.
Recorded for all squares except:

> **W.** 09, 18.
> **E.** 34, 44.
> **S.** 22, 32.

Linton: 'Common and general throughout the county'.

F. arundinacea Schreber   *Tall Fescue*

Native. Streambanks and wet meadows and in rough pastures of the limestone
dales. Occasional; perhaps often overlooked on grazed limestone slopes
where it does not form the tall and large clumps characteristic of places where
water stands in winter.

W. Recorded for 08, Cavendish Golf Course 0473, Peter Dale 1275, Chee
Dale 1273, Miller's Dale 1573.

E. Unthank 3176, Markland Grips 5074, Whitwell Wood 5278, Nethermoor
4464.

S. Ashbourne Hall Pond 1846, Snelston Common 1541, Brailsford Brook
2441, about Derby 33, nr. Stanton-by-Dale 43.

Linton: Burton and Drakelow 21.

F. gigantea (L.) Vill. (*Bromus giganteus* L.)   *Tall Brome Grass*

Native. Damp woods, streamsides and hedgebanks. Locally frequent,
especially in valley-bottom woods.

Recorded for all squares except:

> **W.** 19, 08, 28.
> **E.** 34, 44.
> **S.** 33, 22.

Linton: Mellor 98, Chapel and Hayfield 08, Hathersage 28, Holbrook 34,
Mickleover 33, Stapenhill 22.

F. altissima All. (*F. sylvatica* Vill.)   *Wood Fescue*

Native. Woods on rocky slopes; on both limestone and sandstone. Very
rare.

W. Lovers' Walk, Matlock Bath 2958.

Linton: Stirrup Wood 99, Ashwood Dale 07, Matlock 25.

F. rubra L.   *Red Fescue*

Native. Meadows, pastures, waysides, heaths and hill grasslands; less tolerant
of dry shallow soils and of heavy grazing than *F. ovina*, but often mixed with
it on deeper and moister soils, both calcareous and non-calcareous. Common
and widespread. The common Derbyshire plant is ssp. **rubra.**

Recorded for all squares except:

W. 07.

Ssp. **commutata** Gaud., strongly tufted and lacking the creeping rhizomes of ssp. *rubra*, is a grass of well-drained soils on roadsides, railway banks and waste places.

F. ovina L. *Sheep's Fescue*

Native. In grassland on a wide range of well-drained soils from shallow black limestone soils to strongly acid loams, but sparse or absent on podsolized soils; a characterized dominant grass of limestone grassland and also of the *Festuca-Agrostis* grasslands of the lower slopes of hills on the gritstone and shales. Common and widespread.

Glaucous-leaved forms are frequent on the northern moorlands.

Recorded for all squares except:

E. 48.

Linton: General and abundant'.

F. tenuifolia Sibth. (*F. ovina* var. *capillata* (Lam.) Alef.)

Fine-leaved Sheep's Fescue

Native. Heaths, moorland and open woods on acid soils. Much less common than *F. ovina* but probably overlooked. Recorded from Bunter Sands in neighbouring counties.

S. Bradley Wood 2046.

Linton: Fairbrook, nr. Snake Inn 19, by River Noe, Edale 18, gritstone heaths about Matlock.

X FESTULOLIUM Ascherson & Graebner

F. loliaceum (Hudson) P. Fourn. (*Lolium perenne* × *Festuca pratensis*)

Hybrid Fescue

Native. Waysides, sides of ponds and ditches, meadows, with the parents. Not uncommon.

W. Buxton 07, recorded for 17, Bakewell 26, Washbrook Lane, Thorpe 1651.

E. Pebley Pond 4879, Glapwell 4866.

S. Red House, Thorpe 1748, Bentley Brook 1646, Dale Abbey 4338.

Linton: Tibshelf 46, Alfreton and Shirland 45.

LOLIUM L.

L. perenne L. *Rye-grass*

Native, but much sown for improving meadows and pastures. Old pastures and meadows on good rich soils, roadsides, hedgebanks and waste places. Common and widespread. The common plant is ssp. **perenne.**

Recorded for all squares.

Linton: 'Common. General and plentiful'.

Ssp. multiflorum (Lam.) Husnot (*L. perenne* L. var. *italicum* Braun)

Italian Rye-grass

Introduced. Naturalized on roadsides, field borders, waste places etc. Frequent.

W. Nr. Buxton 0471, recorded for 17, 06, Fenny Bentley 1750, Kniveton 2050.

E. Dore 3981, Killamarsh 4481, Whittington 3974, Markland Grips 5174, Astwith 4464, Allestree 3440.

S. Recorded for all squares except 43.

Linton: Between Birchover and Stanton 26, Renishaw 47, Kelstedge 36, Shirland and Alfreton 45, Sandiacre and Shardlow 43.

L. temulentum L. *Darnel*

Introduced. A very rare weed of arable land and waste places.

No recent records.

Linton: Above Miller's Dale Station 17, nr. Rodsley 24, Repton 32.

VULPIA C. C. Gmelin

V. bromoides (L.) Gray (*Festuca sciuroides* Roth) *Barren Fescue*

Native. Dry waysides, banks, walls. Locally frequent in the south of the county.

S. Roadside, Clifton Cross 1645, Mappleton 1647, Sandybrook 1843, above Bradley Wood 2046, nr. Bradley Knob 2445, Swarkestone 3628.

Linton: Shirland 45, Sawley and Dale 43, Egginton 22.

NARDURUS (Bluff, Nees & Schau.) Reichenb.

N. maritimus (L.) Murbeck

Probably introduced. Waste ground. Very rare.

S. Clifton goods-yard 1644.

Linton: Not recorded.

PUCCINELLIA Parl.

P. distans (Jacq.) Parl. *Reflexed Poa*

Probably introduced. Waste land, rubbish tips etc. Very rare.

E. Plentiful, roadside, Steetley Quarry 57.

Linton: Not recorded.

CATAPODIUM Link

C. rigidum (L.) C. E. Hubbard (*Festuca rigida* L.) *Hard Poa, Fern Grass*

Native. Dry banks, walls and rocky places; also in short open grassland on shallow soils; mainly, but not exclusively, on limestone. Locally frequent.

W. Monsal Head 1871, Middleton Dale 27, The Nabbs, Dovedale 1453, Thorpe Pastures 1551, Thorpe Cloud 1550, Thorpe Quarry 1651, Parwich Hill 1954.

E. Pleasley Vale 5265, railway nr. Black Rocks 3056, Ambergate Station 3551.

S. Birchwood Quarry 1541, Sandybrook 1748, Ashbourne 1746.

Linton: Nr. Bakewell 26, Clowne 57, Sandiacre 43, Ticknall Quarry 32.

(*C. marinum* (L.) C. E. Hubbard has been recorded for the dales of the Wye Valley 17, and for Coombs Dale 2274, but confirmation is needed of the occurrence of this maritime species so far inland.)

POA L.

P. annua L.   *Annual Meadow-grass*

Native. Cultivated land, waysides, waste places, short open grassland etc. Very common and widespread and reaching 1,800 feet.

Recorded for all squares.

Linton: 'Very common'.

P. nemoralis L.   *Wood Meadow-grass*

Native. Woods and shady hedgebanks on a wide range of soils, rocks and walls. Local.

W. Litton Mill 1773, Lathkill 1865, recorded for 26.

E. Ecclesall Wood 3282, Ford Valley 48, nr. Unthank 3176, recorded for 35.

S. Sandybrook nr. Ashbourne 1748, Clifton Cross 1645, Shirley 2745, Recorded for 23, Anchor Church 3327, Grange Wood, Lullington 2714.

Linton: Hartle Dale nr. Bradwell 18, Dovedale 15, Via Gellia 25, Mickleover 33, about Burton 22.

P. compressa L.   *Flattened Meadow-grass*

Native. Short grassland on dry soils, banks, walls and waste ground. Local.

W. Froggatt Wood, 2477, Dovedale 1451.

E. Dore 3081, Elmton and nr. Creswell in 57, Pleasley Vale 5265, Upper Langwith 5169, Crich 3454.

S. Snelston 1541, Spital Lane, Ashbourne 1846, Brailsford 2541, Mackworth 3137.

Linton: Buxton 07, Blackwell Dale 17, Matlock Bridge 26, Via Gellia 25, Barlborough 47, Hazelwood 34, Repton, Swarkestone Bridge and Calke 32.

P. pratensis L.   *Smooth-stalked Meadow-grass*

Native. Meadows, pastures, waysides, cultivated and waste ground, walls. Common and widespread, on a wide range of soils from calcareous to moderately acid but avoiding strongly podsolized and waterlogged soils.

Recorded for all squares.

Linton: 'Common. Generally distributed'.

P. angustifolia L.   *Narrow-leaved Meadow-grass*
Native. Rough hill-grassland, especially on limestone.
Often treated as a subspecies of *P. pratensis*.
W. Chee Dale 1172.
E.   Pebley Pond 4978, Pleasley Vale 5064, Crich limestone inlier 3455.

P. subcaerulea Sm. ('*P. pratensis* var. *subcaerulea* (Sm.)')
Native. Wet pastures and meadows, marshes, streamsides, moist grassy slopes.
Rare, but perhaps overlooked.
W. Pastures above Dovedale 1551.
Linton: Hilton Common 23, Marston Montgomery 13.

P. trivialis L.   *Rough-stalked Meadow-grass*
Native. Meadows and pastures on good moist soil, sides of streams and ponds,
damp waysides and hedgebanks, waste places etc. Common and widespread.
Recorded for all squares.
Linton: 'Common'.

P. palustris L.   *Swamp Meadow-grass*
Probably introduced. Recorded only once.
S.   One plant in a market-garden, Ashbourne 1746.
Linton: Not recorded.

DACTYLIS L.

D. glomerata L.   *Cock's-foot Grass*
Native. Meadows, pastures, rough grassland, roadsides, waste places.
Common and widespread.
Recorded for all squares.
Linton: 'Very common'.

CYNOSURUS L.

C. cristatus L.   *Crested Dog's-tail*
Native. Old grassland on a wide range of soil-types from calcareous to
moderately acid. Common and widespread.
Recorded for all squares except:
> **W.** 07.
> **E.** 44.
Linton: 'General throughout the County'.

C. echinatus L.   *Rough Dog's-tail*
Introduced. Casual on cultivated ground and waste places. Very rare.
S.   Sturston Tip 1946.
Linton: Not recorded.

BRIZA L.

B. media L.   *Quaking Grass*
Native. Meadows, pastures, waysides and rough grassland on good soils;
common on loamy soils over limestone. Frequent and generally distributed.
Recorded for all squares except:

> **W.** 09, 19.
> **E.** 34, 44.

Linton: 'Generally distributed'.

MELICA L.

M. uniflora Retz.   *Wood Melick*
Native. Woods on a wide range of soil-types from moist acid mulls to
calcareous loams; also on cliff-ledges and rock outcrops, especially of limestone.
Common and widespread.

W. Monk's Dale 1374, Cressbrook Dale 1773, Coombs Dale 2774, Lathkill
Dale 1966, Brassington Rocks 2154, Via Gellia 2757.

E. Recorded for all squares except 47, 34, 44.

S. Ashbourne 1745, Mercaston 2642, Yeaveley 1839, Alkmonton 1938,
Dale Abbey 4338.

Linton: 'Occurs in every district though slightly local in its occurrence'.

M. nutans L.   *Mountain Melick*
Native. Rocky woods and scrub, especially on limestone scree, when it is
often associated with *Rubus saxatilis, Geranium sanguineum* and *Convallaria
majalis.* Local and rather uncommon.

W. Chee Tor 1373, Taddington Dale 1670, Monk's Dale 1374, Cressbrook
Dale 1773, Coombs Dale 2274, Lathkill Dale 2066, Via Gellia 2556.

E. Whitwell Woods 5278, Pleasley 5165, nr. Codnor 4250.

Linton: Ashwood Dale 07, Dovedale 15, Norton 38.

ZERNA Panzer

Z. erecta (Hudson) Gray (*Bromus erectus* Hudson)   *Upright Brome-grass*
Native. Limestone pastures, roadsides and dry banks. Local, and on the
Carboniferous Limestone restricted to sunny slopes.

W. Recorded for 08, 07, Miller's Dale 1573, Lathkill Dale 1966.

E. Colliery, Killamarsh 4481, Pebley Pond 4878, Markland Grips 5074,
Pleasley Vale 5265.

S. Sturston 1946.

Linton: Cressbrook Dale 17, Barlborough 47, Bolsover 47, Whitwell 56,
Copse Hill nr. Osmaston 24, Little Eaton 34.

Z. ramosa (Hudson) Lindman (*Bromus ramosus* Hudson)   *Hairy Brome-grass*

Native. Woods, hedgebanks and shady places, especially on limestone but found on a variety of fertile soils. Common and widespread.

Recorded for all squares except:

W. 09, 18, 28.
E. 34, 44.
S. 22, 32, 21.

Linton: 'Generally distributed'.

ANISANTHA C. Koch

A. sterilis (L.) Nevski (*Bromus sterilis* L.)   *Barren Brome*

Native. Gardens, roadsides, hedgebanks, waste places. Frequent in the south, less so in the north and west.

W. West end of Longstone Edge 1972, Coombs Dale 2274, Conksbury Bridge 2165, Alport 2263, Bonsall Moor 2559.

E. Recorded for all squares except 44.

S. Recorded for all squares.

Linton: 'Common. Generally distributed'.

BROMUS L.

B. mollis L.   *Soft Brome-grass*

Native. Fields, meadows, waysides, waste places. Common and widespread.

Recorded for all squares except:

W. 09, 19, 18, 28.
E. 47.

Linton: 'General and abundant'.

B. thominii Hardouin

Probably of hybrid origin, from the cross *B. mollis* × *B. lepidus*. Meadows and waste places. Rare, but probably under-recorded.

W. On soil by roadside, Hognaston Winn 2251.

S. Ashbourne 1646.

Linton: Not recorded.

B. lepidus Holmberg

Doubtfully native. Waste places and roadsides, less commonly in meadows. Uncommon.

W. Roadside soil dump, Hognaston Winn 2251.

E. Markland Grips 5074, Nethermoor 4260.

S. Factory tip Ashbourne 1746, above Bradley Wood 2046, Ednaston 2442, Swarkestone Bridge 3628.

Linton: Not recorded.

B. racemosus L.   *Smooth Brome*
Native. Meadows, less frequently in arable fields. Rather uncommon.
W. Hope 1783.
S.   By River Dove, Ashbourne 1646, Ashbourne 1846, roadside, Clifton
1645, Osmaston 2044, nr. Peathayes 2139.
Linton: Whitwell 57, Shirland 35, Ecclesbourne, above Shottle Station 24,
The Knob, Hulland 24, Yeldersley Lane 24, Calke 32.

B. commutatus Schrader   *Meadow Brome*
Probably native. Meadows, waysides, waste places. Rare, and usually
behaving as a casual.
S.   Sturston, once only 1946, Clifton Goods-yard 1644.
Linton: Between Osmaston and Ladyhole, at roadside 24, by brook at Hilton
23, Ticknall 32.

BRACHYPODIUM Beauv.

B. sylvaticum (Hudson) Beauv. (*B. gracile* Beauv.)   *Slender False-brome*
Native. Woods, hedges, roadsides, banks, chiefly on good fertile soils; a
colonist of abandoned arable fields and of limestone scree-slopes and often
persisting as a woodland relic on limestone slopes. Common and widespread.
Recorded for all squares except:

> **W.** 09, 19, 08.
> **E.**   36.
> **S.**   22.

Linton: 'Common'.

B. pinnatum (L.) Beauv.   *Tor-grass, Heath False-brome*
Native. Calcareous grassland; locally common and forming large pure stands
on the Magnesian Limestone, but infrequent and restricted to sunny slopes on
the Carboniferous Limestone. See also maps (page 93).
W. Recorded for 08, Win Hill, east side 1985, Great Rocks Dale 1172,
Flagg Dale 1173, Burfoot 1672, Grange Mill 2358, Gratton Dale 2059.
E.   Pebley Pond 4878, Markland Grips 5174, Creswell Crags 5374, Whitwell
Wood 5278, Ault Hucknall 4665, Hardwick Ponds 4563/4, Astwith 4464,
Pleasley Vale 5165, Crich 3454.
S.   Osmaston 2043.

AGROPYRON Gaertner

A. caninum (L.) Beauv.   *Bearded Couch-grass*
Native. Woods, hedgebanks and rock-ledges on a wide range of substrata,
including limestone. Frequent and widespread.
W. Recorded for all squares except 09, 19.

E. Beauchief 3381, Brierley Wood 3775, Renishaw 4378, Markland Grips 5074, Quarry, Hate Wood 3463, Pleasley Vale 5765, Ambergate 3551, Little Eaton 3641.

S. Recorded for all squares.

Linton: Stirrup Wood 98, Mosbrough 48, Loscoe 44.

A. repens (L.) Beauv.   *Twitch, Couch-grass, Scutch*

Native. Hedgebanks, roadsides, rough grassy places, waste ground and cultivated ground. Common and widespread; a troublesome weed of fields and gardens.

Recorded for all squares.

Linton: 'General through the County'.

ELYMUS L.

E. arenarius L.   *Lyme-grass*

Introduced. An escape from gardens. Rare.

W. Nr. Froggatt, in a garden 2476.

S. A large colony on the waste from gravel pit, Hulland 2645. Seen 1958, all gone 1961.

Linton: Not recorded.

HORDEUM L.

H. secalinum Schreber   *Meadow Barley*

Native. Meadows, fields, waysides. Uncommon.

W. Meadows by the Wye, Bakewell 26.

E. Markland Grips 5074.

S. Meadow by the Bentley Brook, Ashbourne 1646, roadside nr. Mercaston Brook 2742, Spondon 3935, Scropton 1829, Grove Farm, Drakelow 2318.

Linton: Coxbench, and between Allestree and Quarndon 34, Marston Montgomery 13, Sinfin Moor 33, Sawley, and by the Erewash, nr. Long Eaton 43, between Chellaston and Swarkestone 32.

H. murinum L.   *Wall Barley, Way Bent*

Native. Waysides and waste places, especially near buildings. Common except in the north and west.

W. Recorded for 07, nr. Kirk Ireton 2650.

E. Recorded for all squares except 46, 45.

S. Clifton Goods Yard 1644, by Grange Farm, Osmaston 1944, Brailsford 2541, also Bradley and Mugginton in 24, recorded for 13, very common on waste ground in and about Derby 3537, recorded for 22, Swarkestone 3728, Netherseal 2913.

Linton: Chapel 08, Codnor Park Station 45, Hatton 23, New Stanton and eight other localities in 43.

HORDELYMUS (Jessen) Harz

H. europaeus (L.) Harz (*Hordeum silvaticum* Hudson)   *Wood Barley*

Native. Woods and shady places, especially on calcareous soils. Rare.

W. Monk's Dale 1373/4, nr. Haddon Hall 2366, Lovers' Walk, Matlock Bath 2958.

E. Galleys Wood, Beauchief 3381, nr. Pebley Pond, Barlborough 4880, Spinkhill 4678, Whitwell Wood 5278, Markland Grips 5174, Glapwell Wood 4766, Pleasley Vale 5265.

S. Hermitage Wood, Dale Abbey 4338.

Linton: Stirrup Wood, Charlesworth 99, Bakewell Road, Buxton 07, Great Shacklow Wood, Ashford 16.

KOELERIA Pers.

K. cristata (L.) Pers.   *Crested Hair-grass*

Native. In dry grassland, especially on calcareous soils, and abundant on black limestone soils of the dales. Locally abundant.

W. Cave Dale 1583, recorded for 07, Monk's Dale 1374, Peter Dale 1275, Miller's Dale 1473, Middleton Dale 2176, Longstone Edge 2073, Longdale 1361, Conksbury Bridge 2165, Parwich Hill 1954, Dovedale 1551, Kniveton 2155.

E. Markland Grips 5074, Pleasley Vale 5265.

Linton: 'Very common in the limestone districts'. Crich 35, Calke 32.

TRISETUM Pers.

T. flavescens (L.) Beauv. (*T. pratense* Pers.)   *Yellow Oat*

Native. Pastures, especially on limestone; banks, waysides. Common and widepsread.

Recorded for all squares except:

W. 09.
E. 34, 44.

Linton: 'Common. Generally distributed'.

AVENA L.

A. fatua L.   *Wild Oat*

Introduced. Cultivated and waste ground. A scarce casual, not usually persisting.

W. Monsal Dale 17.

E. Nr. Beauchief Abbey in a fallow field 3381, Pebley Pond 4978, recorded for 36, 35.

S. Ashbourne, one plant 1946, recorded for 13.

Linton: Heanor 44, Brizlincote 22, Calke and Repton 32.

A. sativa L.   *Oat*

Introduced. Common as a relic of cultivation and a frequent roadside weed, not usually persisting.

E.   By colliery, Killamarsh 4481.

S.   Ashbourne, casual 1746, Stores Road, Derby 3637.

Linton: Not recorded.

HELICTOTRICHON Besser

H. pratense (L.) Pilger (*Avena pratensis* L.)   *Meadow Oat*

Native. Grassland, especially on black limestone soils of the dales. Locally abundant.

W.   Cave Dale, Castleton 1583, Cunning Dale 0773, Monk's Dale, Cressbrook Dale and Chee Dale in 17, Calver 2374, Earl Sterndale 0966, Ashford Dale 1869, Biggin Dale 1458, Grange Mill 2358, Brassington 2154.

E.   Scarcliffe Park Wood 5270, Markland Grips 5074, recorded for 36, Pleasley Vale 5265, nr. Denby Common 4046.

Linton: Osmaston-by-Ashbourne 24, Repton 32, Burton 22.

H. pubescens (Hudson) Pilger   *Hairy Oat-grass*

Native. Pastures, moist grassy banks and waysides, chiefly on calcareous soils. Frequent.

W.   Recorded for all squares except 09, 19, 08, 28.

E.   Markland Grips 5074, Pleasley Vale 5265.

S.   Ashbourne 1746, Hulland 2547, recorded for 32.

Linton: Cubley Common 13, Mickleover and Chaddesden 33, Stapenhill, Burton and Willington 22.

ARRHENATHERUM Beauv.

A. elatius (L.) Beauv. ex J. & C. Presl (*A. avenaceum* Beauv.)   *Tall Oat-grass*

Native. Rough grassland, hedgebanks, waysides and waste ground on a wide range of soils; a frequent colonist of bare limestone scree in the dales. Common and widespread.

Recorded for all squares except:

W.   09.

Linton: 'Common. Generally distributed and abundant'.

HOLCUS L.

H. lanatus L.   *Yorkshire Fog*

Native. Meadows and pastures, open woodland, roadsides, margins of moorland flushes, etc. Common and widespread.

Recorded for all squares.

Linton: 'Common. Generally distributed.'

H. mollis L.   *Creeping Soft-grass*

Native. Woodland, scrub and shaded hedgebanks on fertile acid mull soils, where it may dominate extensive areas; a troublesome weed of arable fields on sandy soils. Common and widespread.

Recorded for all squares except:

E.   34, 44.

Linton: 'Common. A troublesome weed.'

DESCHAMPSIA Beauv.

D. caespitosa (L.) Beauv.   *Tufted Hair-grass*

Native. Meadows, pastures and woods, especially where heavy and at least moderately fertile soils are periodically waterlogged. Common and widespread.

Recorded for all squares.

Linton: 'Generally distributed'.

D. flexuosa (L.) Trin.   *Wavy Hair-grass*

Native. Woods, heaths and moorlands on very acid soils with a raw humus layer, and often extensively dominant; a colonist of bare ground and of cliff-ledges and rock outcrops of sandstone or gritstone; also of burnt woodland and heath on acid soils. Common and widespread but avoiding calcareous soils.

Recorded for all squares except:

E.   56, 45.
S.   13, 43.

Linton: Sudbury 13.

AIRA L.

A. praecox L.   *Early Hair-grass*

Native. In dry open situations on shallow light soils, about rocky outcrops and on sandy heaths. More frequent than *A. caryophyllea*, often accompanying it.

W. Chinley 0383, Ringinglow 2983, nr. Buxton 0573, Wye Valley 17, nr. Froggatt Bridge 2476, also Bakewell 2168, White Edge 2677, Farley Moor 2962, Elton 2261, Dovedale 1452.

E. Cutthorpe 3273, Totley Bents 3080, recorded for 36, Hardwick Hall 4563.

S. Bradley Wood 1946, Spinnyford Brook 2445, recorded for 23, Ticknall 3523, Grange Wood 2714.

Linton: Charlesworth Coombs 09, Edale 18, Wessington Common 35, Breadsall Moor 33.

A. caryophyllea L. *Silvery Hair-grass*
Native. On light shallow soils; locally frequent in open situations about limestone outcrops and in dry sandy places.

W. Nr. Buxton 0573, Monsal Head 1871, Lathkill Dale 1966, Heathcote 1560, Thorpe Pasture 1551, Wolfscote Hill 1357, Dovedale 1452, Haven Hill 2051.

E. Nr. Holymoorside 3168, Pleasley Vale 5165.

S. Ashbourne Green 1847, above Bradley Wood 2046, nr. The Knob, Hulland 2445, Scropton 1930, nr. Weston-on-Trent 4128.

Linton: Eyam Moor 27, Hardwick Park 46, Horsley 34, South of Findern 33, Risley 43, Burton 22, Calke and Repton 32.

CALAMAGROSTIS Adanson

C. epigejos (L.) Roth *Bush-grass*
Native. Damp woods, fen scrub, ditches, especially on heavy soils. Rare.

E. Whitwell Wood 5279.

S. Lullington 2714, nr. Yeaveley 1839.

Linton: Scarcliffe Park Wood 57, Gresley 21.

AGROSTIS L.

A. canina L. *Brown Bent-grass, Velvet Bent*
Native. Heaths and moorlands; rarely in wet meadows, swamps and ditches.

W. Nr. Malcoff 0783, nr. Edale 1385, Bamford Moor 2185, nr. Ladmanlow 0371, Cressbrook Dale 1773, Leash Fen 2874, nr. Owler Bar 2978, Axe Edge 0369, nr. Hartington 1561, Friden 1761, Newhaven 1758, Bonsall Moor 2559.

E. Park Bank Wood, Beauchief 3381, Beighton 4483, Cordwell Valley 3176, Hardwick 4563, Morley 3940.

S. Bradley Wood 1946, Shirley Wood 2042, Hulland Moss 2546, Grange Wood, Lullington 2714.

Linton: Walton Wood 36, Aston, Sudbury 13, Calke Park, Scaddow Rocks, Twyford Lane 32.

A. tenuis Sibth. *Common Bent*
Native. Heaths, moorlands, pastures etc., on a wide range of soils; not common on calcareous soils though abundant on the slightly acid loams of the limestone plateau; associated with *Festuca ovina* in the Festuca-Agrostis grasslands of moderately but not strongly acid soils. Common and widespread.

Recorded for all squares except:

E.   45, 34, 44.

A. gigantea Roth *Common Bent-grass*

Native. Damp woodland and grassy places. Insufficiently recorded but probably not infrequent.

W. Chew Wood, Charlesworth 9992, Coombs Dale 2274, Baslow 2572.

E. Killamarsh 4481, Pebley Pond 4878, Markland Grips 5075, Whitwell Wood 5278, Tupton 3866, Hardwick Ponds 4563.

S. Brailsford 2442.

Linton: Via Gellia 25, Matlock and Cromford Canal 35, Calke 32.

A. stolonifera L. (*A. palustris* Huds.) *Marsh Bent, Fiorin*

Native. Roadsides, rough grassland. Generally distributed and common.

Recorded for all squares except:

> **W.** 08.
> **E.** 34, 44.

Linton: East of Allestree Park 34.

PHLEUM L.

P. bertolonii DC. (*P. pratense* var. *nodosum* L.) *Cat's-tail*

Native. Pastures and short rough grassland. Common and widespread.

Recorded for all squares except:

> **W.** 19, 08, 18, 28.
> **E.** 45, 34, 44.
> **S.** 13, 33, 43.

P. pratense L. *Timothy Grass*

Native. Meadows; often sown for cropping as hay. Common and widespread.

Recorded for all squares.

Some records relating to *P. bertolonii* are probably included among the above.

Linton: 'Common. Generally distributed'.

ALOPECURUS L.

A. myosuroides Hudson *Black Twitch*

Native. A weed of gardens and waste ground, usually occasional only, but persistent in fields in some low-lying parts of southern Derbyshire.

W. Windmill Tip 1677, Hartington 1260, Ireton 2650.

E. Windley 3045.

S. Ashbourne 1745/6, Clifton goods yard 1644, Mugginton 2843, Hulland 2446, Bradley 2245, Derby 3537, fields by Swarkestone Bridge 3329.

Linton: Ockbrook 43, Stapenhill 22.

A. pratensis L.   *Meadow Foxtail*
Native. Meadows, pastures and roadsides. General and abundant.
Recorded for all squares.
Linton: 'Common. General and abundant'.

A. geniculatus L.   *Marsh Foxtail*
Native. Marshy pond-sides and wet places in meadows. Common and
generally distributed.
Recorded for all squares except:
> **W.** 19.
> **E.** 57, 34.
Linton: 'Common. Generally distributed'.

A. æqualis Sobol. (*A. fulvus* Sm.)   *Orange Foxtail*
Native. Marshy pond-sides. Local; not recorded from the west and south.
E.   Woodall Pond, Killamarsh 4780, Press Reservoirs 3565, Smith's Pond
3766, Deer Ponds, Hockley 3767, Hardwick Ponds 4563.
Linton: Ponds, Sutton Scarsdale 46.

MILIUM L.

M. effusum L.   *Wood Millet*
Native. In moist shady woods on acid mull soils; common in valley-bottom
woods on acid soils. Local.
W.   Lathkill Dale 1966.
E.   Beauchief Abbey 3381, Ford nr. Eckington 4081, Totley 3079, Cutthorpe
3473, Linacre 3372, Pebley Pond 4878, Clay Cross 3963, nr. Ashover 36,
Ambergate 3551.
S.   Ashbourne 1746, Clifton 1644, Snelston 1543, Cubley 1737, Sudbury
1535, copse east of Alkmonton 1938, Dale Abbey 4338, Melbourne 3825,
Grange Wood 2714.
Linton: Stirrup 98, Strines 28, Bubnell 27, Matlock 35, Abbeydale 38, Salter
Wood, Marehay 34, Loscoe 44.

ANTHOXANTHUM L.

A. odoratum L.   *Sweet Vernal Grass*
Native. Meadows, pastures and moorland. Generally distributed and
abundant.
Recorded for all squares.
Linton: 'Common. Everywhere abundant'.

PHALARIS L.

P. arundinacea L. *Reed Grass*

Native. In periodically flooded places by streams, ponds and ditches and in marshes; usually on inorganic soils rather than peat, and in fens confined to areas accessible to flood-borne silt from streams. Locally common.

Recorded for all squares except:

W. 09, 19, 08.
E. 44.

Linton: 'Common. Generally distributed'.

P. canariensis L. *Canary Grass*

Introduced. A casual of waste ground and rubbish-tips. Occasional.

W. Stoney Middleton 2275, Thorpe 1550.

E. Abbey Lane 3282, Whatstandwell 3254.

S. Ashbourne goods yard 1746, Clifton goods yard 1644, Bradley 2145, various localities in Derby 3636 etc., Chaddesden 3837.

Linton: Burbage, Buxton 07, nr. Matlock and Wirksworth 25, Drakelow 22, Repton and Calke 32.

NARDUS L.

N. stricta L. *Mat Grass*

Native. In grassland on the poorer acid soils with a covering of raw humus or peat, and extensively dominant in moorland at all altitudes; also a serious weed of over-grazed better types of grassland on the lower slopes of the hills. Locally very abundant.

W. Recorded for all squares.

E. Recorded for 48, nr. Astwith 4464, recorded for 35.

S. Bradley Brook 2244, Hulland Moss 2445, nr. the Henmore, Corley 2147, Mugginton 2843.

Linton: Nr. Dore 38, Mickleover 33, nr. Calke Abbey 32, Linton Heath 21.

PANICUM L.

P. miliaceum L. *Proso Millet*

Introduced. A rare casual.

S. Rubbish dump, Stores Road, Derby 3736.

Linton: Not recorded.

SORGHUM Pers.

S. vulgare Pers. *Sorghum*

Introduced. A rare casual.

S. Rubbish dump in Derby 3736.

Linton: Not recorded.

INDEX

The entries in *italics* denote the names used in Linton's 'Flora of Derbyshire' where these differ in their generic names from those used in this Flora.